Birkhäuser

# Frontiers in Mathematics



This series is designed to be a repository for up-to-date research results which have been prepared for a wider audience. Graduates and postgraduates as well as scientists will benefit from the latest developments at the research frontiers in mathematics and at the "frontiers" between mathematics and other fields like computer science, physics, biology, economics, finance, etc. All volumes are online available at SpringerLink.

Christian Budde

# Bi-Continuous Operator Semigroups

## From Theory to Applications

 **Birkhäuser**

Christian Budde 
University of the Free State
Bloemfontein, South Africa

ISSN 1660-8046 ISSN 1660-8054 (electronic)
Frontiers in Mathematics
ISBN 978-3-032-12947-5 ISBN 978-3-032-12948-2 (eBook)
https://doi.org/10.1007/978-3-032-12948-2

Mathematics Subject Classification: 47D03, 37L05, 46A70, 34G10, 93C25, 47D06, 47D09, 47A55



This book is published under the imprint Birkhäuser, www.birkhauser-science.com by the registered company
Springer Nature Switzerland AG
The registered company address is: Gewerbestrasse 11, 6330 Cham, Switzerland

If disposing of this product, please recycle the paper.

*Dedicated to
Lena and Felix*

# Preface

My first systematic exposure to operator semigroups occurred during a masters course in functional analysis that I followed at the Vrije Universiteit Amsterdam. At that stage, it was not evident that this area would become my primary research focus.

Following my studies, I pursued doctoral research under the supervision of Prof. Dr. Bálint Farkas at the Bergische Universität Wuppertal. My doctoral thesis centered on extrapolation spaces and perturbation theory for bi-continuous semigroups.

I continued this research during my postdoctoral fellowship at the North-West University under the supervision of Prof. Dr. Sanne ter Horst. During this period, I observed growing interest in bi-continuous semigroups which reflected in an increasing number of publications. This development motivated me to consolidate existing results, leading to the conception of this book.

I express profound gratitude to my mentor Prof. Dr. Bálint Farkas for his guidance and insightful discussions throughout my academic career. I also acknowledge the broader mathematical community for their contributions to bi-continuous semigroup theory, which have stimulated progress in this field.

I hope this book will serve as both a reference work and motivation for future research. Readers are encouraged to engage critically with the material and explore the mathematical questions it presents.

## What Is in This Book?

This monograph serves two primary purposes. First, it provides a comprehensive treatment of bi-continuous semigroup theory and its applications, ranging from network theory to control theory. The exposition proceeds systematically from fundamental theory through approximation methods, extrapolation techniques as well as perturbation theory, concluding with applications to networks, mean ergodicity and control systems. Each chapter includes illustrative examples to reinforce the theoretical concepts.

Second, this work aims to support emerging researchers and doctoral students by offering a structured foundation for further investigation in operator semigroups. By synthesizing current knowledge, it could provide a reference point for new research projects in this domain.

## What Is Not in This Book?

While self-contained within its specialized domain, this monograph intentionally excludes elementary functional analysis, as we assume readers possess postgraduate-level familiarity with this foundation. The general $C_0$-semigroup theory is discussed only where directly relevant to bi-continuous extensions; for comprehensive treatments, we refer readers to monographs by Davies [80], Goldstein [119], Pazy [198], or Engel and Nagel [101]. Although we mention certain approaches to non-strongly continuous semigroups with some detail, these ultimately fall beyond our primary scope. Representative literature in this direction includes [67, 147, 148, 151, 165, 185, 197, 203].

The holomorphic functional calculus [140, 141], numerical implementations and computational aspects similarly lie outside this book's coverage. We particularly emphasize that while the mixed topology—studied by Wiweger [240] and Cooper [73] and addressed in Appendix B—maintains direct connections to bi-continuous semigroups (which we utilize where appropriate), this monograph adopts the classical presentation framework established by Kühnemund [160] and Farkas [106]. Specifically, we avoid technical complexities arising from abstract locally convex space theory, especially concerning the mixed topology. Interested readers may consult sources such as [73, 142, 153–156] for deeper exploration of these topics.

Bloemfontein, South Africa                                        Prof. Christian Budde

**Acknowledgements** I would like to express my gratitude to the University of the Free State, and in particular, the Department of Mathematics and Applied Mathematics, for their support since I joined the department as a permanent staff member in January 2022. The friendly, open-hearted, and inspiring environment in the department has undoubtedly boosted my productivity and strengthened my ambition to write this book.

I am also deeply appreciative of the continuous communication with my former Ph.D. supervisor, Prof. Dr. Bálint Farkas. His ongoing motivation has fueled my academic journey and encouraged me to engage actively in research within my field of expertise. Additionally, I owe a debt of gratitude to my former postdoctoral supervisor, Prof. Dr. Sanne ter Horst, who granted me the freedom to explore my own scientific style, fostered academic independence, and was consistently available for discussions and support. I extend my heartfelt gratitude to my diverse co-authors for engaging in fruitful collaborations, from which I gained valuable insights and knowledge. I also want to thank my postdoc Dr. Marieme Lasri for the proofreading. A special thanks also goes to the referees who invested a lot of time in order to give constructive feedback which helped in improving this book.

Of course, I also want to thank my family for the continuous encouragement regarding my work. Especially, I want to thank Lena, for her enormous backing and understanding of what mathematics means to me. It also goes without saying that I am very thankful for having Felix in our lives.

Last but not least, I want to acknowledge funding by the *Deutsche Forschungsgemeinschaft* (DFG, German Research Foundation) under grant number 46873678, the *National Research Foundation* (NRF) under grant numbers SRUG220317136, KIC22080247403 and KIC230816143482, as well as the *DSI-NRF Centre of Excellence in Mathematical and Statistical Sciences (CoE-MaSS), South Africa* under grant numbers 2024-031-OPA-Bi-Continuous Semigroups and 2025-013-OPA-Bi-Continous Cosine Families.

Bloemfontein, South Africa      Prof. Christian Budde
March 2026

**Competing Interests** The author has no competing interests to declare that are relevant to the content of this manuscript.

# Introduction

In this introduction we provide a comprehensive overview of the development of bicontinuous operator semigroups, from its origins in $C_0$-semigroup theory to recent advancements. We summarize the evolution of this field, highlighting key milestones that have shaped this research area.

The theory of operator semigroups, and especially $C_0$-semigroups, has been originated by the work of Hille [131] and Yosida [241]. A significant milestone was reached with the publication of the monograph by Hille and Phillips [132]. Since then there has been an enormous research effort in this direction and various academic institutions made substantial contributions in this field, leading to the theory's notable refinement. This progress is well documented in several monographs. For references, we refer for example to the work by Davies [80], Goldstein [119], Pazy [198], Engel and Nagel [101], Arendt, Batty, Hieber and Neubrander [17] or Bátkai, Kramar Fijavž and Rhandi [33], just to mention a few.

Besides the theoretical contributions and the therefore developed functional-analytic toolbox it has evolved into a multifaceted framework. In fact, the landscape of $C_0$-semigroups is characterized by the diverse applications of this theory, extending beyond its traditional domains such as partial differential equations and stochastic processes. In addition, there exist specialized monographs that exclusively focus on the mathematical applications of operator semigroups in the fields of physics and life sciences. For instance, a noteworthy example is [175], which highlights the theory of operator semigroups' significant impact. These dedicated works underscore the strength and relevance of operator semigroup theory in diverse scientific disciplines.

Nevertheless, despite the significant impact of the theory of $C_0$-semigroups, certain limitations are encountered. This becomes evident when examining stochastic differential equations, Ornstein–Uhlenbeck and Feller processes. The operator semigroups arising from these systems give rise to transition semigroups that, as a general rule, do not possess strong continuity. For more information we refer for example to the monographs by Bertoldi and Lorenzi [170] or by Lorenzi and Rhandi [171]. We will briefly recall some of these results in Sect. 1.5. Another motivation arises from the work of Lotz [172]. It was shown that on Grothendieck spaces with the Dunford–Pettis property every $C_0$-semigroup is automatically already uniformly continuous. Typical examples for such spaces are $L^\infty(\Omega, \Sigma, \mu)$ for a positive measure space $(\Omega, \Sigma, \mu)$, $H^\infty(\mathbb{D})$, injective Banach spaces as well as certain $C(\Omega)$ spaces, see for example also [3], [48] or [56].

In order to circumvent those obstacles, many authors developed different approaches to operator semigroups that are not strongly continuous with respect to the norm. The idea is to make use of locally convex topologies. The theory of $C_0$-semigroups on locally convex spaces has, for example, been developed by Miyadera [185], Komatsu [147], Kōmura [148] and Ōuchi [197]. Other approaches are for example semigroups on norming dual pairs of Kunze [165], "C-class" semigroups of Kraaij [151], $\pi$-semigroups of Priola [203] or weakly continuous semigroups of Cerrai [67], to mention a few.

The objective of the theory of bi-continuous semigroups is to establish a comprehensive framework that unifies and contextualizes various individual results. This theory was initiated and developed by Kühnemund [160]. She demonstrated that bi-continuous semigroups possess similar characteristics to $C_0$-semigroups, enabling the development of a systematic theory that encompasses essential results such as a Hille–Yosida generation type theorem [161] and Trotter–Kato approximation theorems [160, Chap. 2], [159]. The versatility and efficacy of this theory have been exemplified through an extensive range of applications, underscoring its flexibility and robustness.

The work of Kühnemund was followed by the research of Farkas [106] in 2003. The objective was to initiate the development of a perturbation theory specifically tailored for bi-continuous semigroups. By taking these initial steps, the foundation for a comprehensive understanding of the behavior and properties of perturbed bi-continuous semigroups has been laid. Farkas covered both bounded-type perturbations [107] as well as perturbations of Miyadera–Voigt type [102, 103, 108].

Subsequently, numerous authors enthusiastically embraced the topic and made valuable contributions to the theory of bi-continuous semigroups, addressing previously unresolved aspects, bridging gaps and refining the existing theory. Their collective efforts have significantly advanced the research of bi-continuous semigroups. Noteworthy publications in this field include works by Albanese and Mangino [8], Albanese, Lorenzi and Manco [7], Jara [140, 141] and Farkas [109].

In 2019 the theory of bi-continuous semigroups was then further developed by Budde [50]. His contributions to the field include the investigation of intermediate and extrapolation spaces [59], the advancement of perturbation theory [53, 60] as well as applications to network theory [61]. Also worth mentioning in that context is the work of Kruse, Schwenninger and Seifert [154, 155, 157, 158], Kunstmann [164] as well as Budde [52, 62].

In conclusion, this book on bi-continuous operator semigroups offers a comprehensive compilation of a rapidly developing field in mathematics. As advancements continue to unfold, our understanding of the evolution equations and operator semigroup deepens. We kindly invite you to join us on this mathematical journey as we delve into the intricacies of bi-continuous operator semigroups and contribute to the ongoing development of this discipline. Welcome to a world of bi-continuous operator semigroups.

# Contents

# Part I
# Bi-continuous Semigroups

# An Excursion on $C_0$-Semigroups

Before we start with the main topic of this book, we will recall some important aspects of the theory of $C_0$-semigroups. There are several monographs on $C_0$-semigroups and for more details we refer the reader from example to the books by Engel and Nagel [101], Goldstein [119], Pazy [197], Arendt, Batty, Hieber and Neubrander [17] or Bátkai, Kramar Fijavž and Rhandi [33], just to mention a few. This chapter is supposed to provide an overview of the theory of $C_0$-semigroups, as this is also crucial to understand the theory of bi-continuous semigroups. However, in order to emphasize the introductory character of this chapter, we will only state the important results without proofs. For more details we then refer to the references mentioned beforehand.

## 1.1 Uniformly Continuous Semigroups and $C_0$-Semigroups

Let $X$ be a Banach space and consider the following Banach space valued initial value problem given by

$$\begin{cases} \dot{u}(t) = Au(t), & t \geq 0, \\ u(0) = x \in X, \end{cases} \tag{ACP}$$

where $(A, D(A))$ is a closed linear operator on $X$. For the sake of simplicity, we will first assume that $A \in \mathscr{L}(X)$, i.e., $A : X \to X$ is a bounded linear operator on $X$. For a linear bounded operator, we can make sense of the expression $e^{tA}$ by means of the exponential series, i.e.,

$$e^{tA} := \sum_{n=0}^{\infty} \frac{t^n A^n}{n!}, \quad t \geq 0.$$

© The Author(s), under exclusive license to Springer Nature Switzerland AG 2026
C. Budde, *Bi-Continuous Operator Semigroups*, Frontiers in Mathematics,
https://doi.org/10.1007/978-3-032-12948-2_1

This series is absolutely convergent due to the boundedness of $A$. Since $X$ is a Banach space the series is therefore also convergent and $e^{tA}$ yields a bounded operator for each $t \geq 0$. Let us summarize some properties, cf. [101, Chap. I, Prop. 3.5] or [33, Prop. 9.2].

**Proposition 1.1** *Let $A \in \mathscr{L}(X)$ and set $T(t) := e^{tA}$, $t \geq 0$. Then the following holds:*

(a) *The functional equation $T(t + s) = T(t)T(s)$ and $T(0) = I$ is satisfied for all $t, s \geq 0$.*
(b) *The function $\mathbb{R}_+ \ni t \mapsto T(t) \in (\mathscr{L}(X), \|\cdot\|)$ is continuous.*
(c) *The map $\mathbb{R}_+ \ni t \mapsto T(t) \in (\mathscr{L}(X), \|\cdot\|)$ is differentiable and satisfies the differential equation*

$$\begin{cases} \frac{\mathrm{d}}{\mathrm{d}t} T(t) = A T(t), & t \geq 0, \\ T(0) = I. \end{cases}$$

The previous result shows that the function $t \mapsto e^{tA}$ solves (ACP) whenever $A \in \mathscr{L}(X)$. It also gives rise to the following definition.

**Definition 1.2** A family of bounded linear operators $(T(t))_{t \geq 0}$ on a Banach space $X$ is called a *uniformly continuous one-parameter operator semigroup* (or *uniformly continuous semigroup* for short) if it satisfies $T(t + s) = T(t)T(s)$ and $T(0) = I$ as well as the condition that $\mathbb{R}_+ \ni t \mapsto T(t) \in \mathscr{L}(X)$ is continuous with respect to the uniform operator topology on $\mathscr{L}(X)$.

Obviously, the family of bounded linear operators given by $(e^{tA})_{t \geq 0}$ for some $A \in \mathscr{L}(X)$ forms a uniformly continuous semigroup by Proposition 1.1. The converse, however, is also true, cf. [101, Chap. I, Thm. 3.7] or [33, Prop. 9.4].

**Theorem 1.3** *Let $(T(t))_{t \geq 0}$ be a uniformly continuous semigroup on a Banach space $X$. Then there exists $A \in \mathscr{L}(X)$ such that $T(t) = e^{tA}$ for all $t \geq 0$.*

**Example 1.4** Let $\Omega$ be a locally compact space and consider the space $C_0(\Omega)$ consisting of all continuous functions on $\Omega$ vanishing at infinity. Recall that a function $f : \Omega \to \mathbb{C}$ vanishes at infinity if for all $\varepsilon > 0$ there exists a compact set $K \subseteq \Omega$ such that $|f(x)| < \varepsilon$ for all $x \in \Omega \setminus K$. Equipped with the $\|\cdot\|_\infty$-norm defined by

$$\|f\|_\infty := \sup_{x \in \Omega} |f(x)|, \quad f \in C_0(\Omega),$$

the space $C_0(\Omega)$ becomes a Banach space. To any continuous function $q : \Omega \to \mathbb{C}$ satisfying the condition

$$\sup_{x\in\Omega} \operatorname{Re}(q(x)) < \infty, \tag{1.1.1}$$

we define

$$T_q(t)f := e^{tq} f, \quad t \geq 0, \ f \in C_0(\Omega).$$

Then (1.1.1) ensures that $T_q(t)$ is a bounded linear operator on $C_0(\Omega)$ for all $t \geq 0$. We call $(T_q(t))_{t\geq 0}$ the *multiplication semigroup*. As a matter of fact, $(T_q(t))_{t\geq 0}$ is uniformly continuous if and only if $q$ is a bounded function, cf. [101, Chap. I, Prop. 4.4].

**Example 1.5**  Let us now consider the space $UC_b(\mathbb{R})$ of bounded and uniformly continuous functions on $\mathbb{R}$. This is also a Banach space, if it is equipped with the $\|\cdot\|_\infty$-norm. We will now give a counterexample of an operator semigroup that is not uniformly continuous. To do so, define

$$(T(t)f)(x) := f(x + t), \quad t \geq 0, \ f \in UC_b(\mathbb{R}), \ x \in \mathbb{R}.$$

For each $t \geq 0$, the operator $T(t)$ is a linear isometry and therefore bounded. Due to the additive structure of $\mathbb{R}$, the family of operators $(T(t))_{t\geq 0}$ also satisfies the semigroup law $T(t + s) = T(t)T(s)$ and $T(0) = I$ for $t, s \geq 0$. However, this operator semigroup fails to be uniformly continuous. To see this, we observe that

$$\|T(t) - I\| = \sup_{\|f\|\leq 1} \ \sup_{x\in\mathbb{R}} |f(x + t) - f(x)|.$$

If the right-hand side would converge to 0 as $t \to 0$ (this is the requirement to be uniformly continuous), then the unit ball of $UC_b(\mathbb{R})$ would be uniformly equicontinuous, which is impossible. However, we observe that the map $t \mapsto T(t)f$ is continuous for each $f \in UC_b(\mathbb{R})$. We call $(T(t))_{t\geq 0}$ the *(left) translation semigroup*.

By Theorem 1.3 and Example 1.5 we see that the assumption of uniform continuity is a too strong requirement. We will now weaken the assumption, which gives rise to the following definition.

**Definition 1.6**  A family of bounded linear operators $(T(t))_{t\geq 0}$ on a Banach space $X$ is called a *strongly continuous (one-parameter) semigroup* (or $C_0$-semigroup for short) if the following properties are satisfied:

(i)   $T(t + s) = T(t)T(s)$ and $T(0) = I$ for all $t, s \geq 0$.
(ii)  $(T(t))_{t\geq 0}$ is strongly continuous, i.e., the orbit map $\mathbb{R}_+ \ni t \mapsto T(t)x \in X$ is continuous for each $x \in X$.

We see that, in comparison with Definition 1.2, we have that the map $\mathbb{R}_+ \ni t \mapsto T(t) \in \mathscr{L}(X)$ is continuous with respect to the strong operator topology on $\mathscr{L}(X)$. By [101, Chap.

I, Prop. 5.3], the second condition of Definition 1.6 can be substituted by $\lim_{t\to 0} T(t)x = x$ for all $x \in X$ or $\lim_{t\to 0} \|T(t)x - x\| = 0$ for all $x \in X$.

**Example 1.7** Let us revise Examples 1.4 and 1.5:

(i) We saw that $(T_q(t))_{t\geq 0}$ is uniformly continuous if and only if $q$ is a bounded function. In general, under the assumption (1.1.1), $(T_q(t))_{t\geq 0}$ is a $C_0$-semigroup, cf. [101, Chap. I, Prop. 4.11]

(ii) As already mentioned in Example 1.5, the translation semigroup $(T(t))_{t\geq 0}$ satisfies that the map $t \mapsto T(t)f$ is continuous for each $f \in \mathrm{UC_b}(\mathbb{R})$. Hence, $(T(t))_{t\geq 0}$ is a $C_0$-semigroup.

(iii) The translation semigroup becomes also strongly continuous on $\mathrm{C_0}(\mathbb{R})$ as well as on the space of $p$-integrable functions $\mathrm{L}^p(\mathbb{R})$ $(1 \leq p < \infty)$.

Due to the uniform boundedness principle, every $C_0$-semigroup is exponentially bounded, cf. [101, Chap. I, Prop. 5.5] or [33, Prop. 9.7].

**Proposition 1.8** *For every $C_0$-semigroup $(T(t))_{t\geq 0}$ on a Banach space $X$, there exist constants $M \geq 1$ and $\omega \in \mathbb{R}$ such that $\|T(t)\| \leq Me^{\omega t}$ for all $t \geq 0$.*

## 1.2    The Generator

We saw that for a uniformly continuous semigroup $(T(t))_{t\geq 0}$ on a Banach space $X$ one has that $\frac{\mathrm{d}}{\mathrm{d}t} T(t) = AT(t)$ for $t \geq 0$, cf. Proposition 1.1. In particular, one can recover the operator $A$ by $A = \frac{\mathrm{d}}{\mathrm{d}t}\big|_{t=0} T(t)$. Therefore, $A$ is—informally speaking—the first derivative of $(T(t))_{t\geq 0}$ at $t = 0$. This gives rise to the following definition.

**Definition 1.9** Let $(T(t))_{t\geq 0}$ be a $C_0$-semigroup on a Banach space $X$. The *(infinitesimal) generator* of $(T(t))_{t\geq 0}$ is defined by

$$Ax := \lim_{t\to 0} \frac{T(t)x - x}{t}, \quad D(A) := \left\{ x \in X : \lim_{t\to 0} \frac{T(t)x - x}{t} \text{ exists} \right\}. \tag{1.2.1}$$

Generally speaking, the (infinitesimal) generator of a $C_0$-semigroup is an unbounded linear operator as the limit in (1.2.1) does not have to exist for each $x \in X$. It is noteworthy that Definition 1.9 works nonetheless nicely together with the definition of uniformly continuous semigroups as by [101, Chap. II, Cor. 1.5] the generator $(A, D(A))$ is a bounded operator if and only if $(T(t))_{t\geq 0}$ is uniformly continuous. Besides that, the generator also has other properties.

**Proposition 1.10** *Let $(T(t))_{t\geq 0}$ be a $C_0$-semigroup on a Banach space $X$ with generator $(A, D(A))$. Then, the following properties hold:*

(a)  *$A : D(A) \to X$ is a linear operator.*

(b)  *If $x \in D(A)$, then $T(t)x \in D(A)$ and $\frac{d}{dt}T(t)x = AT(t)x = T(t)Ax$ for all $t \geq 0$.*

(c)  *For every $t \geq 0$ and $x \in X$ one has*

$$\int_0^t T(s)x\,ds \in X.$$

(d)  *For every $t \geq 0$ one has*

$$T(t)x - x = A \int_0^t T(s)x\,ds \ \textit{if } x \in X,$$
$$= \int_0^t T(s)Ax\,ds \ \textit{if } x \in D(A).$$

(e)  *The generator of a strongly continuous semigroup is a closed and densely defined linear operator that determines the semigroup uniquely.*

For an (unbounded) linear operator $(A, D(A))$ the *spectrum* is defined by

$$\sigma(A) := \{\lambda \in \mathbb{C} : \lambda - A \text{ is not bijective}\}.$$

The *resolvent set* is defined by $\rho(A) := \mathbb{C} \setminus \sigma(A)$. The *resolvent operator* is defined by $R(\lambda, A) := (\lambda - A)^{-1}$ for $\lambda \in \rho(A)$.

Let $(T(t))_{t\geq 0}$ be a $C_0$-semigroup on a Banach space $X$, then by Proposition 1.8 there exist $M \geq 1$ and $\omega \in \mathbb{R}$ such that $\|T(t)\| \leq Me^{\omega t}$ for all $t \geq 0$. Let $(A, D(A))$ be the generator of $(T(t))_{t\geq 0}$. By [101, Chap. II, Thm. 1.10], one has that $\lambda \in \rho(A)$ whenever $\text{Re}(\lambda) > \omega$. Moreover, the resolvent can be represented by the Laplace transform of the semigroup, i.e.,

$$R(\lambda, A)x = \int_0^\infty e^{-\lambda t}T(t)x\,dt, \quad x \in X.$$

We saw, by means of Definition 1.9, that every $C_0$-semigroup $(T(t))_{t\geq 0}$ yields a generator. However, this does not tell you which (unbounded) linear operators are generators of $C_0$-semigroups. This is characterized by the famous Hille–Yosida theorem.

**Theorem 1.11** *Let $(A, D(A))$ be a linear operator on a Banach space $X$ and let $M \geq 1$ and $\omega \in \mathbb{R}$ be constants. The following assertions are equivalent.*

(a)  *$(A, D(A))$ generates a $C_0$-semigroup $(T(t))_{t\geq 0}$ satisfying $\|T(t)\| \leq Me^{\omega t}$ for all $t \geq 0$.*

(b)  $(A, D(A))$ *is closed, densely defined and for every* $\lambda \in \mathbb{C}$ *with* $\mathrm{Re}(\lambda) > \omega$ *one has*
$\lambda \in \rho(A)$ *and*

$$\left\| R(\lambda, A)^n \right\| \leq \frac{M}{(\mathrm{Re}(\lambda) - \omega)^n}, \quad n \in \mathbb{N}.$$

The Hille–Yosida theorem is important as it characterizes those operators on Banach spaces that are generators of $C_0$-semigroups. However, this theorem is not really useful for applications as one, for example, needs to know all natural powers of the resolvent operator explicitly. For contraction semigroups, Theorem 1.11 simplifies to the so-called *Lumer–Phillips theorem*.

**Definition 1.12**  A linear operator $(A, D(A))$ on a Banach space is called *dissipative* if $\|(\lambda - A)x\| \geq \lambda \|x\|$ for all $\lambda > 0$ and $x \in D(A)$

**Theorem 1.13**  *For a densely defined, dissipative operator* $(A, D(A))$ *on a Banach space* $X$ *the following statements are equivalent:*

(a)  *The closure* $\overline{A}$ *of* $A$ *generates a contractive* $C_0$*-semigroup.*
(b)  $\mathrm{Ran}(\lambda - A)$ *is dense in* $X$ *for some (hence all)* $\lambda > 0$.

We turn our attention now to our starting point, the abstract Cauchy problem (ACP). We will show that there is a correspondence between solutions of (ACP) and $C_0$-semigroups. Let us start with a toy-example. Let $\mathrm{UC}_b^1(\mathbb{R})$ be the space of differentiable bounded uniformly continuous functions whose derivatives are also bounded and uniformly continuous. For a given $f \in \mathrm{UC}_b^1(\mathbb{R})$ consider the following partial differential equation

$$\begin{cases} \frac{\partial}{\partial t} u(t, x) = \frac{\partial}{\partial x} u(t, x), & t \geq 0,\ x \in \mathbb{R}, \\ u(0, x) = f(x), & x \in \mathbb{R}. \end{cases} \tag{PDE}$$

This equation can be equivalently rewritten as an (ACP) by taking

$$Af = f', \quad D(A) = \mathrm{UC}_b^1(\mathbb{R}). \tag{1.2.2}$$

Therefore, solutions of (ACP) correspond to solutions of (PDE). Hence, we introduce the concept of (classical) solutions of (ACP).

**Definition 1.14**  A function $u : \mathbb{R}_+ \to X$ is called a *(classical) solution* of (ACP) if $u$ is continuously differentiable with respect to $X$, $u(t) \in D(A)$ for all $t \geq 0$ and (ACP) holds.

There is also a definition of well-posedness for (ACP). In fact, for a "good" solution of (ACP) one expects the existence of an unique solution that depends continuously on the initial data. This idea yields the following definition.

**Definition 1.15** We call (ACP) associated to a closed operator $(A, D(A))$ *well-posed* if $D(A)$ is dense in $X$, for every $x \in D(A)$ there exists a unique solution $u(\cdot, x)$ of (ACP) and for every sequence $(x_n)_{n \in \mathbb{N}}$ in $D(A)$ with $\lim_{n \to \infty} x_n = 0$ one has $\lim_{n \to \infty} u(t, x_n) = 0$ uniformly on compact intervals.

The next result brings well-posedness and $C_0$-semigroups together, cf. [101, Chap. II, Prop. 6.2 and Cor. 6.9].

**Theorem 1.16** *For a closed operator $(A, D(A))$ the associated (ACP) is well-posed if and only if $(A, D(A))$ generates a $C_0$-semigroup $(T(t))_{t \geq 0}$ on $X$. In this case, if one has that $u(0) = x \in D(A)$ the function $u(t) := T(t)x$, $t \geq 0$, is the unique classical solution of (ACP).*

As mentioned in Example 1.7, the translation semigroup on $\mathrm{UC_b}(\mathbb{R})$ is a $C_0$-semigroup. By following Definition 1.9, it is an easy exercise to show that the generator of the translation semigroup actually coincides with the operator defined by (1.2.2). Hence, by Theorem 1.16 the corresponding (ACP) is well-posed and so is (PDE). In particular, from Theorem 1.16 we also obtain the (unique) solution of (ACP) by

$$u(t, x) = (T(t)f)(x) = f(x + t), \quad t \geq 0, \ x \in \mathbb{R}.$$

## 1.3  Perturbations

As mentioned earlier, verifying the conditions of Theorem 1.11 is not an easy task. However, sometimes it is possible to split the operator as a sum of operators. The idea is the following: let $(A, D(A))$ be the generator of a $C_0$-semigroup on a Banach space $X$ and let $(B, D(B))$ be another operator on $X$. Under which assumptions does $A + B$ (on a certain domain) generate again a $C_0$-semigroup on $X$? We will state in this section three common perturbation results for $C_0$-semigroups. We will start with bounded perturbations.

**Theorem 1.17** *Let $(A, D(A))$ be the generator of a $C_0$-semigroup $(T(t))_{t \geq 0}$ on a Banach space $X$ and let $B \in \mathscr{L}(X)$. Then $(A + B, D(A))$ generates also a $C_0$-semigroup $(S(t))_{t \geq 0}$ on $X$. Moreover, the variation of constants formula holds*

$$S(t)x = T(t)x + \int_0^t T(t - s)BS(s)x \, \mathrm{d}s, \quad t \geq 0, \ x \in X.$$

We continue with unbounded perturbations and formulate a version of the so-called *Miyadera–Voigt perturbation theorem*.

**Theorem 1.18** *Let $(A, D(A))$ be the generator of a $C_0$-semigroup $(T(t))_{t\geq 0}$ on a Banach space $X$. Assume that $B \in \mathscr{L}(D(A), X)$ and that there exist $t_0 > 0$ and $q \in (0, 1)$ such that*

$$\int_0^{t_0} \|BT(r)x\| \, \mathrm{d}r \leq q \, \|x\|, \quad x \in D(A).$$

*Then the operator $(A + B, D(A))$ generates a $C_0$-semigroup $(S(t))_{t\geq 0}$ on $X$ which satisfies the variation of constants formula*

$$S(t)x = T(t)x + \int_0^t T(t - s)BS(s)x \, \mathrm{d}s, \quad t \geq 0, \ x \in D(A).$$

The last perturbation result in the framework of $C_0$-semigroups we want to mention is the so-called *Desch–Schappacher perturbation theorem*. For this result, we need the so-called extrapolation spaces. For this, we refer to either Chap. 4 or [101, Chap. II, Sect. 5] or [187].

**Theorem 1.19** *Let $(A, D(A))$ be the generator of a $C_0$-semigroup $(T(t))_{t\geq 0}$ on a Banach space $X$ and let $B \in \mathscr{L}(X, X_{-1})$. Moreover, assume that there exist $t_0 > 0$ and $q \in [0, 1)$ such that*

(i) $\displaystyle\int_0^{t_0} T_{-1}(t_0 - r)Bf(r) \, \mathrm{d}r \in X,$

(ii) $\displaystyle\left\| \int_0^{t_0} T_{-1}(t_0 - r)Bf(r) \, \mathrm{d}r \right\| \leq q \, \|f\|_\infty,$

*for all $f \in C([0, t_0], X)$. Then the operator $(A_{-1} + B)_{|X}$ with domain*

$$D((A_{-1} + B)_{|X}) = \{x \in X : \ A_{-1}x + Bx \in X\},$$

*generates a $C_0$-semigroup $(S(t))_{t\geq 0}$ on $X$ which satisfies the variation of constants formula*

$$S(t)x = T(t)x + \int_0^t T_{-1}(t - s)BS(s)x \, \mathrm{d}s, \quad t \geq 0, \ x \in X.$$

## 1.4   Approximations

Another approach to study complicated operators and their generated $C_0$-semigroups is by means of approximation. As a matter of fact, the proof of the Hille–Yosida theorem, cf. Theorem 1.11, uses approximations. Here, we will state the two so-called *Trotter–Kato approximation theorems* and their consequences.

**Theorem 1.20** *Let $(T_n(t))_{t\geq 0}$ and $(T(t))_{t\geq 0}$ be $C_0$-semigroups on a Banach space $X$ with generators $(A_n, D(A_n))$ and $(A, D(A))$, respectively. Assume all operator semigroups satisfy the estimate $\|T_n(t)\| \leq Me^{\omega t}$ and $\|T(t)\| \leq Me^{\omega t}$ for all $t \geq 0$ and $n \in \mathbb{N}$ and some constants $M \geq 1$ and $\omega \in \mathbb{R}$. Let $D$ be a core for the operator $(A, D(A))$ and consider the following statements:*

(a) *$D \subseteq D(A_n)$ for all $n \in \mathbb{N}$ and $A_n x \to Ax$ for all $x \in D$.*
(b) *For each $x \in D$ there exists $x_n \in D(A_n)$ such that $x_n \to x$ and $A_n x_n \to Ax$.*
(c) *$R(\lambda, A_n)x \to R(\lambda, A)x$ for all $x \in X$ and some/all $\lambda > \omega$.*
(d) *$T_n(t)x \to T(t)x$ for all $x \in X$ uniformly for $t$ in compact sets.*

*Then the implications $(a)\Longrightarrow(b)\Longleftrightarrow(c)\Longleftrightarrow(d)$ hold.*


**Definition 1.21** Let $(A, D(A))$ be a linear operator on a Banach space $X$. A subspace $D$ of the domain $D(A)$ is called a core for $A$ if $D$ is dense in $D(A)$ with respect to the graph norm.


**Theorem 1.22** *Let $(T_n(t))_{t\geq 0}$ be $C_0$-semigroups on a Banach space $X$ with generator $(A_n, D(A_n))$. Assume all operator semigroups satisfy the estimate $\|T_n(t)\| \leq Me^{\omega t}$ for all $t \geq 0$ and $n \in \mathbb{N}$ and some constants $M \geq 1$ and $\omega \in \mathbb{R}$. For some $\lambda_0 > \omega$ consider the following statements:*

(a) *There exists a densely defined operator $(A, D(A))$ such that $A_n x \to Ax$ for all $x$ in a core $D$ of $A$ and such that $\mathrm{Ran}(\lambda_0 - A)$ is dense in $X$.*
(b) *The operators $R(\lambda_0, A_n)$, $n \in \mathbb{N}$, converge strongly to an operator $R \in \mathcal{L}(X)$ with dense range.*
(c) *The semigroups $(T_n(t))_{t\geq 0}$, $n \in \mathbb{N}$, converge strongly (and uniformly for $t \in [0, t_0]$) to a strongly continuous semigroup $(T(t))_{t\geq 0}$ with generator $B$ such that $R = R(\lambda_0, B)$.*

*Then the implications $(a)\Longrightarrow(b)\Longleftrightarrow(c)$ hold. In particular, if $(a)$ holds, then $B = \overline{A}$.*


## 1.5    A Motivating Counterexample: Parabolic Equations on $C_b(\mathbb{R}^N)$

The following example is a great motivation for why one should study bi-continuous semigroups in more detail. On $C_b(\mathbb{R}^N)$ we define the differential operator $(\mathcal{A}, D(\mathcal{A}))$ by

$$\mathcal{A}u(x) = \sum_{i,j=1}^{N} q_{ij}(x)\frac{\partial^2}{\partial x_i \partial x_j}u(x) + \sum_{i=1}^{N} b_i(x)\frac{\partial}{\partial x_i}u(x) + c(x)u(x), \quad x \in \mathbb{R}^N,$$

$$D(\mathcal{A}) = \left\{ u \in C_b(\mathbb{R}^N) \cap \bigcap_{1 \le p < \infty} W_{loc}^{2,p}(\mathbb{R}^N) : \mathcal{A}u \in C_b(\mathbb{R}^N) \right\}.$$

(1.5.1)

For the coefficients of the operator $(\mathcal{A}, D(\mathcal{A}))$ we make the following assumptions.

**Assumption 1.23** (i) $q_{ij} = q_{ji}$ for all $1 \le i, j \le N$ and

$$\sum_{i,j=1}^{N} q_{ij}(x)\xi_i\xi_j \ge \kappa(x)\,|\xi|^2, \quad \kappa(x) > 0, \ \xi, x \in \mathbb{R}^N.$$

(ii) $q_{ij}, b_i, 1 \le i, j \le N$, and $c$ belong to $C_{loc}^\alpha(\mathbb{R}^N)$ for some $\alpha \in (0, 1)$.
(iii) There exists $c_0 \in \mathbb{R}$ such that $c(x) \le c_0$ for all $x \in \mathbb{R}^N$.

Let us now investigate the parabolic case associated to the operator $(\mathcal{A}, D(\mathcal{A}))$. We will see that then abstract Cauchy problems and operator semigroups enter the picture again. For $f \in C_b(\mathbb{R}^N)$ we consider the following (parabolic) partial differential equation

$$\begin{cases} \frac{\partial}{\partial t}u(t, x) = \mathcal{A}u(t, x), & t > 0, \ x \in \mathbb{R}^N, \\ u(0, x) = f(x), & x \in \mathbb{R}^N. \end{cases}$$

(1.5.2)

The following result shows that (1.5.2) admits solutions $u$ which are regular enough in a certain sense.

**Theorem 1.24** *For any $f \in C_b(\mathbb{R}^N)$, there exists a solution*

$$u \in C\left([0, \infty) \times \mathbb{R}^N\right) \cap C_{loc}^{1+\frac{\alpha}{2}, 2+\alpha}\left((0, \infty) \times \mathbb{R}^N\right)$$

*of (1.5.2) and*

$$|u(t, x)| \le e^{c_0 t}\,\|f\|_\infty, \quad t > 0, \ x \in \mathbb{R}^N.$$

The next result shows that the solution we get from Theorem 1.24 actually can be represented by means of operator semigroups.

**Theorem 1.25** *There exists a semigroup of linear operators $(T(t))_{t \ge 0}$ on $C_b(\mathbb{R}^N)$ such that for any $f \in C_b(\mathbb{R}^N)$ the solution of (1.5.2) given by Theorem 1.24 is represented by*

$$u(t, x) = (T(t)f)(x), \quad t \ge 0, \ x \in \mathbb{R}^N.$$

*For any $t > 0$, $T(t)$ satisfies the estimate*

$$\|T(t)f\|_\infty \le e^{c_0 t}\,\|f\|_\infty, \quad f \in C_b(\mathbb{R}^N).$$

*Moreover, there exists a family of Borel measures $p(t, x; dy)$ in $\mathbb{R}^N$ such that*

$$(T(t)f)(x) = \int_{\mathbb{R}^N} f(y)\, p(t, x; dy), \quad t \ge 0,\ x \in \mathbb{R}^N,$$

*and a function $G : (0, \infty) \times \mathbb{R}^N \times \mathbb{R}^N \to \mathbb{R}$ such that*

$$p(t, x; dy) = G(t, x, y)dy, \quad t > 0,\ x, y \in \mathbb{R}^N.$$

*The function $G$ is strictly positive and the functions $G(t, \cdot, \cdot)$ and $G(t, x, \cdot)$ are measurable for any $t > 0$ and any $x \in \mathbb{R}^N$. Further, for almost any fixed $y \in \mathbb{R}^N$, the function $G(\cdot, \cdot, y)$ belongs to $C_{loc}^{1+\frac{\alpha}{2}, 2+\alpha}\left((0, \infty) \times \mathbb{R}^N\right)$, and is a solution of the equation $\frac{\partial}{\partial t}u - \mathcal{A}u = 0$. Finally, if $c_0 \le 0$ then $p(t, x; dy)$ is a stochastically continuous transition function.*

Unfortunately, the semigroup $(T(t))_{t\ge 0}$ does not fit into the framework of $C_0$-semigroups as it is not strongly continuous on neither $C_b(\mathbb{R}^N)$ nor on $UC_b(\mathbb{R}^N)$.

**Proposition 1.26** *Let $f \in C_b(\mathbb{R}^N)$. Then, $\|T(t)f - f\|_\infty \to 0$ as $t \to 0$ if and only if $f \in UC_b(\mathbb{R}^N)$ and*

$$\lim_{t \to 0^+} (f(e^{tB}x) - f(x)) = 0,$$

*uniformly with respect to $x \in \mathbb{R}^N$.*

However, even if Proposition 1.26 shows that $(T(t))_{t\ge 0}$ is not strongly continuous on $C_b(\mathbb{R}^N)$ in general, combining Theorems 1.24 and 1.25 show that $T(t)f$ converges to $f$ as $t \to 0$, locally uniform on $\mathbb{R}^N$. Moreover, there is the following interesting convergence property.

**Proposition 1.27** *Let $(f_n)_{n\in\mathbb{N}}$ be a bounded sequence in $C_b(\mathbb{R}^N)$ such that $f_n \to f$ pointwise as $n \to \infty$ for some $f \in C_b(\mathbb{R}^N)$. Then, $T(\cdot)f_n$ tends to $T(\cdot)f$ locally uniformly in $(0, \infty) \times \mathbb{R}^N$. Further, if $f_n \to f$ uniformly on compact subsets of $\mathbb{R}^N$, then $T(t)f_n \to T(t)f$ locally uniformly in $[0, \infty) \times \mathbb{R}^N$ as $n \to \infty$.*

We will find back most of the properties mentioned here in Chap. 2, when we start the investigation of bi-continuous semigroups. Before this happens, let us emphasize that the operator $(\mathcal{A}, D(\mathcal{A}))$ as well as the associated semigroup $(T(t))_{t\ge 0}$ from Theorem 1.25 are actually motivated by Markov processes and stochastic differential equations.

**Theorem 1.28** *There exists a continuous Markov process $X$ associated with the semigroup $(T(t))_{t\ge 0}$. We have*

$$(T(t)f)(x) = \mathbb{E}_{x\chi t < \tau} f(X_t), \quad t > 0, \ x \in \mathbb{R}^N$$

*for any $f \in \mathrm{B_b}(\mathbb{R}^N)$.*

The connection between the associated Markov process from Theorem 1.28 and the stochastic differential equation associated with $(\mathcal{A}, D(\mathcal{A}))$ is the following. Let $(W_t)_{t\geq 0}$ be a $N$-dimensional Wiener process and let $\{\mathcal{F}_t^W : t \geq 0\}$ be the filtration generated by $(W_t)_{t\geq 0}$. For any $x \in \mathbb{R}^N$ let $\sigma(x) \in \mathscr{L}(\mathbb{R}^N)$ be the unique positive definite matrix such that $Q(x) = \frac{1}{2}\sigma(x)\sigma(x)^*$. We consider the stochastic differential equation

$$\begin{cases} \mathrm{d}\xi_t^x = b(\xi_t^x)\mathrm{d}t + \sigma(\xi_t^x)\mathrm{d}W_t, & t > 0, \\ \xi_0^x \equiv x, \end{cases} \tag{1.5.3}$$

where $x \in \mathbb{R}^N$ is fixed. Let us assume the following.

**Assumption 1.29** The functions $b$ and $\sigma$ are continuous and satisfy

$$\|\sigma(x) - \sigma(y)\|_2^2 + 2\langle b(x) - b(y), x - y\rangle \leq K_R |x - y|^2, \quad x, y \in B_R,$$

and

$$\mathcal{A}\left(1 + |x|^2\right) = \|\sigma(x)\|_2^2 + 2\langle b(x), y\rangle \leq K\left(1 + |x|^2\right), \quad x \in \mathbb{R}^N,$$

where $K, K_R > 0$ are constants and $\|\sigma\|_2^2 = \mathrm{Tr}(\sigma\sigma^*)$.

**Theorem 1.30** *Under Assumption 1.29 there exists a unique (up to equivalence) solution $\xi_x^t$ of (1.5.3) which is equivalent to the Markov process X.*

## Notes on This Chapter

For the classical results on $C_0$-semigroups in this chapter, i.e., Sects. 1.1–1.4 we mainly consulted the monograph by Engel and Nagel [101]. However, we also refer to [17, 33, 119, 197] as additional resources on the standard theory for $C_0$-semigroups. The results in Sect. 1.5 can for example be found in the monograph by Lorenzi and Bertoldi [169]. For further reading we also refer to [170].

# Bi-continuous Semigroups: Preliminaries and Examples

In this first chapter, we introduce the concept of bi-continuous semigroups which was first due to F. Kühnemund. This chapter will serve as an introduction to the basic definition of bi-continuous semigroups and the structure of the underlying space. In fact, we present a lot of examples in order to provide a good understanding of bi-continuous semigroups.

## 2.1   Bi-admissible Spaces

The following assumptions have been proposed by F. Kühnemund, see for example [160, Assump. 1] or [159, Assump. 1.1].

**Assumption 2.1**  Consider a triple $(X, \| \cdot \|, \tau)$ where $X$ is a Banach space, and

1. $\tau$ is a locally convex Hausdorff topology coarser than the norm-topology on $X$, i.e., the identity map $(X, \| \cdot \|) \to (X, \tau)$ is continuous;
2. $\tau$ is sequentially complete on the $\|\cdot\|$-closed unit ball, i.e., every $\| \cdot \|$-bounded $\tau$-Cauchy sequence is $\tau$-convergent;
3. The dual space of $(X, \tau)$ is norming for $X$, i.e.,

$$\|x\| = \sup_{\substack{\varphi \in (X,\tau)' \\ \|\varphi\| \leq 1}} |\varphi(x)|. \tag{2.1.1}$$

In what follows, $(X, \|\cdot\|, \tau)$ satisfying these assumptions will be called a *bi-admissible space*.


C. Budde, *Bi-Continuous Operator Semigroups*, Frontiers in Mathematics,
https://doi.org/10.1007/978-3-032-12948-2_2

**Remark 2.2** Recall that every locally convex topology gives rise to a family of seminorms and vice versa. In this regard (2.1.1) is equivalent to the following: There is a set $\mathcal{P}$ of $\tau$-continuous seminorms defining the topology $\tau$, such that

$$\|x\| = \sup_{p \in \mathcal{P}} p(x). \qquad (2.1.2)$$

Indeed, assume (2.1.1) holds and let $\mathcal{P}$ be the collection of *all* $\tau$-continuous seminorms $p$ such that $p(x) \leq \|x\|$. Then $|\varphi(\cdot)| \in \mathcal{P}$ for each $\varphi \in (X, \tau)'$ with $\|\varphi\| \leq 1$, and (2.1.2) is trivially satisfied. If $q$ is any $\tau$-continuous seminorm, then $q(x) \leq M\|x\|$ for some constant $M$ and for all $x \in X$. So that $q/M \in \mathcal{P}$, proving that $\mathcal{P}$ defines precisely the topology $\tau$. For the converse implication suppose that (2.1.2) holds. Then by the application of the Hahn–Banach theorem we obtain (2.1.1).

**Remark 2.3**  1.  There is the related notion of so-called Saks spaces, see [73]. By definition a *Saks space* is a triple $(X, \|\cdot\|, \tau)$ such that $X$ is a vector space with a norm $\|\cdot\|$ and locally convex topology $\tau$ coarser than the $\|\cdot\|$-topology, but the closed unit ball is $\tau$-complete. In this case, $X$ is a Banach space.
2.  It follows from (2.1.1) that $(X, Y)$ with $Y = (X, \tau)'$ is a norming dual pair, cf. Remark 4.27.
3.  Kraaij puts this setting in the more general framework of locally convex spaces with mixed topologies, see [150, Sec. 4], and also [106, App. A]. You can find more about the mixed topology in Appendix A.

## 2.1.1  The Space of Bounded Continuous Functions

Let $(\Omega, \kappa)$ be an arbitrary topological Hausdorff space and denote by $C_b(\Omega)$ the space of all bounded continuous functions $f : \Omega \to \mathbb{R}$. Equipped with the supremum-norm $\|\cdot\|_\infty$ defined by

$$\|f\|_\infty := \sup_{x \in \Omega} |f(x)|, \quad f \in C_b(\Omega),$$

the space $C_b(\Omega)$ becomes a Banach space. In addition, we consider on $C_b(\Omega)$ the locally convex topology generated by the family of seminorms $\mathcal{P}_{co} := \{p_K : K \subseteq \Omega \text{ compact}\}$ where $p_K(f) := \sup_{x \in K} |f(x)|$, $f \in C_b(\Omega)$. The topology generated by these seminorms is called the *compact-open topology* and will be denoted by $\tau_{co}$. Under certain assumptions on the topology $\kappa$ on $\Omega$ our space $C_b(\Omega)$ satisfies Assumption 2.1:

- If $(\Omega, \kappa)$ is completely regular, then $\tau_{co}$ is Hausdorff.
- If $(\Omega, \kappa)$ is compactly generated, then $\tau_{co}$ is sequentially complete on $\|\cdot\|_\infty$-bounded sets.

Notice that every locally compact or metrizable space has this property. Observe that the point measures are $\tau_{\text{co}}$-continuous and hence one conclude that $(C_b(\Omega), \tau_{\text{co}})$ is norming for $(C_b(\Omega), \|\cdot\|_\infty)$. The details here are left to the reader.

### 2.1.2 The Dual Space

Let $X$ be a Banach space and consider its norm dual $X'$, consisting of all linear functionals $\varphi : X \to \mathbb{C}$ which are continuous with respect to the Banach space norm. It is well-known that $X'$ is again a Banach space by equipping $X'$ with the dual norm $\|\cdot\|_{X'}$ defined by

$$\|\varphi\|_{X'} := \sup_{\substack{x \in X \\ \|x\| \leq 1}} |\langle x, \varphi \rangle|, \quad \varphi \in X',$$

where $\langle \cdot, \cdot \rangle : X \times X' \to \mathbb{C}$ is the (canonical) dual pairing between $X$ and $X'$ defined by $\langle x, \varphi \rangle := \varphi(x)$ for $x \in X$ and $\varphi \in X'$. The family of seminorms $\mathcal{P}_{w^*} := \{p_x : x \in X\}$ where $p_x(\varphi) := |\langle x, \varphi \rangle| = |\varphi(x)|$ yields a locally convex topology, the so-called *weak*-topology* and is denoted by $\tau_{w^*}$ or $\sigma(X', X)$. In fact, it is the coarsest topology making the evaluation maps $\varphi \mapsto \varphi(x)$ continuous. It is left as an exercise to verify that $\tau_{w^*}$ also satisfies Assumption 2.1.

### 2.1.3 The Space of Bounded Linear Operators

Let $(E, \|\cdot\|_E)$ be a Banach space and denote by $\mathscr{L}(E)$ the space of bounded linear operators on $E$. The operator norm $\|\cdot\|_{\mathscr{L}(E)}$ defined by

$$\|T\|_{\mathscr{L}(E)} := \sup_{\|x\|_E \leq 1} \|Tx\|_E, \quad T \in \mathscr{L}(E),$$

makes $\mathscr{L}(E)$ a Banach space as well. The locally convex topology on $\mathscr{L}(E)$ we consider now is generated by the family of seminorms $\mathcal{P}_{\text{sot}} := \{p_x : x \in E\}$ where $p_x(T) := \|Tx\|_E$ and is called the *strong operator topology* which will be denoted by $\tau_{\text{sot}}$. We will show here that $\tau_{\text{sot}}$ satisfies Assumption 2.1. First of all, we observe that $\tau_{\text{sot}}$ is Hausdorff and coarser than the operator norm topology on $\mathscr{L}(E)$. In order to show that $\tau_{\text{sot}}$ is sequentially complete on $\|\cdot\|_{\mathscr{L}(E)}$-bounded set consider $(T_n)_{n \in \mathbb{N}}$ to be a $\|\cdot\|_{\mathscr{L}(E)}$-bounded $\tau_{\text{sot}}$-Cauchy sequence. We conclude that $(T_n x)_{n \in \mathbb{N}}$ is a Cauchy sequence in $E$ and hence convergent. Hence, we define $T : E \to E$ by

$$Tx := \lim_{n \to \infty} T_n x, \quad x \in E.$$

By linearity of the operators $(T_n)_{n \in \mathbb{N}}$ we achieve that $T$ is linear as well. In view of the fact that $(T_n)_{n \in \mathbb{N}}$ was supposed to be bounded with respect to the operator norm, we obtain

$$\|Tx\|_E \le \sup_{n \in \mathbb{N}} \|T_n x\|_E < \infty,$$

and hence $T \in \mathscr{L}(E)$. By construction we attain that $T_n \to T$ with respect to $\tau_{\mathrm{sot}}$. For the norming property we fix sequences $(x_n)_{n \in \mathbb{N}}$ in $E$ and $(\varphi_n)_{n \in \mathbb{N}}$ in $E'$ with $\|x_n\|_E \le 1$ and $\|\varphi_n\|_{E'} \le 1$ for each $n \in \mathbb{N}$ and such that

$$\|T\|_{\mathscr{L}(E)} = \sup_{n \in \mathbb{N}} \|Tx_n\|_E \quad \text{and} \quad \|Tx_n\|_E = |\varphi_n(Tx_n)|.$$

Now define $\Lambda_n \in (\mathscr{L}(E), \tau_{\mathrm{sot}})'$

$$\Lambda_n(T) := \varphi_n(Tx_n)$$

and observe that $\|\Lambda_n\| \le 1$ for each $n \in \mathbb{N}$. Moreover

$$\|T\|_{\mathscr{L}(E)} = \sup_{n \in \mathbb{N}} \|Tx_n\|_E = \sup_{n \in \mathbb{N}} |\varphi_n(Tx_n)| = \sup_{n \in \mathbb{N}} |\Lambda_n(T)| \le \sup_{\substack{\varphi \in (\mathscr{L}(E), \tau_{\mathrm{sot}})' \\ \|\varphi\| \le 1}} |\varphi(T)| \le \|T\|_{\mathscr{L}(E)}$$

which gives us the norming property. We remark that the dual of $(\mathscr{L}(E), \tau_{\mathrm{sot}})$ can be identified with the set $\mathcal{F}(E)$ of finite rank operators on $E$.

## 2.2 Bi-continuous Semigroups

Now we formulate the definition of a bi-continuous semigroup which is originally due to F. Kühnemund, cf. [160, Assumption 1] or [159, Assumption 1.1].

**Definition 2.4** Let $(X, \|\cdot\|, \tau)$ be a bi-admissible space. We call a family of bounded linear operators $(T(t))_{t \ge 0}$ a *bi-continuous semigroup* if it has the following properties.

1. $T(t + s) = T(t)T(s)$ and $T(0) = I$ for all $s, t \ge 0$.
2. $(T(t))_{t \ge 0}$ is strongly $\tau$-continuous, i.e., the map $\varphi_x : [0, \infty) \to (X, \tau)$ defined by $\varphi_x(t) = T(t)x$ is continuous for every $x \in X$.
3. $(T(t))_{t \ge 0}$ is exponentially bounded, i.e., there exist $M \ge 1$ and $\omega \in \mathbb{R}$ such that $\|T(t)\| \le Me^{\omega t}$ for each $t \ge 0$. In this case, we say that the semigroup $(T(t))_{t \ge 0}$ is of type $\omega$.
4. $(T(t))_{t \ge 0}$ is locally-bi-equicontinuous, i.e., if $(x_n)_{n \in \mathbb{N}}$ is a norm-bounded sequence in $X$ which is $\tau$-convergent to 0, then $(T(s)x_n)_{n \in \mathbb{N}}$ is $\tau$-convergent to 0 uniformly for $s \in [0, t_0]$ for each fixed $t_0 \ge 0$.

The following definition will occur frequently in the forthcoming chapters and originates from the theory of strongly continuous semigroups, cf. [101, Chapter I, Definition 5.6].

**Definition 2.5**  For an exponentially bounded semigroup of bounded linear operators $(T(t))_{t \geq 0}$ we define the *growth bound* to be

$$\omega_0(T) := \inf\left\{\omega \in \mathbb{R} : \exists M \geq 1 \; \forall t \geq 0 : \|T(t)\| \leq M e^{\omega t}\right\}.$$

### 2.2.1  The Translation Semigroup

We already saw that $(C_b(\mathbb{R}), \|\cdot\|_\infty, \tau_{co})$ is a bi-admissible space. Now consider the *left-translation semigroup* $(T(t))_{t \geq 0}$ on $C_b(\mathbb{R})$ defined by

$$(T(t)f)(x) := f(x + t), \quad t \geq 0, \; f \in C_b(\mathbb{R}), \; x \in \mathbb{R}.$$

This semigroup actually fails to be strongly continuous with respect to the $\|\cdot\|_\infty$-norm. In fact, take the function $f \in C_b(\mathbb{R})$ defined by $f(x) = \sin(x^2)$. Although $(T(t))_{t \geq 0}$ is not a $C_0$-semigroup, we can show that it is bi-continuous with respect to the compact-open topology. Since $\|T(t)f\|_\infty = \|f\|_\infty$ for each $f \in C_b(\mathbb{R})$, $(T(t))_{t \geq 0}$ becomes a contraction semigroup and therefore it is exponentially bounded. Observe that for a compact subset $K \subseteq \mathbb{R}$ holds that

$$p_K(T(t)f - f) = \sup_{x \in K} |f(x + t) - f(x)|, \quad t \geq 0, \; f \in C_b(\mathbb{R}).$$

Since $f$ is continuous and $K$ is compact, $f_{|K}$ is uniformly continuous, hence for each $\varepsilon > 0$ there exists $\delta > 0$ such that

$$\sup_{x \in K} |f(x + t) - f(x)| < \varepsilon$$

whenever $|t| < \delta$ and we obtain the strong continuity of $(T(t))_{t \geq 0}$ with respect to $\tau_{co}$. For the local bi-equicontinuity let $(f_n)_{n \in \mathbb{N}}$ be a $\|\cdot\|_\infty$-bounded $\tau_{co}$-null sequence. For $t_0 > 0$ and $t \in [0, t_0]$ we obtain

$$p_K(T(t)f_n) = \sup_{x \in K} |f_n(x + t)| \leq \sup_{x \in H} |f_n(x)| = p_H(f_n),$$

where $H := \bigcup_{t \in [0, t_0]} K + t$ is compact. This shows that $T(t)f_n \overset{\tau_{co}}{\to} 0$ uniformly for $t \in [0, t_0]$. We conclude that the left-translation semigroup is indeed a bi-continuous semigroup on $C_b(\mathbb{R})$.

One could think that one can stick to another topology on $C_b(\mathbb{R})$ in order to treat the translation semigroup, namely the topology of pointwise convergence. This topology, which will be denoted by $\tau_p$, is induced by the family of seminorms $\mathcal{P}_p = \{p_x : x \in \mathbb{R}\}$ where $p_x(f) = |f(x)|$, $f \in C_b(\mathbb{R})$. By using the same arguments as in Sect. 2.2.1 one can show that $\tau_p$ satisfies Assumption 2.1. Nevertheless, the translation semigroup does not become bi-continuous with respect to this topology since it is not locally bi-equicontinuous, i.e., define the sequence $(f_n)_{n \in \mathbb{N}}$ in $C_b(\mathbb{R})$ by

$$f_n(x) := \begin{cases} \max\left\{1 - n^2\left|x - \tfrac{1}{n}\right|, 0\right\}, & \text{if } x \in [0, 1], \\ 0, & \text{otherwise.} \end{cases}$$

By construction $(f_n)_{n \in \mathbb{N}}$ is a $\|\cdot\|_\infty$-bounded sequence with pointwise limit 0. Observe that $(T(\tfrac{1}{n}) f_n)(0) = 1$ for all $n \geq 2$ and hence $(T(t))_{t \geq 0}$ fails to be locally bi-equicontinuous.

### 2.2.2  The Diffusion Semigroup

A more involved operator semigroup on $C_b(\mathbb{R})$ is the so-called diffusion semigroup or Gauss–Weierstrass semigroup. This semigroup $(T(t))_{t \geq 0}$ is defined by

$$(T(t)f)(x) := \int_{\mathbb{R}} f(y) \, d\gamma_{x,t}(y) = \frac{1}{\sqrt{4\pi t}} \int_{\mathbb{R}} e^{-\frac{(y-x)^2}{4t}} f(y) \, dy, \quad t \geq 0, \ f \in C_b(\mathbb{R}), \ x \in \mathbb{R},$$

where $\gamma_{x,t}$ denotes the Gaussian measure with mean $x$ and variance $t$ defined via the probability density $g_{x,t}$ on $\mathbb{R}$ defined by

$$g_{x,t}(y) = \frac{1}{\sqrt{4\pi t}} e^{-\frac{(y-x)^2}{4t}}, \quad y \in \mathbb{R}.$$

As a matter of fact, $(T(t))_{t \geq 0}$ is a bi-continuous semigroup on $C_b(\mathbb{R})$. It is easy to verify that $(T(t))_{t \geq 0}$ satisfies the semigroup law. Moreover, we observe that $(T(t))_{t \geq 0}$ is a contraction semigroup, and hence it is exponentially bounded. In order to show that the diffusion semigroup is $\tau_{co}$-continuous fix $f \in C_b(\mathbb{R})$. Since continuous functions on compact sets are known to be automatically uniformly continuous, one can find for arbitrary $\varepsilon > 0$ and a compact set $K \subseteq \mathbb{R}$ a $\delta > 0$ such that from $|y| < \delta$ it follows that $|f(x + y) - f(x)| < \varepsilon$ for all $x \in K$. Therefore, by using the Chebyshev inequality we obtain with $B_\delta := \{y \in \mathbb{R} : |y| < \delta\}$ that

$$\sup_{x \in K} |(T(t)f)(x) - f(x)| \leq \frac{1}{\sqrt{4\pi t}} \sup_{x \in K} \int_{B_\delta} |f(x + y) - f(x)| e^{-\frac{|y|^2}{4t}} \, dy + \frac{2\|f\|_\infty}{\sqrt{4\pi t}} \int_{\mathbb{R} \setminus B_\delta} e^{-\frac{|y|^2}{4t}} \, dy$$

$$< \varepsilon + \frac{2t\|f\|_\infty}{\delta^2},$$

yielding the strong $\tau_{co}$-continuity of $(T(t))_{t \geq 0}$ at 0. That the diffusion semigroup is also locally bi-equicontinuous can be proven as follows: Let $K \subseteq \mathbb{R}$ be compact, $t_0 \geq 0$ and $\varepsilon > 0$. By the regularity properties of the Gaussian measure, there exists a compact subset $L \subseteq \mathbb{R}$ such that $\gamma_{x,t}(L) \geq 1 - \varepsilon$ uniformly for $x \in K$ and $t \in [0, t_0]$. Now, let $(f_n)_{n \in \mathbb{N}}$ be a sequence in $C_b(\mathbb{R})$ which is $\|\cdot\|_\infty$-bounded and $\tau_{co}$-convergent to 0. By definition, there exists $N \in \mathbb{N}$ such that $\sup_{x \in L} |f_n(x)| < \varepsilon$ for all $n \geq N$. Thus

$$\sup_{x \in K} |(T(t)f_n)(x)| \leq \sup_{x \in K} \int_L |f_n(y)| \, d\gamma_{x,t}(y) + \sup_{x \in K} \int_{\mathbb{R} \setminus L} |f_n(y)| \, d\gamma_{x,t}(y) < \varepsilon \left(1 + \|f_n\|_\infty\right),$$

uniformly for $t \in [0, t_0]$. It follows that by Definition 2.4 the semigroup $(T(t))_{t \geq 0}$ is bi-continuous with respect to the compact-open topology.

### 2.2.3  The Adjoint Semigroup

Let $(T(t))_{t\geq 0}$ be a $C_0$-semigroup on a Banach space $X$. Recall that the adjoint of a bounded linear operator $T \in \mathcal{L}(X)$ is defined to be the unique bounded linear operator $T' \in \mathcal{L}(X')$ such that

$$\langle Tx, \varphi \rangle = \langle x, T'\varphi \rangle,$$

for all $x \in X$ and $\varphi \in X'$. The *adjoint semigroup* $(T'(t))_{t\geq 0}$ on the dual Banach space $X'$ consist of all adjoint operators $T(t)'$ on $X'$. We already saw that $(X', \|\cdot\|_{X'}, \tau_{w*})$ is a bi-admissible space. Let us show that the adjoint semigroup is indeed a bi-continuous semigroup on $X'$ with respect to $\tau_{w*}$. Of course, $(T'(t))_{t\geq 0}$ is exponentially bounded as $(T(t))_{t\geq 0}$ is exponentially bounded, i.e., there exists $M \geq 0$ and $\omega \in \mathbb{R}$ such that $\|T(t)\| \leq Me^{\omega t}$ for all $t \geq 0$. Let $\varphi \in X'$ be arbitrary and observe that

$$\|T'(t)\varphi\| = \sup_{\substack{x\in X \\ \|x\|\leq 1}} |\langle x, T'(t)\varphi \rangle| = \sup_{\substack{x\in X \\ \|x\|\leq 1}} |\langle T(t)x, \varphi \rangle| \leq \sup_{\substack{x\in X \\ \|x\|\leq 1}} \|T(t)\|\, \|x\|\, \|\varphi\| \leq Me^{\omega t}\, \|\varphi\|.$$

Hence $(T'(t))_{t\geq 0}$ is exponentially bounded. By the strong continuity of the original $C_0$-semigroup $(T(t))_{t\geq 0}$ we obtain

$$|\langle x, T'(t)\varphi - \varphi \rangle| = |\langle T(t)x - x, \varphi \rangle| \leq \|T(t)x - x\|\, \|\varphi\|$$

for all $x \in X$, $\varphi \in X'$ and $t \geq 0$. Hence $(T'(t))_{t\geq 0}$ is strongly continuous with respect to the weak*-topology $\tau_{w*}$. Finally, let $t_0 > 0$ and $(\varphi_n)_{n\in\mathbb{N}}$ a $\|\cdot\|$-bounded sequence in $X'$ such that $\varphi_n \to 0$ with respect to the weak*-topology $\tau_{w*}$. Then

$$|\langle x, T'(t)\varphi_n \rangle| = |\langle T(t)x, \varphi_n \rangle|$$

for all $x \in X$ which converges to 0 as $n \to \infty$ uniformly on $[0, t_0]$ by the compactness of the set $\{T(t)x : t \in [0, t_0]\}$, showing that $(T'(t))_{t\geq 0}$ is also locally bi-equicontinuous. Therefore, we conclude that the adjoint semigroup $(T'(t))_{t\geq 0}$ is a bi-continuous semigroup with respect to the weak*-topology $\tau_{w*}$.

We will now show that a certain class of bi-continuous semigroups on a dual Banach space are actually automatically implemented by $C_0$-semigroups. To see this, we first need the following result.

**Lemma 2.6** *Let $X$ be a Banach space and $T \in \mathcal{L}(X')$ and assume that $T$ is weak*-continuous on $\|\cdot\|$-bounded sets. Then there exists $S \in \mathcal{L}(X)$ such that $S' = T$.*

**Proof** It suffices to show that $T' \in \mathcal{L}(X'')$ leaves $X$ invariant. To show this we show that $T$ has a weak*-adjoint $T'^{w^*} : X \to X$ which, in this case, coincides with $T'_{|X}$. For $x \in X$ we define $\psi_x : X' \to \mathbb{R}$ by $\psi_x(\varphi) := \langle T\varphi, x \rangle$, $\varphi \in X'$. Then the restriction $\psi_{x|B}$ to the unit ball $B$ in $X'$ is weak*-continuous so $\mathrm{Ker}(\psi_x) \cap B$ is weak*-closed. From the Krein–Šmulian Theorem, cf. [208, Chap. IV, Sect. 6], it follows that $\mathrm{Ker}(\psi_x)$ is weak*-closed, hence $\psi_x$ is weak*-continuous. Therefore, $T$ has a weak*-adjoint $T'^{w^*}$ given by $T'^{w^*}x = \psi_x$. $\qquad\square$

**Theorem 2.7** *Let $(T(t))_{t\geq 0}$ be a bi-continuous semigroup on $X'$ with respect to the weak*-topology such that $T(t)$, $t \geq 0$, is weak*-continuous on $\|\cdot\|$-bounded sets. Then there exists a $C_0$-semigroup $(S(t))_{t\geq 0}$ on $X$ such that $S(t)' = T(t)$ for all $t \geq 0$.*

**Proof** We observe that if $(T(t))_{t\geq 0}$ is a bi-continuous semigroup on $X'$ with respect to the weak*-topology such that $X$ is invariant under the operators $T'(t)$, $t \geq 0$, then $(T(t))_{t\geq 0}$ is the dual of a $C_0$-semigroup on $X$ since $(T'(t)_{|X})_{t\geq 0}$ is weakly and hence strongly continuous. The assertions follow now as an immediate consequence of Lemma 2.6. $\qquad\square$

**Remark 2.8** The assumptions of Theorem 2.7 are satisfied, for example, when $(T(t))_{t\geq 0}$ is a bi-continuous semigroup on $X'$ with respect to the weak*-topology and $X$ is separable. This is due to the fact that in this case the unit ball of $X'$ is metrizable for the weak*-topology. A more general formulation can be found [154, Prop. 4.8].

## 2.2.4 The Implemented Semigroup

Let $E$ be some Banach space and $(T(t))_{t\geq 0}$, $(S(t))_{t\geq 0}$ two $C_0$-semigroups on $E$. We already showed that $(\mathscr{L}(E), \|\cdot\|_{\mathscr{L}(E)}, \tau_{\mathrm{sot}})$ is a bi-admissible space. The semigroup $(\mathcal{U}(t))_{t\geq 0}$ on $\mathscr{L}(E)$ defined by

$$\mathcal{U}(t)L := T(t)LS(t), \quad L \in \mathscr{L}(E), \ t \geq 0,$$

is called the *implemented semigroup*. If $T(t) \equiv I$ or $S(t) \equiv I$, then the corresponding semigroups $(\mathcal{U}_R)(t)_{t\geq 0}$ and $(\mathcal{U}_L)(t)_{t\geq 0}$ are called *right implemented* and *left implemented semigroup*, respectively. These semigroups play an important role in quantum mechanics. To show that $(\mathcal{U}(t))_{t\geq 0}$ is not strongly continuous in general, let $(T(t))_{t\geq 0}$ and $(S(t))_{t\geq 0}$ be two strongly continuous semigroups on the Banach space $E = \mathrm{L}^1(\mathbb{R}, \mathbb{C})$ defined by

$$(T(t)f)(x) := f(x+t), \quad (S(t)f)(x) = \overline{(T(t)f)(s)}, \quad t \geq 0, \ f \in \mathrm{L}^1(\mathbb{R}, \mathbb{C}), \ x \in \mathbb{R}.$$

By considering the operator $L \in \mathscr{L}(E)$, defined by

$$Lf := \mathbf{1}_{[0,1]}f, \quad f \in \mathrm{L}^1(\mathbb{R}, \mathbb{C}),$$

one recognizes that the implemented semigroup on the space $\mathscr{L}(E)$ defined by $\mathcal{U}(t)L := T(t)LS(t), t \geq 0, L \in \mathscr{L}(E)$, is not strongly continuous with respect to the operator norm, i.e., let $(f_n)_{n\in\mathbb{N}}$ be a sequence in $\mathrm{L}^1(\mathbb{R}, \mathbb{C})$ such that $\|f_n\| = 1$ and $\mathrm{supp}(f_n) \subseteq \left[0, \frac{1}{n}\right]$ for all $n \in \mathbb{N}$ and observe that

$$\left\| \mathcal{U}\left(\frac{1}{n}\right) L f_n - L f_n \right\|_1 = \left\| \mathbf{1}_{\left[\frac{1}{n}, \frac{1}{n}+1\right]} f_n - \mathbf{1}_{[0,1]} f_n \right\|_1 = \| f_n \|_1 = 1.$$

Consequently, $(\mathcal{U}(t))_{t \geq 0}$ is not strongly continuous with respect to the norm on $L^1(\mathbb{R}, \mathbb{C})$.

Let us show that $(\mathcal{U}(t))_{t \geq 0}$ is bi-continuous on $\mathscr{L}(E)$ with respect to the strong operator topology $\tau_{\mathrm{sot}}$. Since $(T(t))_{t \geq 0}$ and $(S(t))_{t \geq 0}$ satisfy the semigroup law, also $(\mathcal{U}(t))_{t \geq 0}$ does. Due to the sub-multiplicativity of the operator norm, we conclude that $(\mathcal{U}(t))_{t \geq 0}$ is also exponentially bounded. For the strong continuity with respect to the strong operator topology $\tau_{\mathrm{sot}}$ we just observe that for $L \in \mathscr{L}(E)$ and $x \in E$ one has that

$$\| \mathcal{U}(t) L x - L x \| = \| T(t) L S(t) x - L x \| \leq \| T(t) L (S(t)x - x) \| + \| T(t) L x - L x \|.$$

For the local bi-equicontinuity let $(L_n)_{n \in \mathbb{N}}$ be an operator norm bounded strongly convergent null sequence. For $t_0 > 0$ and $x \in E$ holds

$$\sup_{t \in [0,t_0]} \| T(t) L_n S(t) x \| \leq M \cdot \sup_{t \in [0,t_0]} \| L_n S(t) x \|,$$

where $M := \sup_{t \in [0,t_0]} \| T(t) \|$ which is finite by [101, Chap. I, Prop. 5.5]. Since $(L_n)_{n \in \mathbb{N}}$ converges uniformly on compact sets and both semigroups are strongly continuous on $E$ we conclude that the right-hand side tends to zero, which proves that $(\mathcal{U}(t))_{t \geq 0}$ is locally bi-equicontinuous. Therefore, we conclude that $(\mathcal{U}(t))_{t \geq 0}$ is a bi-continuous semigroup with respect to the strong operator topology $\tau_{\mathrm{sot}}$.

We remark that $(\mathcal{U}(t))_{t \geq 0}$ is a $C_0$-semigroup if and only if both semigroups $(T(t))_{t \geq 0}$ and $(S(t))_{t \geq 0}$ are uniformly continuous meaning that

$$\lim_{t \to 0} \| T(t) - I \| = \lim_{t \to 0} \| S(t) - I \| = 0.$$

### 2.2.5  The Ornstein–Uhlenbeck Semigroup

We consider the Ornstein–Uhlenbeck semigroups on an infinite-dimensional, separable Hilbert space $\mathcal{H}$. These semigroups arise from stochastic differential equations and Ornstein–Uhlenbeck processes, and are in strong connection with Kolmogorov equations. Let us start with the stochastic equation given by

$$\begin{cases} dX(t) = AX(t)dt + Q^{\frac{1}{2}} dW(t), \\ X(0) = x \in \mathcal{H}, \end{cases} \tag{2.2.1}$$

where $(A, D(A))$ is the generator of a $C_0$-semigroup $(S(t))_{t \geq 0}$ on a separable Hilbert space $\mathcal{H}$. Moreover, $W$ is a $\mathcal{H}$-valued $Q$-Wiener process for a self-adjoint, positive and bounded linear operator $Q$ on $\mathcal{H}$. The corresponding transition semigroup or Ornstein–Uhlenbeck semigroup $(P(t))_{t\geq 0}$ on the space $C_b(\mathcal{H})$ is defined by

$$(P(t)f)(x) = \mathbb{E}(f(X(t, x))), \quad t \geq 0, \ f \in C_b(\mathcal{H}), \ x \in \mathcal{H}. \tag{2.2.2}$$

The goal is to show that the Ornstein–Uhlenbeck semigroup on $C_b(\mathcal{H})$ is bi-continuous with respect to $\tau_{co}$. To do so, we define a family $(Q(t))_{t\geq 0}$ of covariance operators on $\mathcal{H}$ by

$$Q(t)x := \int_0^t S(r)QS(r)'x \, dr, \quad t \geq 0, \ x \in \mathcal{H}. \tag{2.2.3}$$

Here $(S(t)')_{t\geq 0}$ denotes the adjoint semigroup of $(S(t))_{t\geq 0}$. In what follows, we have to make an additional assumption on the operator $(Q(t))_{t\geq 0}$.

**Assumption 2.9** Each operator $Q(t)$, $t \geq 0$, is a trace class operator. In particular, let $(e_n)_{n\in\mathbb{N}}$ be any orthonormal basis for $\mathcal{H}$, then $\mathrm{Tr}(Q(t)) = \sum_{n\in\mathbb{N}} \langle e_n, Q(t)e_n \rangle < \infty$ for all $t \geq 0$.

By $\mu_{t,x} := \mathcal{N}(S(t)x, Q(t))$ we denote the Gaussian measure with mean $S(t)x$ and covariance $Q(t)$, which actually exists due to Assumption 2.9, cf. [78, Chap. I, Sect. 2.3.2]. The Ornstein–Uhlenbeck semigroup $(P(t))_{t\geq 0}$ as defined in (2.2.2) can now be expressed as follows

$$(P(t)f)(x) = \mathbb{E}(f(X(t, x))) = \int_{\mathcal{H}} f(\sigma) \, d\mu_{t,x}(\sigma), \quad t \geq 0, \ f \in C_b(\mathcal{H}), \ x \in \mathcal{H}. \tag{2.2.4}$$

The following result shows that the Ornstein–Uhlenbeck semigroup actually is a (non-trivial) example of a bi-continuous semigroup.

**Proposition 2.10** *Under Assumption 2.9 the Ornstein–Uhlenbeck semigroup $(P(t))_{t\geq 0}$ is a bi-continuous contraction semigroup on $C_b(\mathcal{H})$ with respect to $\tau_{co}$.*

***Proof*** We first notice that $(C_b(\mathcal{H}), \|\cdot\|_\infty, \tau_{co})$ is a bi-admissible space. Moreover, it is obvious that $(P(t))_{t\geq 0}$ consists of bounded linear operators such that $\|P(t)\| \leq 1$ for all $t \geq 0$.

Let us now show that $(P(t))_{t\geq 0}$ as defined by (2.2.4) is strongly-$\tau_{co}$-continuous on $C_b(\mathcal{H})$. To do so, let $f \in C_b(\mathcal{H})$, $K \subseteq \mathcal{H}$ compact and $\varepsilon > 0$. Since by assumption $(S(t))_{t\geq 0}$ is a $C_0$-semigroup on $\mathcal{H}$ and every continuous function on $K$ is automatically uniformly continuous, we find $\delta > 0$ and $t_0 \geq 0$ such that for all $\|y\| < \delta$ and $t \in [0, t_0]$ one has that

$$\sup_{x\in K} |f(y + S(t)x) - f(y)| < \frac{\varepsilon}{2}.$$

Using this, there exists $\tilde{t} \leq t_0$ such that for all $t \leq \tilde{t}$:

$$\sup_{x \in K} |(P(t)f)(x) - f(x)| \leq \sup_{x \in K} \int_{\|y\| < \delta} |f(\sigma + S(t)x) - f(\sigma)| \, \mathrm{d}\mu_{0,t}(\sigma) + 2\|f\|_\infty \int_{\|y\| \geq \delta} \mathrm{d}\mu_{0,t}(\sigma)$$

$$\leq \frac{\varepsilon}{2} + \frac{2\|f\|_\infty}{\delta^2} \mathrm{Tr}(Q(t)) < \varepsilon.$$

For the second last estimate we make use of the Chebyshev inequality. The last step follows from Assumption 2.9 since this implies that $\mathrm{Tr}(Q(t)) \to 0$ as $t \to 0$.

Finally, we have to show that $(P(t))_{t \geq 0}$ is also locally bi-equicontinuous. Let $K \subseteq \mathcal{H}$ be compact, $t_0 \geq 0$ and $\varepsilon > 0$. Furthermore, let $(f_n)_{n \in \mathbb{N}}$ be a $\|\cdot\|_\infty$-bounded $\tau_{co}$-null-sequence in $C_b(\mathcal{H})$. It is left as an exercise to show that the family of measures $\{\mu_{x,t} : t \in [0, t_0], x \in K\}$ is tight, i.e., there exists a compact set $K' \in \mathcal{H}$ such that $\mu_{x,t}(\mathcal{H} \setminus K') < \infty$ for all $t \in [0, t_0]$ and $x \in \mathcal{K}$. Therefore, there exists $N \in \mathbb{N}$ such that for all $n \geq N$ and $t \in [0, t_0]$ one has that

$$\sup_{x \in K} |(P(t)f_n)(x)| \leq \sup_{x \in K} \int_{K'} |f_n(\sigma)| \, \mathrm{d}\mu_{x,t}(\sigma) + \sup_{x \in K} \int_{X \setminus K'} |f_n(\sigma)| \, \mathrm{d}\mu_{x,t}(\sigma) < \varepsilon + \varepsilon \|f_n\|_\infty.$$

This concludes the proof and shows that the Ornstein–Uhlenbeck semigroup $(P(t))_{t \geq 0}$ is indeed bi-continuous with respect to the compact-open topology $\tau_{co}$. $\quad\square$

### 2.2.6 Semigroups on $C_b(\Omega)$

In this section, we want to restrict ourself to the space $X = C_b(\Omega)$, where $\Omega$ is a Polish space. From Sect. 2.1.1 we know that $(C_b(\Omega), \|\cdot\|_\infty, \tau_{co})$ is a bi-admissible space. The aim of this section is to investigate (arbitrary) bi-continuous semigroups on $C_b(\Omega)$.

We notice that as a Banach space the dual of $C_b(\Omega)$ is isomorphic to the space $M(\beta\Omega)$ consisting of all bounded complex Borel measures on the Stone–Cech compactification $\beta\Omega$ of $\Omega$. By $\mathcal{J}$ we denote the set of $\|\cdot\|_\infty$-continuous linear functionals on $C_b(\Omega)$ which are also $\tau_{co}$-continuous on $\|\cdot\|_\infty$-bounded sets.

**Lemma 2.11** *$\mathcal{J}$ is a $\|\cdot\|_\infty$-closed subspace of $M(\beta\Omega)$.*

***Proof*** Obviously, $\mathcal{J}$ is a linear subspace. To show that $\mathcal{J}$ is indeed $\|\cdot\|_\infty$-closed, let $(\varphi_n)_{n \in \mathbb{N}}$ be a sequence in $\mathcal{J}$ such that $\varphi \to \varphi$ in the norm dual of $C_b(\Omega)$. Moreover, let $(f_\alpha)_{\alpha \in A}$ be a $\|\cdot\|_\infty$-bounded net in $C_b(\Omega)$ which $\tau_{co}$-converges to some $f \in C_b(\Omega)$. It suffices to show that $\varphi(f_\alpha) \to \varphi(f)$. To see this, observe that

$$|\varphi(f_\alpha - f)| \leq |\varphi(f - f_\alpha) - \varphi_n(f - f_\alpha)| + |\varphi_n(f - f_\alpha)|$$

$$\leq K \|\varphi - \varphi_n\| + |\varphi_n(f - f_\alpha)|.$$

The first term tends to 0 for $n \to \infty$ by the assumption $\varphi_n \to \varphi$ with respect to the dual norm. In particular, for given $\varepsilon > 0$ there exists $N \in \mathbb{N}$ such that $\|\varphi - \varphi_n\| < \frac{\varepsilon}{2K}$ for $n \geq N$. By fixing this $N \in \mathbb{N}$ and the continuity assumptions on $\varphi_n, n \in \mathbb{N}$, we obtain $|\varphi_n(f - f_\alpha)| < \frac{\varepsilon}{2}$ as $f_\alpha \to f$. $\hspace{2cm}$ $\square$

The next result helps us to identify the space $\mathcal{J}$ explicitly.

**Proposition 2.12** *The space $\mathcal{J}$ can be identified with* $M(\Omega)$ *inside* $M(\beta\Omega)$.

**Proof** By assumption, $\Omega$ is a Polish space so in particular a topologically complete space. Hence, $\Omega$ is a $G_\delta$-set in $\beta\Omega$, see for example [236, Chap. 1]. Therefore, we are able to identify $M(\Omega)$ with a subspace of $M(\beta\Omega)$ by the map $\iota : M(\Omega) \to M(\beta\Omega)$ defined by

$$(\iota(\nu))(B) := \nu(\Omega \cap B), \quad \nu \in M(\Omega), \ B \subseteq \beta\Omega \text{ Borel set.}$$

This map $\iota$ yields an injection with range

$$\mathrm{Ran}(\iota) = \{\mu \in M(\beta\Omega) : \ \mu(\beta\Omega \setminus \Omega) = 0\}$$

and $\iota(M(\Omega)) \subseteq \mathcal{J}$. For the converse inclusion, let $\mu \in \mathcal{J}$ be arbitrary and suppose $\mu$ is positive. For the sake of a contradiction, assume that $\mu \notin \iota(M(\Omega))$. Hence, we have that $\mu(\beta\Omega \setminus \Omega) > 0$. As observed beforehand, $\Omega$ is a $G_\delta$-set in $\beta\Omega$ so that $\beta\Omega \setminus \Omega$ is a $F_\sigma$-set. Therefore, there exists a closed set $F \subseteq \beta\Omega$ with $F \cap \Omega = \varnothing$ and $\mu(F) > 0$. Furthermore, let $K \subseteq \Omega$ be compact. Then, $K$ is closed in $\beta\Omega$. Therefore, there exists a continuous function $f_{K,F} : \beta\Omega \to [0, 1]$ such that $f_{K,F}(x) = 0$ for $x \in K$ and $f_{K,F}(x) = 1$ for $x \in F$. This yields that

$$\langle f_{F,K}, \mu \rangle \geq \mu(F) > 0,$$

contradicting that $K$ was chosen to be arbitary and the fact that $\mu$ is $\tau_{\mathrm{co}}$-continuous. $\hspace{1cm}$ $\square$

**Lemma 2.13** *Let* $T : [0, \infty) \to \mathscr{L}(C_b(\Omega))$ *be a* $\tau_{\mathrm{co}}$-*strongly continuous function an let* $\mathcal{K} \subseteq M(\Omega)$ *be a weak**-*compact set. Then the map*

$$[0, \infty) \times \mathcal{K} \ni (t, \nu) \mapsto T'(t)\nu \in M(\beta\Omega)$$

*is continuous if we equip both $\mathcal{K}$ and $M(\beta\Omega)$ with the weak**-*topology* $\sigma(M(\beta\Omega), C_b(\Omega))$.

**Proof** Fix $t \in [0, \infty)$ and $\nu \in M(\Omega)$ and observe that for $f \in C_b(\Omega)$ one obtains

$$\left| \langle f, T'(t)\nu - T'(s)\mu \rangle \right| = |\langle T(t)f, \nu \rangle - \langle T(s)f, \mu \rangle|$$

$$\leq |\langle T(t)f, \nu - \mu \rangle| + |\langle T(t)f - T(s)f, \mu \rangle| < \frac{\varepsilon}{2} + \frac{\varepsilon}{2},$$

where the last inequality follows by the bi-continuity assumptions on $T$ and by an application of Prokhorov's theorem, cf. [216, Sect. II.2], for $s \in U$ and $\mu \in V$ for appropriate neighborhoods $U$ and $V$ of $t$ and $\mu$, respectively. $\hspace{1cm}$ $\square$

As $[0, t_0] \times \mathcal{K}$ is compact for the product topology the following result is a straightforward consequence of Lemma 2.13.

**Corollary 2.14** *Let $T : [0, \infty) \to \mathscr{L}(C_b(\Omega))$ be a $\tau_{co}$-strongly continuous semigroup. For a weak*-compact set $\mathcal{K} \subseteq M(\Omega)$ and $t_0 > 0$ the set of measures given by*

$$\left\{ T'(t)\nu : \ t \in [0, t_0], \ \nu \in \mathcal{K} \right\}$$

*is $\sigma(M(\beta\Omega), C_b(\Omega))$-compact.*

Recall that a family $\mathcal{M} \subseteq M(\Omega)$ is called *tight* if for all $\varepsilon > 0$ there exists $K \subseteq \Omega$ compact such that $|\mu| (\Omega \setminus K) < \varepsilon$ for all $\mu \in \mathcal{M}$, see also [216, Sect. II.2].

**Corollary 2.15** *Let $T : [0, \infty) \to \mathscr{L}(C_b(\Omega))$ be a $\tau_{co}$-strongly continuous semigroup consisting of operators that are $\tau_{co}$-continuous on $\|\cdot\|$-bounded sets. For a $\|\cdot\|$-bounded and weak*-compact set $\mathcal{K} \subseteq M(\Omega)$ and $t_0 > 0$ the set of measures given by*

$$\left\{ T'(t)\nu : \ t \in [0, t_0], \ \nu \in \mathcal{K} \right\}$$

*is tight.*

***Proof*** We first notice that if $T$ is a $\|\cdot\|$-bounded operator on $C_b(\Omega)$ which is also $\tau_{co}$-continuous on $\|\cdot\|$-bounded sets then $T' \in \mathscr{L}(M(\beta\Omega))$ leaves $M(\Omega)$ invariant by means of Proposition 2.12. The assertion now follows by Corollary 2.14. $\qquad\square$

Our final result of this section will show that certain bi-continuous semigroups on $C_b(\Omega)$ satisfy a property which we will later call *local*, cf. Definition 7.3. This property will be important in the framework of perturbation theory in Chap. 7.

**Theorem 2.16** *Let $(T(t))_{t \geq 0}$ be a bi-continuous semigroup on $C_b(\Omega)$ and assume that all operators $T(t)$, $t \geq 0$, are $\tau_{co}$-continuous on $\|\cdot\|$-bounded sets. Then for all $t_0 \geq 0$, $K \subseteq \Omega$ compact and $\varepsilon > 0$ there exists $M > 0$ and $K' \subseteq \Omega$ compact such that*

$$\sup_{x \in K} |T(t)f(x)| \leq M \cdot \sup_{x \in K'} |f(x)| + \varepsilon \|f\|_\infty, \tag{2.2.5}$$

*for all $t \in [0, t_0]$ and $f \in C_b(\Omega)$.*

***Proof*** Let $\varepsilon > 0$, $t_0 > 0$ and $K \subseteq \Omega$ compact be arbitrary. By following Corollary 2.15 or [67], we are able to find a compact set $K' \subseteq \Omega$ such that $(T'(t)\delta_x)(\Omega \setminus K') \leq \varepsilon$ for all $t \in [0, t_0]$. Hence, we obtain

$$\sup_{x \in K} |T(t)f(x)| = \sup_{x \in K} \left| \int_\Omega f \, \mathrm{d} T'(t)\delta_x \right|$$

$$\leq \sup_{x \in K} \int_{K'} |f| \, \mathrm{d} \left| T'(t)\delta_x \right| + \sup_{x \in K} \int_{\Omega \setminus K'} |f| \, \mathrm{d} \left| T'(t)\delta_x \right|$$

$$\leq \sup_{t \in [0,t_0]} \|T(t)\| \cdot \sup_{x \in K'} |f(x)| + \varepsilon \|f\|_\infty \, .$$

$\square$

**Remark 2.17**  In terms of the seminorm system $\mathcal{P}_{\mathrm{co}}$ that generates the compact-open topology $\tau_{\mathrm{co}}$, see Sect. 2.1.1, we can reformulate (2.2.5) as follows: for every $\varepsilon > 0$, $t_0 > 0$ and $p \in \mathcal{P}_{\mathrm{co}}$, there exists $M > 0$ and $q \in \mathcal{P}_{\mathrm{co}}$ such that

$$p(T(t)f) \leq M q(f) + \varepsilon \|f\|_\infty$$

for all $t \in [0, t_0]$ and $f \in C_{\mathrm{b}}(\Omega)$. This property will be later on the motivation for Definition 7.3.

### 2.2.7  Adjoints of Bi-continuous Semigroups and Semigroups on $M(\Omega)$

Given a $C_0$-semigroup $(T(t))_{t \geq 0}$ on a Banach space $X$, we saw in Sect. 2.2.3 that the adjoint semigroup $(T'(t))_{t \geq 0}$ on the dual Banach space $X'$ forms a bi-continuous semigroup with respect to the weak*-topology. By Theorem 2.7 every semigroup on $X'$ which is bi-continuous with respect to the weak*-topology is of this form if $X$ is separable. Given a bi-continuous semigroup $(T(t))_{t \geq 0}$ we would like to define its adjoint in such a way that it is again a bi-continuous semigroup. This affiliates to the theory of sun-dual semigroups, cf. [101, Chap. II, Sect. 2.6]. In order to fit into the theory of bi-continuous semigroups we need to find a suitable space that satisfies Assumption 2.1. For this purpose, for $X$ a Banach space with an additional locally convex topology $\tau$ satisfying Assumption 2.1, let $X^\circ$ be the subspace of $X'$ consisting of all $\|\cdot\|$-bounded linear functionals which are $\tau$-sequentially continuous on $\|\cdot\|$-bounded sets of $X$.

**Proposition 2.18**  *Let $X$ be a Banach space equipped with an additional locally convex topology satisfying Assumption 2.1. Then $X^\circ$ is a closed linear subspace of $X'$, hence it is a Banach space.*

**Proof**  Obviously, $X^\circ$ is a linear space of $X'$. To show that $X^\circ$ is closed, let $(\varphi_n)_{n \in \mathbb{N}}$ be a sequence in $X^\circ$ such that $\|\varphi_n - \varphi\| \to 0$ as $n \to \infty$ for some $\varphi \in X'$. The claim is that $\varphi \in X^\circ$. Hence, let $(x_n)_{n \in \mathbb{N}}$ be a $\|\cdot\|$-bounded $\tau$-null sequence in $X$. Then

$$|\varphi(x_n)| \le |\varphi(x_n) - \varphi_k(x_n)| + |\varphi_k(x_n)|$$
$$\le K \, \|\varphi - \varphi_k\| + |\varphi_k(x - x_n)| \, .$$

By taking $k \in \mathbb{N}$ sufficiently large the first summand becomes arbitrarily small by the assumption that $\|\varphi_n - \varphi\| \to 0$ as $n \to \infty$. Fixing $k \in \mathbb{N}$ and using the continuity assumptions on the elements of the sequence $(\varphi_n)_{n \in \mathbb{N}}$ the second summand also can be made arbitrarily small. Hence, we conclude that $\varphi(x_n) \to 0$ so that $\varphi \in X^\circ$. $\qquad\square$

The space $X^\circ$ can easily be equipped with the dual norm inherited from $X'$ as it is a closed subspace by Proposition 2.18. Furthermore, we equip $X^\circ$ also with the weak topology $\tau^\circ := \sigma(X^\circ, X)$. Of course, the topology $\tau^\circ$ is Hausdorff as $X$ separates the points of $X^\circ$ and also $X$ is norming. Moreover, $\tau^\circ$ is weaker than the $\|\cdot\|$-topology on $X'$. It therefore remains to show that also the second condition of Assumption 2.1 holds true. However, this is in general not true. Therefore, we will make this an assumption.

**Assumption 2.19** For $B$ the open unit ball of $X'$ the space $X^\circ \cap \overline{B}$ is sequentially complete for $\sigma(X^\circ, X)$.

The next example shows that Assumption 2.19 does not follow automatically from our general Assumption 2.1.

**Example 2.20** Let $E$ be a non-reflexiv Banach space and $X = E'$ its norm dual and $\tau = \sigma(X', X)$ the weak*-topology. Suppose that $E'$ is separable so that the topology $\sigma(X'', X)$ is metrizable on $\|\cdot\|$-bounded sets. Then $X^\circ = (X, \tau)' = E$ and $\tau^\circ = \sigma(E, E')$, the weak*-topology. Now, let $y \in E'' \setminus E$ be arbitrary with $\|y\|_{E'} \le 1$. By Goldstine's Theorem, there exists a sequence $(y_n)_{n \in \mathbb{N}}$ in $E$ with $\|y_n\|_E \le 1$ converging to $y$ with respect to $\sigma(E'', E')$. This shows that $\tau^\circ = \sigma(E, E')$, as it is the restriction of $\sigma(E'', E')$ to $E$, is not complete on the closed unit ball of $X^\circ$. To be more concrete, take $E = c_0$, $E' = \ell^1$, $E'' = \ell^\infty$, $y = \mathbf{1}$ the constant 1 sequence and $y_n$, $n \in \mathbb{N}$, the sequence with first $n$ entries 1 and 0 otherwise. So $X = \ell^1$ with $\tau = \sigma(\ell^1, c_0)$ and $y_n \to y$ in $\ell^\infty$ with respect to $\sigma(\ell^\infty, \ell^1)$.

The framework of Assumption 2.1 is clear now. The following result is straightforward and is left to the reader.

**Proposition 2.21** *An operator $B \in \mathscr{L}(X)$ which is $\tau$-sequentially continuous on $\|\cdot\|$-bounded sets the adjoint operator $B' \in \mathscr{L}(X')$ leaves $X^\circ$ invariant.*

In view of Proposition 2.21 we denote, for a linear operator $B \in \mathscr{L}(X)$ that is $\tau$-sequentially continuous on $\|\cdot\|$-bounded sets, by $B^\circ$ the restriction of $B' \in \mathscr{L}(X')$ to $X^\circ$. Now, let $(T(t))_{t \ge 0}$ be a bi-continuous semigroup on a Banach space $X$ with respect to a locally convex topology $\tau$. The family of operators $(T^{\circ}(t))_{t\geq 0}$ is obviously strongly continuous with respect to $\tau^{\circ}$ by definition as well as exponentially bounded. For the local bi-equicontinuity we need to make another assumption.

**Assumption 2.22** Every $\|\cdot\|$-bounded $\tau^{\circ}$-null sequence $(\varphi_n)_{n\in\mathbb{N}}$ in $X^{\circ}$ is $\tau$-equicontinuous on $\|\cdot\|$-bounded sets.

**Remark 2.23** In order to check if Assumptions 2.19 and 2.22 are valid we refer to the work by Kruse and Schwenninger [155]. In particular, they showed that both assumptions are fullfilled if the space $(X, \gamma)$, where $\gamma$ denotes the mixed topology (see Appendix A) is a Mackey–Mazur space, cf. [155, Thm. 3.8]. For the actual proof of this result as well as the consequences we refer to the corresponding work of Kruse and Schwenninger [155]. They also show that this criterion covers all known examples from Sect. 2.1, see also [155, Example 3.9].

Under all assumptions, we are able to prove the following theorem.

**Proposition 2.24** *Let $(T(t))_{t\geq 0}$ be an operator semigroup on a Banach space $X$ which is bi-continuous with respect to $\tau$ and suppose that Assumptions 2.19 and 2.22 are satisfied. Then $(T^{\circ}(t))_{t\geq 0}$ is an operator semigroup on $X^{\circ}$ which is bi-continuous with respect to $\tau^{\circ}$.*

*Proof* As mentioned before Assumption 2.22, it is sufficient to show that $(T^{\circ}(t))_{t\geq 0}$ is locally bi-equicontinuous. To do so, let $(\varphi_n)_{n\in\mathbb{N}}$ be a $\|\cdot\|$-bounded $\tau^{\circ}$-null sequence and $x \in X$. For $t_0 > 0$ the set $\{T(t)x : t \in [0, t_0]\}$ is $\tau$-compact. By Assumption 2.22 the sequence $(\varphi_n)_{n\in\mathbb{N}}$ is $\tau$-equicontinuous on $\|\cdot\|$-bounded sets so that by an adaptation of [208, Chap. III, Thm. 4.5] one has

$$(T^{\circ}(t)\varphi_n)(x) = \varphi_n(T(t)x) \to 0,$$

uniformly for $t \in [0, t_0]$. This proves the claim. $\qquad\square$

The following result by F. Sentilles allows us to identify $C_b(\Omega)^{\circ}$ with $M(\Omega)$, cf. [212]. Sentilles makes use of the strict topology $\beta_0$ on $C_b(\mathbb{R})$ which actually coincides with the mixed topology $\gamma$ if $\Omega$ is at least a completely regular Hausdorff space, see [213, Thm. 2.4]. For more information on the mixed topology, we refer to Appendix A. As a matter of fact, if $\Omega$ is a completely regular Hausdorff space the mixed topology and the submixed topology coincide on $C_b(\Omega)$, see Example B.7 and Proposition B.6.

**Theorem 2.25** *Let $\Omega$ be either a $\sigma$-compact, locally compact or Polish space. Then the following assertions are true:*

*(a) $\beta_0 = \mu(C_b(\Omega), M(\Omega))$, the Mackey topology.*

*(b) A linear operator $T : C_b(\Omega) \to C_b(\Omega)$ is $\beta_0$-continuous if and only if it is $\beta_0$-sequentially continuous. The same holds for all linear functionals.*

*(c) Every $\sigma(M(\Omega), C_b(\Omega))$-null sequence $(\varphi_n)_{n \in \mathbb{N}}$ in $(C_b(\Omega), \beta_0)'$ is $\beta_0$-equicontinuous.*

In Sect. 2.2.6 we discussed bi-continuous semigroups on $C_b(\Omega)$. We will now link those to bi-continuous semigroups on $M(\Omega)$. First of all, let us mention that if $\Omega$ is a Polish space then $M(\Omega)$ satisfies Assumption 2.19, see for example [10], as well as Assumption 2.22 by Theorem 2.25(c). In particular, the topology $\tau^\circ = \sigma(C_b(\Omega), M(\Omega))$ is suitable for our framework. With the knowledge of this section, we are able to formulate the following result.

**Proposition 2.26** *Let $\Omega$ be a Polish space and $(T(t))_{t \geq 0}$ a bi-continuous semigroup on $C_b(\Omega)$ with respect to $\tau_{co}$. Then the semigroup $(T^\circ(t))_{t \geq 0}$ defined by $T^\circ(t) := T(t)'_{|M(\Omega)}$, $t \geq 0$, is a bi-continuous semigroup on $M(\Omega)$ with respect to $\tau^\circ$.*

Surprisingly, the converse of Proposition 2.26 also holds true.

**Theorem 2.27** *Let $\Omega$ be a Polish space. Let $(S(t))_{t \geq 0}$ be a bi-continuous semigroup on $M(\Omega)$ with respect to $\tau^\circ$. Then there exists a bi-continuous semigroup $(T(t))_{t > 0}$ on $C_b(\Omega)$ with respect to $\tau_{co}$ such that $T^\circ(t) = S(t)$, $t \geq 0$.*

***Proof*** For $f \in C_b(\Omega)$ we set

$$(T(t)f)(x) := \int_\Omega f \, d(S(t)\delta_x), \quad t \geq 0, \ f \in C_b(\Omega), \ x \in \Omega,$$

where $\delta_x$ denotes the Dirac measure at a point $x \in \Omega$. We first show that $T(t)f \in C_b(\Omega)$ for all $t \geq 0$. The boundedness follows as

$$\sup_{x \in \Omega} |(T(t)f)(x)| \leq \|S(t)\| \cdot \|f\|_\infty.$$

For the continuity, let $(x_n)_{n \in \mathbb{N}}$ be a sequence in $\Omega$ such that $x_n \to x$ for $n \to \infty$. Then, $\delta_{x_n} \to \delta_x$ in $M(\Omega)$ with respect to $\tau^\circ$. Since by assumption $S(t)$, $t \geq 0$, is $\tau^\circ$-continuous on $M(\Omega)$, we have $S(t)\delta_{x_n} \to S(t)\delta_x$ and the continuity of $T(t)f$ follows. Hence, we have $T(t) \in \mathscr{L}(C_b(\Omega))$ for all $t \geq 0$. Of course, $(T(t))_{t \geq 0}$ is an exponentially bounded semigroup. It suffices to show that for each fixed $t \geq 0$ the operator is $\tau_{co}$-sequentially continuous on $\|\cdot\|_\infty$-bounded sets and that the semigroup $(T(t))_{t \geq 0}$ is $\tau_{co}$-strongly continuous. Then by construction $T^\circ(t) = S(t)$ for $t \geq 0$.

For the first part, fix $t > 0$ and assume the contrary, i.e., there exists a $\|\cdot\|_\infty$-bounded sequence $(f_n)_{n \in \mathbb{N}}$ such that $f_n \to 0$ with respect to $\tau_{co}$ but there exists a compact set $K \subseteq \Omega$ and $\varepsilon > 0$ such that

$$\sup_{x \in K} |(T(t)f_n)(x)| \geq \varepsilon,$$

for all $n \in \mathbb{N}$. For each $n \in \mathbb{N}$ take $x_n \in K$ such that $|(T(t)f_n)(x_n)| \geq \varepsilon$. By compactness, we can assume without loss of generality that $x_n \to x$ for some $x \in K$. Then $\delta_{x_n} \to \delta_x$ with respect to $\tau^\circ$ and we obtain that $S(t)\delta_{x_n}$, $t \geq 0$, is a $\tau^\circ$-convergent sequence. By Theorem 2.25(c) this sequence in $\beta_0$-equicontinuous. By [208, Chap. III, Sect. 4.5] we can therefore deduce that

$$|(T(t)f_n)(x_n)| \leq \sup_{m \in \mathbb{N}} \left| \int_\Omega f_n \, d(S(t)\delta_{x_m}) \right| \to 0,$$

for $n \to \infty$. Contradiction.

For the $\tau_{\mathrm{co}}$-strong continuity, let $K \subseteq \Omega$ be compact. Assume for the sake of a contradiction that there exist $f \in C_b(\Omega)$, $\varepsilon > 0$ and $t_n \in [0, 1]$, $n \in \mathbb{N}$, with $t_n \to 0$ for $n \to \infty$ such that

$$\sup_{x \in K} |(T(t_n)f)(x) - f(x)| \geq \varepsilon,$$

for all $n \in \mathbb{N}$. For each $n \in \mathbb{N}$ take $x_n \in K$ such that

$$\left| \int_\Omega f \, d(S(t)\delta_{x_n} - \delta_{x_n}) \right| = |(T(t_n)f)(x_n) - f(x_n)| \geq \varepsilon.$$

By compactness we can pass to a subsequence and assume that $x_n \to x$ as $n \to \infty$ for some $x \in K$. This means that $\delta_{x_n} \to \delta_x$ with respect to $\tau^\circ$. By the fact that $(S(t))_{t \geq 0}$ is locally bi-equicontinuous with respect to $\tau^\circ$ we get

$$\sup_{s \in [0,1]} \left| \int_\Omega f \, d(S(s)\delta_{x_n} - \delta_{x_n}) \right| \to 0,$$

for $n \to \infty$, which is again a contradiction. $\qquad\square$

## Notes on This Chapter

This chapter serves as introduction to bi-continuous semigroups. As mentioned already, the theory of bi-continuous semigroups was initiated by the work of F. Kühnemund. Big parts of both Sects. 2.1 and 2.2 arose from Kühnemund's work, see [159, Sect. 1.1] and [160]. However, also later work has been considered, e.g., from Farkas [106, Chap. 2] or Scirrat [211]. For Sects. 2.2.7 and 2.2.6 we also consulted [109]. It is also worth mentioning that recently there was work done of the sun-dual theory of bi-continuous semigroups [155].

It is noteworthy that the implemented semigroups introduced in Sect. 2.2.4 have also applications to quantum physics. The question whether automorphisms and operators are implemented arises naturally if one studies the quantum mechanical aspects in mathematical physics. Typically, issues of symmetries apply here. For example Ludwig's Theorem states that on every Hilbert space $\mathcal{H}$ with $\dim(\mathcal{H}) \geq 3$ each order-lattice automorphism $\alpha$ of the effect algebra, defined by $\mathcal{E} := \{T \in \mathscr{L}(\mathcal{H}) : 0 \leq T \leq 1\}$, is unitarily implemented, i.e., there exists a unitary operator $U \in \mathscr{L}(\mathcal{H})$ such that $\alpha(T) = UTU^*$ for each $T \in \mathcal{E}$, see for example [63]. Another important symmetry phenomenon in the operator algebraic setting of quantum mechanics is the so-called Wigner symmetry [166, Chap. 5]. In this context one also sees implemented semigroups appearing. As it is stated in [49, Chap. 3, Sect. 2], all continuous one-parameter groups of Wigner symmetries are implemented by strongly continuous one-parameter groups of unitary operators. The most common setting in quantum mechanics is the case of Hilbert spaces. Implemented semigroups on Hilbert spaces also appear in other resources. In particular, in [93, Sects. 4.4, 6.1, 7.1 and 7.2] Eisner discusses stability of implemented operators as well as of implemented semigroups with applications to Lyapunov's equations. In [19, Chap. D-IV] the point of interest is the asymptotic behavior of positive implemented semigroups on $C^*$-algebras and von Neumann algebras. The implemented semigroups also come to light if one considers the following operator equation

$$AX + XB = Y, \ X \in \mathscr{L}(F, E),$$

where $A$ and $B$ are generators of strongly continuous semigroups on Banach spaces $E$ and $F$. This operator-valued equation has been considered by several authors [20, 125, 198–200]. We now consider extrapolation spaces and Desch–Schappacher perturbations of implemented semigroup on Banach spaces. In particular, we round out the work of J. Alber [9], where extrapolation spaces for these semigroups are studied but only with respect to the space of strong continuity. In Chap. 8 we will come back to implemented semigroups to investigate Desch–Schappacher perturbations of implemented semigroups and show that the domains of these semigroups have a module structure.

The Ornstein–Uhlenbeck semigroup from Sect. 2.2.5 has a close relation to the differential operator $(\mathcal{A}, D(\mathcal{A}))$ introduced in Sect. 1.5 and therefore also to Markov processes as well as stochastic differential equations, cf. [169, Chap. 9]. Beyond the framework of bi-continuous semigroups, the analysis and applications of the Ornstein–Uhlenbeck semigroup attracted a lot of attention in the past, especially within the Italian operator semigroup community, see for example [68, 75, 77, 203], but also beyond [58, 70, 71, 120, 204], just to mention a few. For more references we refer again to [169, 170].

# The Generator of Bi-continuous Semigroups

**3**

In this chapter, we introduce the notion of the generator of a given bi-continuous semigroup. Moreover, we show that this generator has nice properties as they are also known from the strongly continuous case. For that, we also show that the resolvent of the generator can be represented by the Laplace transform of the semigroup.

## 3.1 The Generator and Its Basic Properties

As in the case of $C_0$-semigroups one can define a generator for a bi-continuous semigroup in the following way.

**Definition 3.1** Let $(T(t))_{t\geq 0}$ be a bi-continuous semigroup on a bi-admissible space $(X, \|\cdot\|, \tau)$. The *(infinitesmial) generator* $(A, D(A))$ is defined by

$$Ax := \tau\lim_{t\to 0} \frac{T(t)x - x}{t}$$

with the domain

$$D(A) := \left\{ x \in X : \tau\lim_{t\to 0} \frac{T(t)x - x}{t} \text{ exists and } \sup_{t\in(0,1]} \frac{\|T(t)x - x\|}{t} < \infty \right\}.$$

This definition of the generator leads to a couple of important properties. Recall from Sect. 1.2 that for a linear operator $(A, D(A))$ on a Banach space $X$ the *resolvent set* $\rho(A)$ consists of $\lambda \in \mathbb{C}$ such that $\lambda - A$ is invertible, i.e., there exists a bounded operator $B$ with $Bx \in D(A)$ for each $x \in X$ such that $(\lambda - A)Bx = x$ for each $x \in X$ and $B(\lambda - A)x = x$ for each $x \in D(A)$. The inverse of $\lambda - A$ is often denoted by $R(\lambda, A) := (\lambda - A)^{-1}$. The complement $\sigma(A) := \mathbb{C} \setminus \rho(A)$ is called the *spectrum*. Before we analyze the properties of the generator, we need the following technical result, which on the one hand side is a first consequence of the definition of bi-continuous semigroups, cf. Definition 2.4, and on the other side allows us to Riemann integrate these semigroups which finally leads to the Laplace transform representation of the resolvent operators of the generator.

**Proposition 3.2** *Let $(T(t))_{t\geq0}$ be a bi-continuous semigroup of type $\omega$ on a bi-admissible space $(X, \|\cdot\|, \tau)$. Then, for every $t \geq 0$ and $\lambda \in \mathbb{C}$, there exists the operator $R_t(\lambda) : X \to X$ defined by*

$$R_t(\lambda)x = \int_0^t e^{-\lambda s} T(s)x \, ds,$$

*for all $x \in X$. The integral has to be understood as a $\tau$-Riemann integral.*

**Proof** Fix $\lambda \in \mathbb{C}$ and $t \geq 0$, then the function $\xi_x(s) := e^{-\lambda s} T(s)x, s \in [0, t]$, is a uniformly $\tau$-continuous $X$-valued function for all $x \in X$. Let $P = \{0 = s_0 \leq s_1' \leq s_1 \leq \cdots \leq s_n' \leq s_n = t\}$ be a partition of the interval $[0, t]$ with the corresponding Riemann sum $S(\xi_x(\cdot), P)$ given by

$$S(\xi_x(\cdot), P) = \sum_{m=1}^{n} \xi_x(s_m')(s_m - s_{m-1}).$$

This is a $\tau$-Cauchy net. Notice that an equivalent $\|\cdot\|$-bounded $\tau$-Cauchy sequence is given by the Riemann sum according to an equidistant partition $P_n, n \in \mathbb{N}$, i.e.,

$$S(\xi_x(\cdot), P_n) = \frac{t}{n} \sum_{m=1}^{n} \xi_x\left(\frac{tm}{n}\right).$$

Making use of the assumption that $X$ is sequentially complete on $\|\cdot\|$-bounded sets, we conclude that $(S(\xi_x(\cdot), P_n))_{n\in\mathbb{N}}$ actually converges. By an application of [196, Prop. 1.1] we may conclude that $t \mapsto \xi_x(t)$ is Riemann integrable for all $x \in X$. $\qquad\square$

Now we summarize some properties the generator of a bi-continuous semigroup.

**Theorem 3.3** *Let $(T(t))_{t\geq0}$ be a bi-continuous semigroup on a bi-admissible space $(X, \|\cdot\|, \tau)$ with generator $(A, D(A))$. Then the following hold:*

(a) *For $x \in D(A)$ we have $T(t)x \in D(A)$, $t \geq 0$, $T(t)x$ is continuously differentiable in $t$ with respect to the topology $\tau$ and $T(t)Ax = AT(t)x$ for all $t \geq 0$.*
(b) *An element $x \in X$ belongs to $D(A)$ and $Ax = y$ if and only if*

$$T(t)x - x = \int_0^t T(s)y \, ds,$$

*for all $t \geq 0$.*

(c) *The operator $A$ is bi-closed, i.e., whenever $(x_n)_{n \in \mathbb{N}}$ is a sequence in $D(A)$ such that $x_n \xrightarrow{\tau} x$ and $Ax_n \xrightarrow{\tau} y$ and both sequences are $\| \cdot \|$-bounded, then $x \in D(A)$ and $Ax = y$.*

**Proof** (a)   Assume that $x \in D(A)$, then for $t \geq 0$ one has that

$$T(t)Ax = \tau\lim_{h \to 0} \frac{T(t+h)x - T(t)x}{h} = \tau\lim_{h \to 0} \frac{(T(h) - I)T(t)x}{h},$$

showing that $T(t)x \in D(A)$ and that the right derivative $\frac{d^+}{dt}T(t)x$ exists. In particular, one has

$$\frac{d^+}{dt}T(t)x = AT(t)x = T(t)Ax.$$

In order to show that $T(t)x$ is also differentiable, let $\varphi \in (X, \tau)'$. Then

$$\frac{d^+}{dt}\varphi(T(t)x) = \varphi\left(\frac{d^+}{dt}T(t)x\right) = \varphi(T(t)Ax),$$

which implies that the map $t \mapsto \frac{d^+}{dt}\varphi(T(t)x)$ is continuous. From this we conclude that the map $t \mapsto \varphi(T(t)x)$ is differentiable. To see this, we recall from [241, p. 239], that if one of the Dini derivatives of a continuous real-valued function is finite and continuous, then this function is differentiable and the derivative coincides with the corresponding Dini derivative. In particular, in our case we obtain that

$$\frac{d}{dt}\varphi(T(t)x) = \varphi(T(t)Ax).$$

Now using that $(T(t))_{t \geq 0}$ is bi-continuous, we know that the integral $\int_0^t T(t)Ax \, dt$ in $X$ exists and one obtains

$$\varphi(T(t)x - x) = \int_0^t \frac{d}{ds}\varphi(T(s)x) \, ds = \int_0^t \varphi(T(s)Ax) \, ds = \varphi\left(\int_0^t T(s)Ax \, ds\right)$$

showing that in fact

$$T(t)x - x = \int_0^t T(s)Ax \, ds \tag{3.1.1}$$

for all $t \geq 0$ and that $t \mapsto T(t)x$ is differentiable with

$$\frac{d}{dt}T(t)x = T(t)Ax.$$

(b)   Let $x \in D(A)$ and $y = Ax$, then (3.1.1) yields the first implication from the left to the right of the assertion. For the converse, assume that for $x \in X$ the equality $T(t)x - x =$

$\int_0^t T(s)y \, ds$ holds for all $t \geq 0$. From this one obtains

$$\tau\lim_{t \to 0} \frac{T(t)x - x}{t} = \tau\lim_{t \to 0} \frac{1}{t} \int_0^t T(s)y \, ds = y,$$

which actually shows that $x \in D(A)$ and $Ax = y$.

(c) Let $(x_n)_{n \in \mathbb{N}}$ be a $\|\cdot\|$-bounded sequence in $D(A)$ which converges to $x \in X$. Furthermore, assume that also $(Ax_n)_{n \in \mathbb{N}}$ is $\|\cdot\|$-bounded and $\tau$-convergent with limit $y \in X$. By assertion (b) of this theorem, one obtains

$$T(t)x_n - x_n = \int_0^t T(s)Ax_n \, ds,$$

for all $t \geq 0$. Since by definition the semigroup is locally bi-equicontinuous, one has

$$T(t)x - x = \int_0^t T(s)y \, ds,$$

for all $t \geq 0$, which implies by using assertion (b) again, that $x \in D(A)$ and $Ax = y$.

$\square$

### 3.1.1   Examples of Generators

The following result identifies the generator of the translation semigroup as the first derivative on an appropriate domain.

**Proposition 3.4** *Let $(T(t))_{t \geq 0}$ be the left-translation semigroup on $C_b(\mathbb{R})$. Then its generator $(A, D(A))$ is given by*

$$Af = f', \quad D(A) = C_b^1(\mathbb{R}),$$

*where $C_b^1(\mathbb{R})$ denotes the space of differentiable functions with bounded and continuous derivative.*

**Proof** Let $f \in D(A)$, then the limit $\tau\lim_{t \to 0} \frac{T(t)f - f}{t}$ exists and converges to $Af$ in $C_b(\mathbb{R})$ as $t \to 0$. In particular, one has that $\lim_{t \to 0} \frac{T(t)f(x) - f(x)}{t} = (Af)(x)$. Hence, $f$ is differentiable and $Af = f' \in C_b(\mathbb{R})$ which shows that $f \in C_b^1(\mathbb{R})$ and hence $D(A) \subseteq C_b^1(\mathbb{R})$. For the converse inclusion, assume that $f \in C_b^1(\mathbb{R})$. The fundamental theorem of calculus yields

$$\frac{f(x+t) - f(x)}{t} = \frac{1}{t} \int_0^t f'(x+s) \, ds, \quad x \in \mathbb{R}, \ t > 0.$$

Let $\varepsilon > 0$ be arbitrary. By assumption that $f \in C_b^1(\mathbb{R})$, we actually have that $f'$ is continuous. Therefore, $f'$ becomes uniformly continuous on compact subsets. Hence we find $\delta > 0$ such that $\left|f'(x) - f'(y)\right| < \varepsilon$ whenever $|x - y| < \delta$. Then

$$\left|\frac{f(x+t) - f(x)}{t} - f'(x)\right| \leq \frac{1}{t}\int_0^t \left|f'(x+s) - f'(s)\right|\, \mathrm{d}s < \varepsilon$$

for every $x \in \mathbb{R}$ and $t \in (0, \delta)$. Hence, the expression $\frac{T(t)f-f}{t}$ converges to $f'$ uniformly on compact sets to $f'$ for $t \to 0$. This shows that $C_b^1(\mathbb{R}) \subseteq D(A)$. $\qquad\square$

Next, we consider the adjoint semigroup as introduced during the last section in Example 2.2.3. In order to identify the generator of the adjoint semigroup, we recall the following definition of the adjoint of an unbounded operator.

**Definition 3.5** For a densely defined operator $(A, D(A))$ on $X$, the *adjoint operator* $(A', D(A'))$ on $X'$ is defined by

$$D(A') := \left\{x' \in X' : \exists y' \in X' \,\forall x \in D(A) : \langle Ax, x'\rangle = \langle x, y'\rangle\right\}, \quad A'x' := y'.$$

The following result actually shows that the generator of the adjoint semigroup is just the adjoint of the generator of the corresponding $C_0$-semigroup, see also [101, Chap. II, Sect. 2.5] or [226, Thm. 1.2.3].

**Proposition 3.6** *Let $(T(t))_{t\geq 0}$ be a strongly continuous semigroup on $X$. The generator $(B, D(B))$ of the adjoint semigroup $(T'(t))_{t\geq 0}$ on $X'$ is given by $D(B) = D(A')$ and $Bx' := A'x'$.*

The following result yields a characterization for the generator of adjoints of bi-continuous semigroups, see Sect. 2.2.7, by means of adjoints of unbounded operators. For obvious reasons, it is strongly related to Proposition 3.6.

**Lemma 3.7** *Let $(T(t))_{t\geq 0}$ be a bi-continuous semigroup on a bi-admissible space $(X, \|\cdot\|, \tau)$ and let $(A, D(A))$ be the generator of $(T(t))_{t\geq 0}$. Suppose that Assumptions 2.19 and 2.22 on $X^\circ$ and $\tau^\circ$ are satisfied. Let us denote the generator of $(T^\circ(t))_{t\geq 0}$ by $(A^\circ, D(A^\circ))$. Then*

$$D(A^\circ) = \left\{x' \in X^\circ : \exists y' \in X^\circ \,\forall x \in D(A) : \langle Ax, x'\rangle = \langle x, y'\rangle\right\}, \quad A^\circ x' = y'.$$

***Proof*** In order to prove the assertion of the lemma, we introduce an operator $(B, D(B))$ on $X^\circ$ by

$$D(B) := \left\{x' \in X^\circ : \exists y' \in X^\circ \,\forall x \in D(A) : \langle Ax, x'\rangle = \langle x, y'\rangle\right\}, \quad Bx' := y'.$$

We have to show that $D(A^\circ) = D(B)$ and $A^\circ x' = Bx'$. To do so, we first observe that the following holds for arbitrary $x \in D(A)$ and $x' \in X^\circ$:

$$\left\langle Ax, \int_0^t T^\circ(s)x' \, \mathrm{d}s \right\rangle = \int_0^t \langle Ax, T^\circ(s)x' \rangle \, \mathrm{d}s = \int_0^t \langle T(s)Ax, x' \rangle \, \mathrm{d}s = \left\langle \int_0^t T(s)Ax \, \mathrm{d}s, x' \right\rangle$$

$$= \left\langle A \int_0^t T(s)x \, \mathrm{d}s, x' \right\rangle = \langle T(t)x - x, x' \rangle = \langle x, T^\circ(t)x' - x' \rangle,$$

showing that $B \int_0^t T^\circ(s)x' \, \mathrm{d}s = T^\circ(t)x' - x'$ for all $x' \in X^\circ$. Furthermore, one obtains for $x \in D(A)$ and $x' \in D(B)$ that

$$\left\langle x, B \int_0^t T(s)^\circ x' \, \mathrm{d}s \right\rangle = \int_0^t \langle T(s)Ax, x' \rangle \, \mathrm{d}s = \left\langle x, \int_0^t T^\circ(s)Bx' \, \mathrm{d}s \right\rangle.$$

Therefore, one concludes that $B \int_0^t T^\circ(s)x' \, \mathrm{d}s = \int_0^t T^\circ(s)Bx' \, \mathrm{d}s$ for $x' \in D(B)$. We will use this in what follows. Firstly, fix $x' \in D(B)$. By the previous observations we obtain

$$\lim_{t \to 0} \frac{1}{t} \langle x, T^\circ(t)x' - x' \rangle = \lim_{t \to 0} \frac{1}{t} \left\langle x, B \int_0^t T^\circ(s)x' \, \mathrm{d}s \right\rangle = \lim_{t \to 0} \frac{1}{t} \int_0^t \langle x, T^\circ(s)Bx' \rangle \, \mathrm{d}s = \langle x, Bx' \rangle.$$

Hence, the limit $\lim_{t \to 0} \frac{1}{t}(T'(t)x' - x')$ exists in $X^\circ$ and is equal to $Bx'$. We conclude that $D(B) \subseteq D(A^\circ)$ and $Bx' = A^\circ x'$ hold. This shows that $B \subseteq A^\circ$. For the converse, fix $x' \in D(A^\circ)$, let $x \in D(A)$ be arbitrary and take notice of the following equality

$$\langle x, A^\circ x' \rangle = \lim_{t \to 0} \frac{1}{t} \langle x, T'(t)x' - x' \rangle = \langle Ax, x' \rangle.$$

This shows that $x' \in D(B)$ and $A^\circ x' = Bx'$, showing that $A^\circ \subseteq B$. $\qquad\square$

Last but not least, we determine the generator of the diffusion semigroup, cf. Example 2.2.2. We want to show that the generator of the diffusion semigroup on $C_b(\mathbb{R})$ is, as one expects, the Laplace operator on a certain domain. It is left to the reader to show that the maximal distributional Laplacian operator $\Delta_0$ on the domain $D(\Delta_0) = \{f \in C_0(\mathbb{R}) : \Delta_0 f \in C_0(\mathbb{R})\}$ generates a $C_0$-semigroup on the Banach space $C_0(\mathbb{R})$. Now let $(A, D(A))$ be the generator of the diffusion semigroup. Since on $C_0(\mathbb{R})$ the diffusion semigroup $(T(t))_{t \geq 0}$ is strongly continuous with respect to the $\| \cdot \|_\infty$-norm, we have that $D(\Delta_0) \subseteq D(A)$. Moreover, $D(\Delta_0)$ is invariant under $(T(t))_{t \geq 0}$. Since $D(\Delta_0)$ is $\|\cdot\|$-dense in $C_0(\mathbb{R})$ and $C_0(\mathbb{R})$ bi-dense in $C_b(\mathbb{R})$ we conclude that $D(\Delta_0)$ is bi-dense in $C_b(\mathbb{R})$. In order to identify the generator $(A, D(A))$ we need the following definition, generalizing the definition of a core of a linear operator to the bi-continuous setting.

**Definition 3.8** Let $(A, D(A))$ be a linear operator on a bi-admissible space $(X, \| \cdot \|, \tau)$. A subspace $D \subseteq D(A)$ is called a *bi-core* for $A$ if for all $x \in D(A)$ there exists a sequence $(x_n)_{n \in \mathbb{N}}$ in $D$ such that both $(x_n)_{n \in \mathbb{N}}$ and $(Ax_n)_{n \in \mathbb{N}}$ are $\|\cdot\|$-bounded and $x_n \xrightarrow{\tau_A} x$, where $\tau_A$ is the locally convex topology determined by the family of seminorms $\mathcal{P}_A := \{p(\cdot) + p(A\cdot) : \ p \in \mathcal{P}\}$ on $D(A)$.

Similar to [101, Chap. II, Prop. 1.7] we can formulate and prove the following result.

**Proposition 3.9** *Let $(A, D(A))$ be the generator of a bi-continuous semigroup $(T(t))_{t \geq 0}$ on a bi-admissible space $(X, \|\cdot\|, \tau)$. Let $D \subseteq D(A)$ be a bi-dense subset which is invariant under the semigroup $(T(t))_{t \geq 0}$. Then $D$ is a bi-core for $A$, i.e., for all $x \in D(A)$ there exists $(x_n)_{n \in \mathbb{N}}$ in $D$ such that $x_n \xrightarrow{\tau} x$ and $Ax_n \xrightarrow{\tau} Ax$ and both sequences are $\|\cdot\|$-bounded.*

***Proof*** Let $x \in D(A)$ be arbitrary. By assumption, there exists a sequence $(x_n)_{n \in \mathbb{N}}$ in $D$ which is $\tau$-convergent to $x$ such that both $(x_n)_{n \in \mathbb{N}}$ and $(Ax_n)_{n \in \mathbb{N}}$ are $\|\cdot\|$-bounded. By Theorem 3.3(a) for each $n \in \mathbb{N}$ the map $t \mapsto T(t)x_n \in D$ is $\tau_A$-continuous (see Definition 3.8). Hence, we conclude that $\int_0^t T(s)x_n \, \mathrm{d}s$, which is defined as an Riemann integral, belongs to the closure of $\overline{D}^{\tau_A}$. Similarly, the $\tau_A$-continuity of the map $t \mapsto T(t)x$ for $x \in D(A)$ as well as Theorem 3.3(b) yields for $\widehat{p} \in \mathcal{P}_A$ that

$$\widehat{p}\left(\frac{1}{t}\int_0^t T(s)x \, \mathrm{d}s - x\right) = p\left(\frac{1}{t}\int_0^t T(s)x \, \mathrm{d}s - x\right) + p\left(\frac{1}{t}\int_0^t T(s)Ax \, \mathrm{d}s - Ax\right),$$

which converges to 0 whenever $t \to 0$ and

$$\widehat{p}\left(\frac{1}{t}\int_0^t T(s)x_n \, \mathrm{d}s - \frac{1}{t}\int_0^t T(s)x \, \mathrm{d}s\right),$$

also converges to 0 for $n \to \infty$ for each $t > 0$. In summary, for every $\varepsilon > 0$ there exists $t > 0$ and $n \in \mathbb{N}$ such that

$$\widehat{p}\left(\frac{1}{t}\int_0^t T(s)x_n \, \mathrm{d}s - x\right) < \varepsilon.$$

Therefore, we conclude that $x \in \overline{D}^{\tau_A}$. $\qquad\qquad\square$

Our previous observations show that $D(\Delta_0)$ is indeed a bi-core for $(A, D(A))$. Now define $(\Delta, D(\Delta))$ on the domain

$$D(\Delta) := \{f \in C_b(\mathbb{R}) : \ \Delta f \in C_b(\mathbb{R})\},$$

where the Laplace operator $\Delta$ is understood in the distributional sense. It is not hard to see that $(\Delta, D(\Delta))$ is bi-closed, cf. Theorem 3.3. Since $C_c^\infty(\mathbb{R}) \subseteq D(\Delta_0)$ and every $f \in D(\Delta)$ can be approximated within the compact-open topology by a $\|\cdot\|$-bounded sequence $f_n \in C_c^\infty(\mathbb{R})$ with $Af_n = \Delta_0 f_n \xrightarrow{\tau_{co}} \Delta f$, we see that $D(\Delta_0)$ is a bi-core for $(\Delta, D(\Delta))$. Moreover, $A$

coincides with $\Delta$ on $D(\Delta_0)$, hence we conclude that $(A, D(A))$ actually is given by the maximal distributional Laplacian, i.e.,

$$Af = \Delta f, \quad D(A) = \{f \in C_b(\mathbb{R}) : \Delta f \in C_b(\mathbb{R})\} \, .$$

## 3.2 Standard Constructions

In this section, we consider several constructions of bi-continuous semigroups by given ones. Basic constructions for $C_0$-semigroups can be for example found in [101, Chap. II, Sect. 2(a)]. Those constructions will be important in later chapters.

### 3.2.1 Similar Semigroups

Let $(X, \|\cdot\|_X, \tau_X)$ and $(Y, \|\cdot\|_Y, \tau_Y)$ be bi-admissible spaces and $(T(t))_{t\geq 0}$ a bi-continuous semigroup on $X$. Assume that $B \in \mathscr{L}(X, Y)$ is invertible and that the operators $B$ and $B^{-1}$ are sequentially $\tau_X$-complete and sequentially $\tau_Y$-complete on $\|\cdot\|_X$-bounded and $\|\cdot\|_Y$-bounded sets, respectively. Then we can define an operator semigroup $(S(t))_{t\geq 0}$ on $Y$ by $S(t) := BT(t)B^{-1}$, $t \geq 0$. We show that this in fact defines a bi-continuous semigroup on $Y$ and that its generator is given by $(BAB^{-1}, BD(A))$. Indeed, by construction $(S(t))_{t\geq 0}$ is both $\tau_Y$-strongly continuous and exponentially bounded. Let us show that $(S(t))_{t\geq 0}$ is also locally bi-equicontinuous. To do so, take $t_0 > 0$ and let $(y_n)_{n\in\mathbb{N}}$ be a $\|\cdot\|_Y$-bounded sequence such that $y_n \to 0$ with respect to $\tau_Y$. Then, the sequence $(B^{-1}y_n)_{n\in\mathbb{N}}$ is a sequence in $X$ which is $\|\cdot\|_X$-bounded and $\tau_X$-convergent to 0. Hence, by the local bi-equicontinuity of $(T(t))_{t\geq 0}$ we conclude that $T(t)B^{-1}y_n \to 0$ with respect to $\tau_X$ uniformly for $t \in [0, t_0]$. For the fact that also $BT(t)B^{-1}y_n \to 0$ with respect to $\tau_Y$ uniformly for $t \in [0, t_0]$ we refer to reader to follow the arguments given in the proof of Lemma 7.2. By simply calculating the generator of $(S(t))_{t\geq 0}$ one obtains the desired result.

### 3.2.2 Renorming

It is a known fact that for bounded $C_0$-semigroups one can find an equivalent norm making it a contraction semigroup. For the case of bi-continuous semigroups, we have to take care of the geometric properties as we deal with two topologies. Let us start with a bi-continuous semigroup on a bi-admissible space $(X, \|\cdot\|, \tau)$. Without loss of generality we may assume that $(T(t))_{t\geq 0}$ is bounded. In fact, one observes that if $(T(t))_{t\geq 0}$ is a bi-continuous semigroup with generator $(A, D(A))$, then $(e^{\omega t}T(t))_{t\geq 0}$ is a bi-continuous semigroup with generator $(A + \omega, D(A))$. On $X$ we define a new norm given by

$$\||x\|| := \sup_{t\geq 0} \|T(t)x\|, \quad x \in X.$$

We observe that $\||\cdot\||$ is a norm equivalent to $\| \cdot \|$. In fact, one has that $\|x\| \leq \||x\|| \leq M\|x\|$, for all $x \in X$, where $M \geq 0$ comes from the boundedness of the semigroup $(T(t))_{t \geq 0}$. By construction, $(T(t))_{t \geq 0}$ becomes a contraction semigroup. We have to show that $(X, \||\cdot\||, \tau)$ is a bi-admissible space. We only prove the norming property here, as it is the least straight-forward one. For $x \in X$ we have

$$\||x\|| = \sup_{t \geq 0} \|T(t)x\| = \sup_{t \geq 0} \sup_{\substack{\varphi \in (X,\tau)' \\ \|\varphi\| \leq 1}} |\varphi(T(t)x)| \leq \sup_{t \geq 0} \sup_{\substack{\varphi \in (X,\tau)' \\ \||\varphi\|| \leq 1}} |\varphi(T(t)x)| \leq \||x\||,$$

showing that equality has to hold at all places. Finally, we observe that $(T(t))_{t \geq 0}$ is indeed a bi-continuous semigroup.

## 3.3 The Laplace Transform

Before we prove some other properties of the generator of a bi-continuous semigroup in Theorem 3.15, we want to study the resolvent of a generator of a bi-continuous semigroup. Actually we want to show that the resolvent can be expressed as a Laplace transform as in the case of strongly continuous semigroups. To do so, we start with the following result. Recall, how for a semigroup $(T(t))_{t \geq 0}$ the growth bound $\omega_0 = \omega_0(T)$ in Definition 2.5 and the operators $R_t(\lambda)$, $t \geq 0$, $\lambda \in \mathbb{C}$, in Proposition 3.2 have been defined.

**Lemma 3.10** *Let $(T(t))_{t \geq 0}$ be a bi-continuous semigroup on a bi-admissible space $(X, \| \cdot \|, \tau)$ with growth bound $\omega_0 \in \mathbb{R}$. For $\omega \in \mathbb{R}$, we denote $\mathbb{C}_{>\omega} := \{\lambda \in \mathbb{C} : \operatorname{Re}(\lambda) > \omega\}$. Then the following properties hold.*

(i) *Let $t \geq 0$ and $\lambda \in \mathbb{C}$, then $R_t(\lambda) \in \mathscr{L}(X)$ and*

$$R(\lambda) = \tau\!\lim_{t \to \infty} R_t(\lambda),$$

*exists for $\lambda \in \mathbb{C}_{>\omega_0}$ with respect to the operator norm and satisfies the estimate*

$$\|R(\lambda)\| \leq \frac{M}{\operatorname{Re}(\lambda) - \omega},$$

*for all $\lambda \in \mathbb{C}_{>\omega}$, $\omega > \omega_0$, and some constant $M \geq 1$.*

(ii) *For all $x \in X$ one has*

$$\tau\!\lim_{\substack{\lambda \to \infty \\ \lambda > \omega}} \lambda R(\lambda)x = x.$$

***Proof*** (i) To prove the first part of the result, we use the norming property of $(X, \tau)'$. Recall that by definition, there exist $M \geq 1$ and $\omega \in \mathbb{R}$ with $\omega > \omega_0$ such that $\|T(t)\| \leq M\mathrm{e}^{\omega t}$

for all $t \geq 0$. Using these constants, one obtains

$$
\begin{aligned}
\|R_t(\lambda)x\| &= \sup_{\substack{\varphi \in (X,\tau)' \\ \|\varphi\| \leq 1}} \left| \varphi \left( \int_0^t e^{-\lambda s} T(s)x \, ds \right) \right| \\[2mm]
&= \sup_{\substack{\varphi \in (X,\tau)' \\ \|\varphi\| \leq 1}} \left| \int_0^t e^{-\lambda s} \varphi(T(s)x) \, ds \right| \\[2mm]
&\leq M \|x\| \int_0^t e^{-(\mathrm{Re}(\lambda) - \omega)s} \, ds \\[2mm]
&\leq \frac{M}{\mathrm{Re}(\lambda) - \omega} \|x\|.
\end{aligned}
$$

(ii) Let $\mathcal{P}$ be the family of seminorms generating the locally convex topology $\tau$. Take $\varepsilon > 0$, $x \in X$ and $p \in \mathcal{P}$ arbitrarily. Since $(T(t))_{t \geq 0}$ is $\tau$-continuous there exists $\delta > 0$ such that $p(T(t)x - x) < \varepsilon$ whenever $t \in [0, \delta)$. From this one gets

$$
\begin{aligned}
p(\lambda R(\lambda)x - x) &= p \left( \int_0^\infty \lambda e^{-\lambda t} T(t)x \, dt - \int_0^\infty \lambda e^{-\lambda t} x \, dt \right) \\[2mm]
&\leq p \left( \int_0^\delta \lambda e^{-\lambda t} (T(t)x - x) \, dt \right) + p \left( \int_\delta^\infty \lambda e^{-\lambda t} (T(t)x - x) \, dt \right) \\[2mm]
&\leq \int_0^\delta \lambda e^{-\lambda t} p(T(t)x - x) \, dt + \sup_{\substack{\varphi \in (X,\tau)' \\ \|\varphi\| \leq 1}} \left| \varphi \left( \int_\delta^\infty \lambda e^{-\lambda t} (T(t)x - x) \right) \right| \\[2mm]
&\leq \varepsilon \int_0^\infty \lambda e^{-\lambda t} \, dt + \sup_{\substack{\varphi \in (X,\tau)' \\ \|\varphi\| \leq 1}} \int_\delta^\infty \lambda e^{-\lambda t} \varphi(T(t)x - x) \, dt \\[2mm]
&\leq \varepsilon + \|x\| \underbrace{\left( \frac{M \lambda e^{(\omega - \lambda)\delta}}{\lambda - \omega} + e^{-\lambda \delta} \right)}_{\to 0 \text{ for } \lambda \to \infty}.
\end{aligned}
$$

$\square$

Now we show that the resolvent of an generator $(A, D(A))$ of a bi-continuous semigroup can actually be expressed as a Laplace transform of the corresponding semigroup $(T(t))_{t \geq 0}$. Recall that for $\Lambda \subseteq \mathbb{C}$ a family of bounded linear operators $\mathcal{J} = \{J(\lambda) : \lambda \in \Lambda\}$ on a Banach space $X$ is called a *pseudoresolvent* if the following equality holds for all $\lambda, \mu \in \Lambda$:

$$
J(\lambda) - J(\mu) = (\mu - \lambda) J(\lambda) J(\mu).
$$

The following result is a classical exercise consisting of standard computations and is therefore left to the reader.

**Proposition 3.11** *The family of operators* $(R(\lambda))_{\lambda \in \mathbb{C}_{>\omega_0}}$ *is a pseudoresolvent.*

That the family of operators $(R(\lambda))_{\lambda \in \mathbb{C}_{>\omega_0}}$ gives rise to an unbounded and closed operator $(A, D(A))$ and one can express the resolvent of the generator of a bi-continuous semigroup by the Laplace transform is a conclusion from the following result.

**Lemma 3.12** *For each* $\lambda \in \mathbb{C}_{>\omega_0}$, *the operator* $R(\lambda)$ *is injective.*

Now we have almost reached our goal. In fact, combining Proposition 3.11 and Lemma 3.12 we may conclude from [101, Chap. III, Prop. 4.6] that there exists a closed operator $(A, D(A))$ such that $\mathbb{C}_{>\omega_0} \subseteq \rho(A)$ and $J(\lambda) = R(\lambda, A)$ for all $\lambda \in \mathbb{C}_{>\omega_0}$. This gives rise to the following definition.

**Definition 3.13** The *generator* $(A, D(A))$ of a bi-continuous semigroup $(T(t))_{t \geq 0}$ on a bi-admissible space $(X, \| \cdot \|, \tau)$ is the unique operator on $X$ such that its resolvent $R(\lambda, A)$ is given by

$$R(\lambda, A)x = \int_0^\infty e^{-\lambda t} T(t)x \, dt, \tag{3.3.1}$$

for all $\lambda \in \mathbb{C}_{>\omega_0}$ and all $x \in X$.

This definition seems on the first glance to be confusing since we already introduced the generator by means of Definition 3.1. However, the following result solves this problem, i.e., Definitions 3.1 and 3.13 give rise to the same operator.

**Theorem 3.14** *Let* $(T(t))_{t \geq 0}$ *be a bi-continuous semigroup on a bi-admissible space* $(X, \| \cdot \|, \tau)$ *with generator* $(A, D(A))$ *as defined in Definition 3.1 and with generator* $(B, D(B))$ *as defined in Definition 3.13. Then* $A = B$.

***Proof*** Let $x \in X$ and $\lambda \in \mathbb{C}_{>\omega_0}$ and observe that

$$\frac{T(h)R(\lambda, B)x - R(\lambda, B)x}{h} = \frac{e^{\lambda h} - 1}{h} \int_0^\infty e^{-\lambda t} T(t)x \, dt - \frac{e^{\lambda h}}{h} \int_0^h e^{-\lambda t} T(t)x \, dt.$$

By taking $h \to 0$ the expression converges to $\lambda R(\lambda, B)x - x = BR(\lambda, B)x$ showing that $B \subseteq A$. For the converse inclusion, let $x \in D(A)$ and defined $y := (\lambda - A)x$. By making use of Theorem 3.3 one has

$$A \int_0^\infty e^{-\lambda t} T(t)x \, dt = \int_0^\infty e^{-\lambda t} T(t) Ax \, dt.$$

Therefore,

$$R(\lambda, B)y = (\lambda - A) \int_0^\infty e^{-\lambda t} T(t)x \, dt = (\lambda - A)R(\lambda, B)x = x,$$

showing that $A \subseteq B$. $\qquad\qquad\qquad\qquad\qquad\qquad\qquad\qquad\qquad\qquad\qquad\square$

## 3.4  Resolvent Estimates

As promised above, we now can prove some other properties of generators $(A, D(A))$ of bi-continuous semigroups.

**Theorem 3.15** *Let $(T(t))_{t\geq 0}$ be a bi-continuous semigroup on a bi-admissible space $(X, \|\cdot\|, \tau)$ with generator $(A, D(A))$. Then the following properties hold.*

(a) *The domain $D(A)$ is bi-dense in $X$, i.e., for each $x \in X$ there exists a $\|\cdot\|$-bounded sequence $(x_n)_{n\in\mathbb{N}}$ in $D(A)$ which $\tau$-converges to $x$.*
(b) *$(A, D(A))$ is a Hille–Yosida operator, i.e., for all $\omega > \omega_0$ there exists $M \geq 1$ such that for all $n \in \mathbb{N}$ and $\lambda \in \mathbb{C}_{>\omega}$ the following estimate holds*

$$\left\| R(\lambda, A)^n \right\| \leq \frac{M}{(\mathrm{Re}(\lambda) - \omega)^n}.$$

(c) *Let $(x_n)_{n\in\mathbb{N}}$ be a $\|\cdot\|$-bounded sequence in $X$ which $\tau$-converges to $0$. Let $\alpha > \omega_0$. Then*

$$\tau\lim_{n\to\infty} (\lambda - \alpha)^k R(\lambda, A)^k x_n = 0,$$

*uniformly for $k \in \mathbb{N}$ and $\lambda > \alpha$.*
(d) *$\tau\lim_{\lambda\to\infty} \lambda^k R(\lambda, A)^k x = x$ for all $x \in X$ and $k \in \mathbb{N}$.*

***Proof*** (a) Let $x \in X$ and define a sequence $(x_n)_{n\in\mathbb{N}}$ of elements in $X$ by $x_n := nR(\lambda, A)x$ for $n > \omega_0$ and $x_n := 0$ otherwise. By construction, one has that $x_n \in D(A)$ for each $n \in \mathbb{N}$. Moreover, by Lemma 3.10 the sequence $(x_n)_{n\in\mathbb{N}}$ is a $\|\cdot\|$-bounded sequence such that $x_n \overset{\tau}{\to} x$ for $n \to \infty$.

(b) Let $x \in X, \omega > \omega_0$ and $\lambda \in \mathbb{C}_{>\omega}$ be arbitrary. Making use of the assumption that $(X, \tau)'$ is norming for $X$ one obtains the following estimate

$$\left\| \frac{R(\mu, A)x - R(\lambda, A)x}{\mu - \lambda} + \int_0^\infty t e^{-\lambda t} T(t)x \, dt \right\|$$

$$\leq \sup_{\substack{\varphi \in (X,\tau)' \\ \|\varphi\| \leq 1}} \int_0^\infty \left| \frac{e^{-\mu t} - e^{-\lambda t}}{\mu - \lambda} + t e^{-\lambda t} \right| \cdot |\varphi(T(t)x)| \, dt$$

$$\leq M \|x\| \int_0^\infty \left| \frac{e^{-\mu t} - e^{-\lambda t}}{\mu - \lambda} + t e^{-\lambda t} \right| e^{\omega t} \, dt.$$

We observe that as a consequence of Lebesgue's dominated convergence theorem the expression on the left-hand side tends to zero if $\mu$ tends to $\lambda$. As an interim result one concludes that by induction the following holds

$$\frac{d^n}{d\lambda} R(\lambda, A)x = (-1)^n \int_0^\infty t^n e^{-\lambda t} T(t)x \, dt, \tag{3.4.1}$$

for all $x \in X$, $n \in \mathbb{N}$ and $\lambda \in \mathbb{C}_{>\omega_0}$. Furthermore, by following the proof of Proposition 3.11 as well as [101, Chap. IV, Prop. 1.3] one gets that the map $\lambda \mapsto R(\lambda, A)$ is holomorphic in $\mathbb{C}_{>\omega_0}$ such that

$$\frac{d^n}{d\lambda} R(\lambda, A)x = (-1)^n n! R(\lambda, A)^{n+1} x, \tag{3.4.2}$$

for each $x \in X$ and $n \in \mathbb{N}$. Putting (3.4.1) and (3.4.2) together yields the desired Hille–Yosida estimate, i.e., for all $x \in X$, $n \in \mathbb{N}$ and $\lambda \in \mathbb{C}_{>\omega}$

$$\left\| R(\lambda, A)^n x \right\| \leq \sup_{\substack{\varphi \in (X,\tau)' \\ \|\varphi\| \leq 1}} \frac{1}{(n-1)!} \int_0^\infty t^{n-1} \left| e^{-\lambda t} \varphi(T(t)x) \right| dt$$

$$\leq \frac{M \|x\|}{(n-1)!} \int_0^\infty t^{n-1} e^{(\omega - \mathrm{Re}(\lambda))t} \, dt$$

$$= \frac{M}{(\mathrm{Re}(\lambda) - \omega)^n} \|x\|.$$

(c) Let $\mathcal{P}$ be the family of seminorms that generate $\tau$. Let $\varepsilon > 0$, $p \in \mathcal{P}$ and $\alpha > \omega_0$ be arbitrary. It is left to the reader to check that in fact the semigroup $(e^{-\alpha t} T(t))_{t \geq 0}$ is globally bi-equicontinuous, i.e., for each $\|\cdot\|$-bounded $\tau$-convergent sequence $(x_k)_{k \in \mathbb{N}}$ with limit $x \in X$ one has that $e^{-\alpha t} T(t)x_n \xrightarrow{\tau} e^{-\alpha t} T(t)x$ uniformly for $t \geq 0$ for $n \to \infty$. Again by using (3.4.1) and (3.4.2), there exists $K \in \mathbb{N}$ such that for all $k \geq K$ one has

$$p\left((\lambda - \alpha)^n R(\lambda, A)^n (x_k - x)\right) \leq \frac{(\lambda - \alpha)^n}{(n-1)!} \int_0^\infty t^{n-1} e^{-(\lambda - \alpha)t} p(e^{-\alpha t} T(t)(x_k - x)) \, dt < \varepsilon,$$

uniformly for $n \in \mathbb{N}$ and $\lambda > \alpha$.

(d) Without loss of generality we suppose that the semigroup $(T(t))_{t \geq 0}$ is $\|\cdot\|$-bounded, i.e., there exists $M \geq 1$ such that $\|T(t)\| \leq M$ for all $t \geq 0$. By $\mathcal{P}$ we again denoted the generating family of seminorms of the locally convex topology $\tau$. Let $x \in X$, $n \in \mathbb{N}$, $\varepsilon > 0$ and $p \in \mathcal{P}$ be arbitrary. Since $(T(t))_{t \geq 0}$ is strongly $\tau$-continuous, we find $\delta > 0$ such that for $t \in [0, \delta)$ holds that $p(T(t)x - x) < \frac{\varepsilon}{2}$. Using (3.4.1) and (3.4.2) yields $\lambda_0 > \omega_0$ such that for all $\lambda \geq \lambda_0$ the following holds

$$
\begin{aligned}
& p(\lambda^n R(\lambda, A)^n x - x) \\
& \leq p\left( \frac{\lambda^n}{(n-1)!} \int_0^\delta t^{n-1} e^{-\lambda t} (T(t)x - x)\, dt \right) + p\left( \frac{\lambda^n}{(n-1)!} \int_\delta^\infty t^{n-1} e^{-\lambda t} (T(t)x - x)\, dt \right) \\
& < \frac{\varepsilon}{2} \frac{\lambda^n}{(n-1)!} \int_0^\delta t^{n-1} e^{-\lambda t}\, dt + (1+M)\,\|x\| \frac{\lambda^n}{(n-1)!} \int_\delta^\infty t^{n-1} e^{-\lambda t}\, dt \\
& \leq \frac{\varepsilon}{2} + (1+M)\,\|x\| \left( \frac{\lambda^n}{(n-1)!} \delta^{n-1} e^{-\lambda \delta} + \cdots + \lambda \delta e^{-\lambda \delta} + e^{-\lambda \delta} \right) \\
& \leq \varepsilon.
\end{aligned}
$$

$\square$

As Hille–Yosida operators are important all over, it deserves a separate definition.

**Definition 3.16** A linear operator $(A, D(A))$ on a Banach space $X$ is called a *Hille–Yosida operator* (of type $\omega$), if there exists $M \geq 1$ such that for all $n \in \mathbb{N}$ and $\lambda \in \mathbb{C}_{>\omega}$ the following estimate holds

$$
\left\| R(\lambda, A)^n \right\| \leq \frac{M}{(\mathrm{Re}(\lambda) - \omega)^n}. \tag{3.4.3}
$$

## 3.5   A Hille–Yosida Generation Type Theorem

We have seen in Theorem 3.15 that the resolvent of generators of bi-continuous semigroups satisfy certain properties. However, the converse question, which operator is a generator of a bi-continuous semigroup, is not answered yet. The answer to this question is actually given by the famous Hille–Yosida generation theorem for bi-continuous semigroups. Here we will discuss the original proof given by F. Kühnemund [160, Thm. 16] which uses integrated semigroups. In a later chapter, we will give another proof by using extrapolation spaces. In order to prove the Hille–Yosida theorem in the way Kühnemund did it, we need a result about a connection between Hille–Yosida operators, see Definition 3.16, and integrated semigroups, cf. [17, Sect. 3.3] and [129].

**Proposition 3.17** *Let $(A, D(A))$ be a Hille–Yosida operator of type $\omega$ and let $\underline{X} := \overline{D(A)}^{\|\cdot\|}$. Then there exists an integrated semigroup $(F(t))_{t \geq 0}$ on $X$ with the following properties:*

(a) *The map $\mathbb{R}_+ \ni t \mapsto F(t)x \in X$ is continuously differentiable for all $x \in D(A)$ with respect to the norm. The operator family $(F'(t)_{|\underline{X}})_{t\geq 0}$ is a $C_0$-semigroup on $\underline{X}$.*
(b) *The integrated semigroup $(F(t))_{t\geq 0}$ is given by*

$$F(t) = \lim_{k\to\infty} (-1)^k (k+1) \int_{\frac{k}{t}}^{\infty} s^k R(s, A)^{k+2} \, ds,$$

*for all $t > 0$, $F(0) = 0$ and $\|F(t+h) - F(t)\| \leq M \int_t^{t+h} e^{\omega r} \, dr$ for all $t, h \geq 0$.*

Now, we are able to state and prove the Hille–Yosida theorem for bi-continuous semigroups.

**Theorem 3.18** *Let $(X, \|\cdot\|, \tau)$ be a bi-admissible space. For a linear operator $(A, D(A))$ on $X$, the following assertions are equivalent:*

(a) *The operator $(A, D(A))$ is the generator of a bi-continuous semigroup $(T(t))_{t\geq 0}$ satisfying $\|T(t)\| \leq Me^{\omega t}$ for all $t \geq 0$ and some $\omega \in \mathbb{R}$ and $M \geq 1$.*
(b) *The operator $(A, D(A))$ is a Hille–Yosida operator of type $(M, \omega)$, i.e.,*

$$\|R(\lambda, A)^n\| \leq \frac{M}{(\lambda - \omega)^n}$$

*for all $n \in \mathbb{N}$ and for all $\lambda > \omega$. Moreover, $(A, D(A))$ is bi-densely defined and the family*

$$\left\{(\lambda - \alpha)^n R(\lambda, A)^n : n \in \mathbb{N}, \ \lambda \geq \alpha\right\} \tag{3.5.1}$$

*is bi-equicontinuous for each $\alpha > \omega$, meaning that for each norm bounded $\tau$-null sequence $(x_k)_{k\in\mathbb{N}}$ one has $(\lambda - \alpha)^n R(\lambda, A)^n x_k \to 0$ in $\tau$ uniformly for $n \in \mathbb{N}$ and $\lambda \geq \alpha$ as $n \to \infty$.*
(c) *There exi $(F(t))_{t\geq 0}$ satisfying the following conditions:*

  (i) *$F(\cdot)x \in C^1(\mathbb{R}_+, X)$ for all $x \in D(A)$ and $(F'(t)_{|\underline{X}})_{t\geq 0}$ exists and is a $C_0$-semigroup on $\underline{X}$.*
  (ii) *$F(\cdot)x \in C^1(\mathbb{R}_+, (X, \tau))$ for all $x \in X$.*
  (iii) *The family of operator $(F'(t))_{t\geq 0}$ is locally bi-equicontinuous.*
  (iv) *$(F'(t))_{t\geq 0}$ is exponentially bounded on $X$.*
  (v) *For all $\lambda > \omega$ and $x \in X$ we have*

$$R(\lambda, A)x = \int_0^{\infty} e^{-\lambda t} F'(t)x \, dt.$$

***Proof*** We prove the implications (a)$\Rightarrow$(b), (b)$\Rightarrow$(c), and (c)$\Rightarrow$(a) in order to show that the statements are all equivalent.

(a)$\Rightarrow$(b): If $(A, D(A))$ is the generator of a bi-continuous semigroup $(T(t))_{t\geq 0}$, then assertion (b) follows directly from Theorems 3.3 and 3.15.

(b)$\Rightarrow$(c): Firstly, we assume that $\alpha = 0$. By Proposition 3.17 we know that there exists an integrated semigroup $(F(t))_{t\geq 0}$ satisfying conditions (i). For property (ii) let $x \in X$ and $t \geq 0$ and define a sequence $(D_m(x, t))_{m\in\mathbb{N}}$ in $X$ by

$$D_m(x, t) := \frac{\left(F(t + \frac{1}{m}) - F(t)\right) x}{\frac{1}{m}}, \quad m \in \mathbb{N}.$$

Again by Proposition 3.17, this sequence in $\|\cdot\|$-bounded. We show that $(D_m(x, t))_{m\in\mathbb{N}}$ actually defines a $\tau$-Cauchy sequence. To do so, let $\varepsilon > 0$, $x \in X$ and $p \in \mathcal{P}$ be arbitrary. By the assumptions of (b) there exists a $\|\cdot\|$-bounded and $\tau$-convergent sequence $(x_n)_{n\in\mathbb{N}}$ in $D(A)$, say with limit $x$, and $N_1 \in \mathbb{N}$ such that for $n \geq N_1$

$$p(s^k R(s, A)^k (x - x_n)) < \frac{\varepsilon}{3},$$

uniformly for $k \in \mathbb{N}$ and $s \geq 0$. Using Proposition 3.17 we obtain the following estimate by using the previous estimate

$$p(D_m(x - x_n, t))$$

$$= p\left(\lim_{k\to\infty} m\left[(k+1)\int_{\frac{k}{t}}^{\infty} s^k R(s, A)^{k+2}(x - x_n)\, ds - (k+1)\int_{\frac{k}{t+\frac{1}{m}}}^{\infty} s^k R(s, A)^{k+2}(x - x_n)\, ds\right]\right)$$

$$= p\left(m \lim_{k\to\infty} (k+1)\left[\int_{\frac{k}{t+\frac{1}{m}}}^{\frac{k}{t}} s^k R(s, A)^{k+2}(x - x_n)\, ds\right]\right) < \frac{\varepsilon}{3}m \lim_{k\to\infty}\left[-\frac{1}{s}\right]_{\frac{k}{t+\frac{1}{m}}}^{\frac{k}{t}}$$

$$= \frac{\varepsilon}{3} \lim_{k\to\infty}\left(1 + \frac{1}{k}\right) = \frac{\varepsilon}{3}$$

$$\tag{3.5.2}$$

for all $n \geq N_1$ and uniformly for $m \in \mathbb{N}$ and $t \geq 0$. Since (i) is valid, $(D_m(x_n, t))_{m\in\mathbb{N}}$ is a $\tau$-Cauchy sequence for all $n \in \mathbb{N}$. In particular, there exists $N_2 \in \mathbb{N}$ such that for all $m, k \geq N_2$

$$p(D_m(x, t) - D_l(x, t))$$
$$\leq p(D_m(x - x_{N_2}, t)) + p(D_m(x_{N_2}, t) - D_l(x_{N_2}, t)) + p(D_l(x - x_{N_2}, t)) < \varepsilon.$$

Since $(X\tau)$ is sequentially complete on $\|\cdot\|$-bounded sets, the map $t \mapsto F(t)x$ is differentiable with respect to $\tau$ for all $x \in X$.

Before proving that its derivative is continuous, we show assertion (iii). Clearly, by (3.5.2) we conclude that the operator family $\left\{F'(t) : t \geq 0\right\}$ is globally bi-equicontinuous.

Let us show that $F'(\cdot)x$ is $\tau$-continuous for all $x \in X$ to conclude assertion (ii). Let $x \in X, \varepsilon > 0, t \geq 0$ and $p \in \mathcal{P}$ be arbitrary. Furthermore, let $(x_n)_{n \in \mathbb{N}}$ be a $\|\cdot\|$-bounded and $\tau$-convergent sequence in $D(A)$ with limit $x \in X$. Since $F'(\cdot)x$ is $\tau$-continuous for each $x \in D(A)$ we can find $\delta > 0$ (depending on $n$) such that

$$p(F'(t)x_n - F'(s)x_n) < \frac{\varepsilon}{3}$$

whenever $|t - s| < \delta$. Again by (3.5.2) we can find $N \in \mathbb{N}$ such that $|t - s| < \delta$ implies that

$$p(F'(t)x - F'(s)x) \leq p(F'(t)(x - x_N)) + p(F'(t)x_N - F'(s)x_N) + p(F'(s)(x_N - x)) < \varepsilon,$$

showing that indeed (ii) holds.

Property (iv) again follows by Proposition 3.17 and the norming property of $(X, \tau)'$. In fact, for all $x \in X$ we have

$$
\begin{aligned}
\|F'(t)x\| &= \sup_{\substack{\varphi \in (X, \tau)' \\ \|\varphi\| \leq 1}} \left| \varphi \left( \tau\text{-}\lim_{m \to \infty} D_m(x, t) \right) \right| \\
&= \sup_{\substack{\psi \in (X, \tau)' \\ \|\varphi\| \leq 1}} \left| \tau\text{-}\lim_{m \to \infty} \varphi(D_m(x, t)) \right| \\
&\leq M \|x\| \lim_{m \to \infty} m \int_t^{t + \frac{1}{m}} e^{\omega s} \, ds = M e^{\omega t} \|x\|.
\end{aligned}
$$

For (v) we observe that

$$\int_0^a e^{-\lambda t} F'(t)x \, dt,$$

as a $\tau$-integral exists for all $a \geq 0$, $\lambda > \omega$ and $x \in X$, due to the sequential completeness of $(X, \tau)$ on $\|\cdot\|$-bounded sets. By the norming property of $(X, \tau)'$ and by following the lines of the proof of Lemma 3.10 we conclude that

$$\int_0^\infty e^{-\lambda t} F'(t)x \, dt,$$

exists as $\tau$-improper integral. It remains to prove that

$$R(\lambda, A)x = \int_0^\infty e^{-\lambda t} F'(t)x \, dt$$

for all $x \in X$ and $\lambda > \omega$. For this, let $x \in X$ and $\varepsilon > 0$ be arbitrary. There exists a $\|\cdot\|$-bounded $(x_n)_{n \in \mathbb{N}}$ in $D(A)$ and $N \in \mathbb{N}$ such that $p(x - x_n) < \frac{\varepsilon}{2}$ and hence $p(R(\lambda, A)(x - x_n)) < \frac{\varepsilon}{2}$ for all $n \geq N$. By construction $(F(t))_{t \geq 0}$ is an integrated semigroup and by (iii) one obtains

$$p\left(R(\lambda, A)x - \int_0^\infty e^{-\lambda t} F'(t)x \, dt\right)$$

$$\leq p(R(\lambda, A)x - x_N) + p\left(\int_0^\infty e^{-\lambda t} F'(t)(x_N - x) \, dt\right) < \varepsilon,$$

showing that assertion (v) holds as well. By a rescaling argument, we obtain the desired properties for arbitrary $\alpha > \omega$.

(c)$\Rightarrow$(a): Let $(F(t))_{t\geq 0}$ be the given integrated semigroup. The goal is to construct a bi-continuous semigroup $(T(t))_{t\geq 0}$. For this goal define

$$T(t)x := F'(t)x = \tau\lim_{h\to 0} \frac{F(t+h)x - F(t)x}{h}, \quad x \in X, \ t \geq 0.$$

By the assumption it remains to show that the semigroup law is satisfied for $(T(t))_{t\geq 0}$ on $X$. By (i) the semigroup law already holds on $\underline{X}$. By the same arguments as in Lemma 3.10 and equality (iv) one sees that $\tau\lim_{\lambda\to\infty} \lambda R(\lambda, A)x = x$. Hence, there exists a $\|\cdot\|$-bounded sequence $(x_n)_{n\in\mathbb{N}}$ in $D(A)$ which is $\tau$-convergent to $x$. Since $x_n \in D(A)$ for all $n \in \mathbb{N}$ we have $T(t+s)x_n = T(t)T(s)x_n$ for all $t, s \geq 0$ and $n \in \mathbb{N}$ and

$$\tau\lim_{n\to\infty} T(t+s)x_n = T(t+s)x.$$

Since $(T(s)x_n)_{n\in\mathbb{N}}$ is $\|\cdot\|$-bounded and $\tau$-convergent we also obtain

$$\tau\lim_{n\to\infty} T(t)T(s)x_n = T(t)T(s)x.$$

Since the locally convex topology $\tau$ is supposed to be Hausdorff limits are unique. Hence the semigroup law holds for $(T(t))_{t\geq 0}$. By the same way we obtains $T(0) = I$. $\qquad\square$

As it is already the case of $C_0$-semigroups, the Hille–Yosida theorem is not easy to apply when it comes to explicit examples. For this reason, we will present a Lumer–Phillips type generation theorem for bi-continuous semigroups later in Sect. 6.4.

### 3.5.1 An Application to Semigroups Induced by Flows

Around 1970, J.W. Neuberger [193, 194] and D.L. Lovelady [172] started a theory of non-linear, jointly continuous semigroups (flows) on a metric space $\Omega$ in terms of their Lie generator. A certain analog to ideas of S. Lie, they define an associated linear semigroup, the semigroup induced by the flow, on the Banach space $X$ of bounded, continuous functions on $\Omega$. This semigroup, in general, is not strongly continuous for the supremum norm. However, there exists a locally convex topology on $X$ such that strong continuity holds. 20 years later, in joint work by J.R. Dorroh and J.W. Neuberger [90, 91] a complete characterization of the Lie generator of a jointly continuous flow on a complete, separable, metric space $\Omega$

was established. The main step towards these results was, using a result by F.D. Sentilles [213], to introduce the "strict" locally convex topology $\beta$ on the space $C_b(\Omega)$ of bounded, continuous functions on $\Omega$, for which the associated linear semigroup becomes strongly continuous.

**Definition 3.19**  A *jointly continuous flow* on a topological space $\Omega$ is a map $\varphi : \mathbb{R}_+ \times \Omega \to \Omega$, $(t, x) \mapsto \varphi_t(x)$, satisfying

(i)   $\varphi_0 = I$, $\varphi_{t+s} = \varphi_t \varphi_s$ for all $t, s \geq 0$.
(ii)  The mapping $(t, x) \mapsto \varphi_t(x)$ is jointly continuous.

The *Lie generator* of a jointly continuous flow $\varphi$ is the linear operator $(A, D(A))$ on $C_b(\Omega)$ consisting of all pairs $(f, g)$ such that $f, g \in C_b(\Omega)$ and

$$g(x) = \lim_{t \to 0} \frac{f(\varphi_t(x)) - f(x)}{t}$$

for all $x \in \Omega$. A linear operator $(A, D(A))$ on $C_b(\Omega)$ is called a *derivation* if $f, g \in D(A)$ implies $fg \in D(A)$ and $A(fg) = fA(g) + A(f)g$. The linear semigroup $(T(t))_{t \geq 0}$ on $C_b(\Omega)$ defined by

$$T(t)f = f \circ \varphi_t, \quad f \in C_b(\Omega), \; t \geq 0,$$

is called the *semigroup induced by the flow* $\varphi$.

Let us summarize some results that follow immediately from the work of Dorroh and Neuberger. The first observation is that the semigroup induced by a given jointly continuous flow $\varphi$ is bi-continuous with respect to the compact-open topology, cf. [90, Thm. 2.1 and 2.2].

**Proposition 3.20**  *Let $\varphi$ be a jointly continuous on $\Omega$ and $(T(t))_{t \geq 0}$ the semigroup induced by the flow $\varphi$ on $C_b(\Omega)$. Then $(T(t))_{t \geq 0}$ is a bi-continuous contraction semigroup with respect to the compact-open topology $\tau_{co}$.*

Every bi-continuous semigroup yields a generator as defined by Definition 3.1. As a matter of fact, this generator coincides with the Lie generator of the jointly continuous flow $\varphi$, cf. [91, Prop. 2.4].

**Proposition 3.21**  *Let $\varphi$ be a jointly continuous flow on $\Omega$ and let $f, g \in C_b(\Omega)$ and $\tau = \tau_{co}$. Then the following assertions are equivalent:*

(i)   $g(x) = \lim_{t \to 0} \dfrac{f(\varphi_t(x)) - f(x)}{t}$ *for all $x \in \Omega$.*

(ii)  $g = \tau\lim\limits_{t \to 0} \dfrac{f \circ \varphi_t - f}{t}$

We can now apply Theorem 3.18 to obtain a generation theorem for bi-continuous contraction semigroups induced by jointly continuous flows. Moreover, by Proposition 3.21, this yields a characterization of the Lie generator of a jointly continuous flow thereby reproving the results in [91, Thm. 3.1 and 3.2].

**Theorem 3.22**  *Let $(A, D(A))$ be a linear operator on $C_b(\Omega)$. The following are equivalent:*

(a)  *The operator $(A, D(A))$ is the Lie generator of a jointly continuous flow $\varphi$ on $\Omega$.*

(b)  *$(A, D(A))$ is a derivation and generates a bi-continuous contraction semigroup $(T(t))_{t \geq 0}$ on $C_b(\Omega)$ with respect to $\tau_{co}$ which is induced by a jointly continuous flow $\varphi$.*

(c)  *$(A, D(A))$ is a bi-densely defined Hille–Yosida operator of type 0 and a derivation, and the family $\left\{ (s - \alpha)^k R(s, A)^k : k \in \mathbb{N}, \; s \geq \alpha \right\}$ is bi-equicontinuous for every $\alpha > 0$.*

***Proof*** (a)$\Rightarrow$(b): It is easy to see that the Lie generator of a jointly continuous flow $\varphi$ on $\Omega$ is a derivation. Define $T(t)f := f \circ \varphi_t$, $f \in C_b(\Omega)$, $t \geq 0$. By Proposition 3.20, we see that $(T(t))_{t \geq 0}$ is a bi-continuous contraction semigroup with respect to the compact-open topology $\tau_{co}$ on $C_b(\Omega)$. Let $(B, D(B))$ its generator. By combining Proposition 3.21 with Definition 3.1 we have $A = B$.

(b)$\Rightarrow$(c): The Hille–Yosida theorem, cf. Theorem 3.18, immediately yields assertion (c).

(c)$\Rightarrow$(a): By Theorem 3.18 there exists a semigroup $(T(t))_{t \geq 0}$ which is bi-continuous with respect to the compact-open topology $\tau_{co}$ on $C_b(\Omega)$. Let us show that each operator is an algebra homomorphism, i.e.,

$$T(t)(fg) = T(t)f \cdot T(t)g, \quad f, g \in C_b(\Omega), \; t > 0. \tag{3.5.3}$$

Indeed, for $f, g \in D(A)$ we define a function $\eta : [0, t] \to C_b(\Omega)$ by

$$\eta(s) := T(t - s)\left[T(s)f \cdot T(s)g\right].$$

This map is differentiable with respect to $\tau_{co}$ by Theorem 3.3. Moreover, $\eta(0) = T(t)(fg)$ and $\eta(t) = T(t)f \cdot T(t)g$. Since $(A, D(A))$ is a derivation, we have $\eta'(s) = 0$ for $s \in [0, t]$. Hence $\eta(0) = \eta(t)$ and $T(t)(fg) = T(t)f \cdot T(t)g$. The bi-density of $D(A)$ yields (3.5.3) for all $f, g \in C_b(\Omega)$. Since $(T(t))_{t \geq 0}$ consists of $\tau_{co}$-continuous algebra homomorphisms, for each $x \in \Omega$ there exists a unique $y \in \Omega$ such that $T(t)f(x) = f(y)$ for all $f \in C_b(\Omega)$, cf. [91, Prop. 2.1]. Therefore, $(T(t))_{t \geq 0}$ is a contraction semigroup and for every $t \geq 0$ one can define a flow $\varphi$ on $\Omega$ by

$$f(\varphi_t(x)) = T(t)f(x), \quad x \in \Omega, \; f \in C_b(\Omega).$$

This flow is jointly continuous by [90, Thm. 3.3]. Furthermore, by Theorem 3.14, the generator of $(T(t))_{t\geq 0}$ is given by

$$Af = \tau\!\lim_{t\to 0} \frac{f \circ \varphi_t - f}{t}.$$

By Proposition 3.21, this is equivalent to the fact that $(A, D(A))$ is the Lie generator of the flow $\varphi$. This concludes the proof of the equivalent assertions. $\qquad\square$

## Notes on This Chapter

In this chapter we introduced the generator of bi-continuous semigroups in two different ways; one which is related to the classical definition for $C_0$-semigroups (Definition 3.1) and one by means of resolvents and the Laplace transform (Definition 3.13). Both operators actually yield the same operator. Both approaches were investigated by Kühnemund, see [159, Sect. 1.2] or [160, Sect. 3]. The proof of the Hille–Yosida type generation theorem for bi-continuous semigroups presented here, cf. Theorem 3.18, via integrated semigroups is also due to Kühnemund, see [159, Thm. 1.28] or [160, Thm. 16]. We will present another more straightforward proofs by means of extrapolation spaces later in Sect. 5.1.

# General Intermediate and Extrapolated Spaces  4

In this chapter, we delve into the construction of general intermediate and extrapolated spaces, essential tools in the analysis of linear operators on Banach spaces. The focus is on developing a framework that extends the classical Sobolev and extrapolation spaces for boundedly invertible operators, which will also serve as a foundation for the treatment of bi-continuous semigroups in the subsequent chapter. While many of the results presented here are well known, we aim to provide a comprehensive exposition to address cases involving non-densely defined operators, introducing new techniques and concepts where necessary. These developments are critical for the further exploration of operator theory and its applications.

## 4.1 Sobolev and Extrapolation Spaces for Invertible Operators

In this section, we construct abstract Sobolev (Hölder) and extrapolation spaces (the so-called Sobolev scale) for a boundedly invertible linear operator defined on a Banach space. Some of the results are well known and even standard, but we chose to include them here for the sake of completeness and also because they are needed for the construction of spaces when we deal with not densely defined operators. The emphasis will be, however, on this latter case, when the construction is new, see Sect. 4.1.2. We also note that everything what follows is also valid for operators on Fréchet spaces.

The theory presented in this chapter approaches the theory of intermediate and extrapolation spaces and will be a preparation for Chap. 5 where we eventually also discuss intermediate and extrapolated spaces in the framework of bi-continuous semigroups.

The following is a standing assumption in this chapter.

© The Author(s), under exclusive license to Springer Nature Switzerland AG 2026    57
C. Budde, *Bi-Continuous Operator Semigroups*, Frontiers in Mathematics,
https://doi.org/10.1007/978-3-032-12948-2_4

**Assumption A** We suppose that $A : D(A) \to X$ is a (not necessarily densely defined) linear operator on a Banach space $X$ with 0 in the resolvent set $\rho(A)$.

As a matter of fact, it is only for convenience to suppose $0 \in \rho(A)$ instead of $\rho(A) \neq \emptyset$. Indeed, if $\lambda \in \rho(A)$ we may consider $A - \lambda$ and carry out the constructions for this new operator satisfying $0 \in \rho(A - \lambda)$. The arising spaces will not depend on $\lambda \in \rho(A)$ (up to isomorphism).

### 4.1.1 Abstract Sobolev Spaces

The material presented here is standard, see Nagel [186], Nagel, Nickel, Romanelli [188] or Engel, Nagel [101, Sect. II.5], and some parts are valid even for operators on locally convex spaces, when one has to argue with a family of generating seminorms instead of one norm. We set $X_1 := D(A)$ which becomes a Banach space if endowed with the graph norm

$$\|x\|_A := \|x\| + \|Ax\|. \tag{4.1.1}$$

An equivalent norm is given by $\|x\|_{X_1} := \|Ax\|$ since we have assumed $0 \in \rho(A)$. Then we have the isometric isomorphism

$$A : X_1 \to X \quad \text{with inverse} \quad A^{-1} : X \to X_1.$$

**Definition 4.1** Recall the assumption that $0 \in \rho(A)$, and take $n \in \mathbb{N}, n \geq 1$.

(a) We define
$$X_n := D(A^n) \quad \text{and} \quad \|x\|_{X_n} := \|A^n x\| \quad \text{for } x \in X_n.$$

If we want to stress the dependence on $A$, then we write $X_n(A)$ and $\| \cdot \|_{X_n(A)}$.

(b) Let
$$X_\infty(A) := \bigcap_{n \in \mathbb{N}} X_n,$$

often abbreviated as $X_\infty$.

(c) We further set
$$\underline{X} := \overline{D(A)}, \quad \underline{A} := A|_{\underline{X}},$$

the part of $A$ in $\underline{X}$, i.e.:

$$D(\underline{A}) = \left\{ x \in D(A) : Ax \in \underline{X} \right\}.$$

Moreover, we let
$$\underline{X}_n := D(\underline{A}^n), \quad \|x\|_{\underline{X}_n} := \|\underline{A}^n x\|.$$

To be specific about the underlying operator $A$ we write $\underline{X}_n(A)$ and $\|x\|_{\underline{X}_n(A)}$.

(d) For $n \in \mathbb{N}$ we set $A_n := A|_{X_n}$, the part of $A$ in $X_n$, in particular $A_0 = A$. Similarly, we let $\underline{A}_n := \underline{A}|_{\underline{X}_n}$, for example, $\underline{A}_0 = \underline{A}$. By this notation we also understand implicitly that the surrounding space is $X_n(A)$, respectively, $\underline{X}_n(A)$ with its norm, see Remark 4.2.

**Remark 4.2** 1. By "underlining" we always indicate an object which is in some sense smaller than the one without underlining. The space $\underline{X}(A)$ is connected with the domain of $D(A)$, and the whole issue of distinguishing between $X$ and $\underline{X}$ becomes relevant only if $A$ is not densely defined but its part $\underline{A}$ *is* (cf. Remark 4.5). We keep to the notation $\underline{A}$ for the part of the operator $A$ instead of $A|_{\underline{X}}$.

2. If $A$ is densely defined, then $X_n(A) = \underline{X}_n(A)$ for each $n \in \mathbb{N}$. In particular, if $\underline{X}_1(A) = D(\underline{A})$ is dense in $\underline{X}(A)$, then $\underline{X}_n(A) = X_n(A)$ for each $n \in \mathbb{N}$.

3. For $n \in \mathbb{N}$ we evidently have $X_1(A^n) = X_n(A)$. Also $\underline{X}_1(A^n) = \underline{X}_n(A)$ holds, because $D(\underline{A}^n) = D(\underline{A^n})$. Indeed, the inclusion "$D(\underline{A^n}) \subseteq D(\underline{A}^n)$" is trivial. While for $x \in D(\underline{A}^n)$ we have $x \in \underline{X}$ and $A^n x \in \underline{X}$, implying $A^{n-1}x \in D(\underline{A})$, and then recursively $x \in D(\underline{A^n})$.

4. For $x \in D(A_n) = D(A^{n+1})$ we have $\|x\|_{X_1(A_n)} = \|A_n x\|_{X_n(A)} = \|A^{n+1}x\| = \|x\|_{X_{n+1}(A)}$. Similarly $D(\underline{A}_n) = D(\underline{A}^{n+1})$.

**Proposition 4.3** *Suppose $\underline{A}$ is densely defined in $\underline{X}$.*

(a) *For $n \in \mathbb{N}$ the mappings $A^n : X_n \to X$ and $\underline{A}^n : \underline{X}_n \to \underline{X}$ are isometric isomorphisms.*

(b) *For $n \in \mathbb{N}$ the operators $A_n : X_{n+1} \to X_n$ and $\underline{A}_n : \underline{X}_{n+1} \to \underline{X}_n$ are isometric isomorphisms that intertwine $A_{n+1}$ and $A_n$, respectively, $\underline{A}_{n+1}$ and $\underline{A}_n$.*

(b) *$X_\infty$ is dense in $\underline{X}_n$ for each $n \in \mathbb{N}$. As a consequence, $\underline{X}_m$ is dense in $\underline{X}_n$ for each $m, n \in \mathbb{N}$ with $m \geq n$.*

*Proof* The statements (a) and (b) are trivial by construction.

(c) This is [18, Thm. 6.2] due to Arendt, El-Mennaoui, and Kéyantuo, because $\underline{A}$ is densely defined in $\underline{X}$. $\qquad\square$

**Remark 4.4** We note that the proof of the assertion (c) in [18, Thm. 6.2] is based on a Mittag-Leffler-type result due to Esterle [104] which is valid in complete metric spaces. Hence the statements (a), (b), and (c) all remain true for Fréchet spaces with verbatim the same proof as in [18].

Henceforth, another standing assumption will be the following (though not everywhere needed).

**Assumption B** The operator $\underline{A} := A|_{\underline{X}} : D(\underline{A}) \to \underline{X}$ is densely defined, i.e.:

$$\overline{D(\underline{A})} = \underline{X}.$$

**Remark 4.5** The condition of $D(\underline{A})$ being dense in $\underline{X}$ holds, for example, if there are $M, \omega > 0$ such that $(\omega, \infty) \subseteq \rho(A)$ and

$$\|\lambda R(\lambda, A)\| \le M \quad \text{for all } \lambda > \omega. \tag{4.1.2}$$

Indeed, in this case, we have for $x \in D(A)$

$$\|\lambda R(\lambda, A)x - x\| = \|R(\lambda, A)Ax\| \le \frac{M\|Ax\|}{\lambda} \to 0 \quad \text{for } \lambda \to \infty.$$

Hence $D(A^2) \subseteq D(\underline{A})$ is dense in $D(A)$ for the norm of $X$, and this implies the density of $D(\underline{A})$ in $\underline{X}$. An operator $A$ satisfying (4.1.2) is often said to have a *ray of minimal growth*, see, e.g., [175, Chap. 3], and also Sect. 4.2. Another term used is *"weak Hille–Yosida operator."*

**Proposition 4.6** *If $T \in \mathscr{L}(X)$ is a linear operator commuting with $A^{-1}$, then the spaces $X_n$ and $\underline{X}_n$ are $T$-invariant, and $T \in \mathscr{L}(X_n)$ for $n \in \mathbb{N}$.*

**Proof** The condition means that $Tx \in D(A)$ for each $x \in D(A)$ and for such $x$ we have $ATx = TAx$. This implies the invariance of $X_1$ and that $\|Tx\|_{X_1(A)} \le \|T\|\|x\|_{X_1(A)}$. Using the boundedness assumption we see that $\underline{X}_1$ remains invariant under $T$. For general $n \in \mathbb{N}$ we may argue by recursion, or simply invoke Remark 4.2.  $\square$

### 4.1.2  Extrapolation Spaces

The construction for the extrapolation spaces here is standard if $A$ is densely defined, or if $A$ is a Hille–Yosida operator, see, e.g., [190].

For $x \in X$ we define $\|x\|_{\underline{X}_{-1}(A)} := \|A^{-1}x\|$. Then the surjective mapping

$$A : (D(A), \|\cdot\|) \to (X, \|\cdot\|_{\underline{X}_{-1}(A)})$$

becomes isometric, and hence has a uniquely continuous extension

$$\underline{A}_{-1} : (\underline{X}, \|\cdot\|) \to (\underline{X}_{-1}, \|\cdot\|_{\underline{X}_{-1}(\underline{A})}),$$

which is an isometric isomorphism, where $(\underline{X}_{-1}, \|\cdot\|_{\underline{X}_{-1}(\underline{A})})$ denotes the completion of $(\underline{X}, \|\cdot\|_{\underline{X}_{-1}(A)})$. By construction we obtain immediately:

**Proposition 4.7** *The space $X$ is continuously and densely embedded in $\underline{X}_{-1}$. If $\underline{A}$ is densely defined in $\underline{X}$, then also $X_\infty$ is dense in $\underline{X}_{-1}$. As a consequence $(\underline{X}_{-1}, \|\cdot\|_{\underline{X}_{-1}(\underline{A})})$ is the completion of $(\underline{X}, \|\underline{A}^{-1} \cdot \|)$.*

***Proof*** The space $X$ is dense in $\underline{X}_{-1}$ by construction. For $x \in X$ we have

$$\|x\|_{\underline{X}_{-1}(\underline{A})} = \|AA^{-1}x\|_{\underline{X}_{-1}(\underline{A})} = \|\underline{A}_{-1}A^{-1}x\|_{\underline{X}_{-1}(\underline{A})} \leq \|\underline{A}_{-1}\| \cdot \|A^{-1}x\| \leq \|\underline{A}_{-1}\| \cdot \|A^{-1}\| \cdot \|x\|,$$

showing the continuity of the embedding. The last assertion follows since $X_\infty$ is dense in $D(A)$ with respect to $\|\cdot\|$. $\qquad\qquad\square$

Of course one can iterate the whole procedure and obtain the following chain of dense and continuous embeddings

$$\underline{X}_0 \hookrightarrow \underline{X}_{-1} \hookrightarrow \underline{X}_{-2} \hookrightarrow \cdots \hookrightarrow \underline{X}_{-n} \quad \text{for } n \in \mathbb{N},$$

where for $n \geq 1$ the space $\underline{X}_{-n}$ is a completion of $\underline{X}_{-n+1}$ with respect to the norm $\|\cdot\|_{\underline{X}_{-n}(\underline{A})}$ defined by $\|x\|_{\underline{X}_{-n}(\underline{A})} = \|\underline{A}_{-n+1}^{-1}x\|_{\underline{X}_{-n+1}(\underline{A})}$ and

$$\underline{A}_{-n} : \underline{X}_{-n+1} \to \underline{X}_{-n}$$

is a unique continuous extension of $\underline{A}_{-n+1} : D(\underline{A}_{-n+1}) \to \underline{X}_{-n+1}$ to $\underline{X}_{-n}$.

These spaces, just as well the ones in the next definition, are called *extrapolation spaces* for the operator $A$, see, e.g., [190] or [101, Sect. II.5] for the case of semigroup generators. The spaces $\underline{X}_{-1}$, $\underline{X}_{-2}$ and the operator $\underline{A}_{-2}$ will be used to define the extrapolation space $X_{-1}(A)$. To this purpose we identify $X$ with a subspace of $\underline{X}_{-1}$ and of $\underline{X}_{-2}$.

**Definition 4.8** Consider $X$ as a subspace of $\underline{X}_{-2}$, and define

$$X_{-1} := \underline{A}_{-2}(X) := \left\{ \underline{A}_{-2}x : x \in X \right\} \quad \text{and} \quad \|x\|_{X_{-1}} := \|\underline{A}_{-2}^{-1}x\|.$$

Furthermore, we set $D(A_{-1}) := X$ and for $x \in X$ we define $A_{-1}x := \underline{A}_{-2}x$. To indicate the dependence on the operator $A$ we write $X_{-1}(A)$ and $\|\cdot\|_{X_{-1}(A)}$.

**Remark 4.9** It is easy to see that the operator $A_{-1}$ is the part of $\underline{A}_{-2}$ in $X_{-1}$.

In what follows, we will define higher order extrapolation spaces and prove that all these spaces line up in a scale, where one can switch between the levels with the help of (a version) of the operator $A$ (or $A_{-1}$).

**Proposition 4.10** *The operator $A_{-1}$ is an extension of $\underline{A}_{-1}$, $(X_{-1}, \|\cdot\|_{X_{-1}})$ is a Banach space, the norms of $\underline{X}_{-1}$ and $X_{-1}$ coincide on $\underline{X}_{-1}$, and $\underline{X}_{-1}$ is a closed subspace of $X_{-1}$. The mapping $A_{-1} : X \to X_{-1}$ is an isometric isomorphism.*

***Proof*** The first assertion is true because $\underline{A}_{-2}$ is an extension of $\underline{A}_{-1}$. That $X_{-1}$ is a Banach space is immediate from the definition. Since $\underline{A}_{-2}^{-1}\underline{A}_{-1} = I$ on $X$, we have $\underline{A}_{-1}^{-1}x \in \underline{X} \subseteq X$

for $x \in \underline{X}_{-1}$, so that $\|\underline{A}_{-2}^{-1}x\| = \|\underline{A}_{-2}^{-1}\underline{A}_{-1}\underline{A}_{-1}^{-1}x\| = \|\underline{A}_{-1}^{-1}x\| = \|x\|_{\underline{X}_{-1}}$. This establishes that the norms coincide. Since $\underline{X}_{-1}$ is a Banach space (with its own norm), it is a closed subspace of $X_{-1}$. That $A_{-1}$ is an isometric isomorphism follows from the definition. $\qquad\square$

**Remark 4.11** By construction we have $\underline{X}_{-1}(\underline{A}_{-n}) = \underline{X}_{-(n+1)}(\underline{A})$ and $X_{-1}(A_{-n}) = X_{-(n+1)}(A)$ for each $n \in \mathbb{N}$.

**Proposition 4.12** *For $n \in \mathbb{Z}$ the operators $A_n : X_{n+1} \to X_n$ and $\underline{A}_n : \underline{X}_{n+1} \to \underline{X}_n$ are isometric isomorphisms that intertwine $A_{n+1}$ and $A_n$, respectively, $\underline{A}_{n+1}$ and $\underline{A}_n$.*

***Proof*** For $n \in \mathbb{N}$ this is Proposition 4.3. So we assume $n < 0$. For $n = -1$ the statement about isometric isomorphisms is just the definition, and the intertwining property is also evident. By recursion we obtain the validity of the assertion for general $n \le -1$ and for the operator $\underline{A}_n$. By Remark 4.11 it suffices to prove that $A_{-1}$ intertwines $A_{-1}$ and $A_0 = A$. For $x \in D(A_0) = D(A)$ we have $A_{-1}x \in X = D(A_{-1})$ and $Ax = A_{-1}^{-1}A_{-1}A_{-1}x$. $\qquad\square$

Thus for $n \in \mathbb{N}$ we have the following chain of embeddings (continuous and dense, denoted by $\hookrightarrow$) and inclusions as closed subspaces (denoted by $\subseteq$):

$$\cdots \hookrightarrow \underline{X}_n \subseteq X_n \hookrightarrow \underline{X} \subseteq X \hookrightarrow \underline{X}_{-1} \subseteq X_{-1} \hookrightarrow \underline{X}_{-2} \subseteq X_{-2} \hookrightarrow \cdots \underline{X}_{-n} \subseteq X_{-n} \hookrightarrow \cdots ,$$

where in general the inclusions are strict (see the examples in Sect. 5.2). We also have the following chain of isometric isomorphisms

$$\cdots \longrightarrow \underline{X}_{n+1} \xrightarrow{\underline{A}_n^{-1}} \underline{X}_n \longrightarrow \cdots \longrightarrow \underline{X}_1 \xrightarrow{\underline{A}_0^{-1}} \underline{X}_0 \xrightarrow{\underline{A}_{-1}^{-1}} \underline{X}_{-1} \longrightarrow \cdots \longrightarrow \underline{X}_{-n+1} \xrightarrow{\underline{A}_{-n}^{-1}} \underline{X}_{-n} \longrightarrow \cdots$$

and

$$\cdots \longrightarrow X_{n+1} \xrightarrow{A_n^{-1}} X_n \longrightarrow \cdots \longrightarrow X_1 \xrightarrow{A_0^{-1}} X_0 \xrightarrow{A_{-1}^{-1}} X_{-1} \longrightarrow \cdots \longrightarrow X_{-n+1} \xrightarrow{A_{-n}^{-1}} X_{-n} \longrightarrow \cdots .$$

**Proposition 4.13** (a) $\underline{X}_1(\underline{A}_{-1}) = \underline{X}$ and $X_1(A_{-1}) = X$ *with the same norms.*
(b) $\underline{X}_{-1}(\underline{A}_1) = \underline{X}$ *with the same norms.*
(c) $(\underline{A}_1)_{-1} = \underline{A}$.
(c) $X_{-1}(A_1) = X$ *with the same norms, and $(A_1)_{-1} = A$.*

***Proof*** (a) By definition $X_1(A_{-1}) = D(A_{-1}) = X$ with the graph norm of $A_{-1}$. Since $A_{-1}$ extends $A$, for $x \in X$ we have $\|A_{-1}x\|_{X_{-1}(A)} = \|Ax\|_{X_{-1}} = \|A^{-1}Ax\| = \|x\|$. The first statement then follows, because $\underline{X}_1(\underline{A}_{-1}) = X_1(\underline{A}_{-1}) = \overline{D(A)} = \underline{X}$ with the same norms.

(b) For $x \in \underline{X}_1(\underline{A}) = D(\underline{A}^2)$ we have

$$\|x\|_{\underline{X}_{-1}(\underline{A}_1)} = \|\underline{A}_1^{-1}x\|_{\underline{X}_1(\underline{A})} = \|\underline{A}\,\underline{A}_1^{-1}x\| = \|x\|,$$

which can be extended by density for all $x \in \underline{X}$, showing the equality of the spaces $\underline{X}_{-1}(\underline{A}_1) = \underline{X}$ (with the same norm).

(c) By construction the operator $(\underline{A}_1)_{-1} : \underline{X}_1(A) \to \underline{X}_{-1}(\underline{A}_1)$ is the unique continuous extension of

$$\underline{A}_1 : D(\underline{A}_1) = D(\underline{A}^2) \to \underline{X}_1(A),$$

and $(\underline{A}_1)_{-1}$ is an isometric isomorphism. For $x \in \underline{X}_1(A)$ we have $\|x\|_{\underline{X}_{-1}(\underline{A}_1)} = \|\underline{A}_1^{-1}x\|_{\underline{X}_1(A)} = \|x\|$. But then it follows that $(\underline{A}_1)_{-1} = \underline{A} : D(\underline{A}) \to \underline{X}$.

(d) The space $X_{-1}(A_1)$ is defined by

$$X_{-1}(A_1) := (\underline{A}_1)_{-2}(X_1(A)) = ((\underline{A}_1)_{-1})_{-1}(X_1(A)) = \underline{A}_{-1}(X_1(A)) = AX_1(A) = X,$$

by part (c). For the norm equality let $x \in X$. Then

$$\|x\| = \|AA^{-1}x\| = \|A^{-1}x\|_{X_1(A)} = \|\underline{A}_{-1}^{-1}x\|_{X_1(A)} = \|(\underline{A}_1)_{-2}^{-1}x\|_{X_1(A)} = \|x\|_{X_{-1}(A_1)}.$$

For the last assertion we note that $(A_1)_{-1} = (\underline{A}_1)_{-2}|_{X_1(A)} = A$  $\square$

Recall the standing assumption that $\underline{A} = A|_{\underline{X}}$ is densely defined in $\underline{X} = \overline{D(A)}$. The following proposition plays the key role for the extension of operators to the extrapolation spaces, particularly for the construction of extrapolated semigroups in Sect. 4.3.

**Proposition 4.14** (a)  *Let $n \in \mathbb{N}$. If $T \in \mathscr{L}(X)$ is a linear operator commuting with $A^{-1}$, then the operator $T$ has a unique continuous extension to $\underline{X}_{-n}$ denoted by $\underline{T}_{-n}$. The operator $\underline{T}_{-n}$ is the restriction of $\underline{T}_{-n-1}$. The space $X_{-n}$ is invariant under $\underline{T}_{-n-1}$, whose restriction is denoted by $T_{-n}$, for which $T_{-n} \in \mathscr{L}(X_{-n})$. For $k, n \in -\mathbb{N}$ the operators $\underline{T}_n, \underline{T}_k$ are all similar; the same holds for $T_n$ and $T_k$.*
(b)  *Let $\underline{T} \in \mathscr{L}(\underline{X})$ such that it leaves $D(A)$ invariant and commutes with $\underline{A}^{-1} = A^{-1}|_X$. Then $\underline{T}_{-1}x = A\underline{T}A^{-1}x$ for each $x \in X$, and as a consequence, $\underline{T}_{-1} : \underline{X}_{-1} \to \underline{X}_{-1}$ leaves $X$ invariant (and, of course, extends $\underline{T}$).*

***Proof*** (a) For $x \in X$ we have

$$\|Tx\|_{X_{-1}(A)} = \|A^{-1}Tx\| = \|TA^{-1}x\| \le \|T\| \cdot \|A^{-1}x\| = \|T\| \cdot \|x\|_{X_{-1}(A)}.$$

Therefore $T : (X, \|\cdot\|_{X_{-1}(A)}) \to (X, \|\cdot\|_{X_{-1}(A)})$ is continuous, and hence has a unique continuous extension $\underline{T}_{-1}$ to $\underline{X}_{-1}$. This extension commutes with $\underline{A}_{-1}^{-1}$, because $T$ commutes with $A^{-1}$ and $\underline{A}_{-1}^{-1}$ is the unique continuous extension of $A^{-1}$. By iteration we obtain the continuous extensions $\underline{T}_{-n}$ onto $\underline{X}_{-n}$, which then all commute with the corresponding $\underline{A}_{-n}^{-1}$. By construction $\underline{T}_{-n}$ is a restriction of $\underline{T}_{-n-1}$. We prove that $X_{-1}$ is invariant under $\underline{T}_{-2}$. Let

$x \in X_{-1}$, hence $x = \underline{A}_{-2}y$ for some $y \in X$. Then $Ty = \underline{T}_{-2}y = \underline{T}_{-2}\underline{A}_{-2}^{-1}x = \underline{A}_{-2}^{-1}\underline{T}_{-2}x$, hence $\underline{T}_{-2}x = \underline{A}_{-2}Ty \in X_{-1}$, i.e., the invariance of $X_{-1}$ is proved. We have for $x \in X_{-1}$ that $\|\underline{T}_{-1}x\|_{X_{-1}} = \|\underline{A}_{-2}^{-1}\underline{T}_{-1}x\| = \|\underline{A}_{-2}^{-1}\underline{T}_{-2}x\| = \|\underline{T}_{-2}\underline{A}_{-2}^{-1}x\| \leq \|\underline{T}_2\| \cdot \|\underline{A}_{-2}^{-1}x\| = \|\underline{T}_2\| \cdot \|x\|_{X_{-1}}$, therefore $\underline{T}_{-1} \in \mathscr{L}(X_{-1})$. The assertion about $T_{-n}$ follows by recursion.

It is enough to prove the similarity of $T_0 = T$ and $T_{-1}$, and the similarity of $\underline{T}_0$ and $\underline{T}_{-1}$. The latter assertions can be proved as follows: For $x \in D(A)$ we have

$$\underline{A}_{-1}^{-1}\underline{T}_{-1}\underline{A}_{-1}x = \underline{A}_{-1}^{-1}\underline{T}_{-1}Ax = \underline{A}_{-1}^{-1}TAx = \underline{A}_{-1}^{-1}ATx = \underline{A}_{-1}^{-1}\underline{A}_{-1}Tx = \underline{T}x,$$

then by continuity and denseness the equality follows even for $x \in \underline{X}$. For the similarity of $T$ and $T_{-1}$ take $x \in X$. Then

$$A_{-1}^{-1}T_{-1}A_{-1}x = \underline{A}_{-2}^{-1}\underline{T}_{-2}\underline{A}_{-2}x = \underline{T}_{-1}x = Tx.$$

(b) Let $x \in X \subseteq \underline{X}_{-1}$. Then there is a sequence $(x_n)$ in $\underline{X}$ with $x_n \to x$ in $\underline{X}_{-1}$ (see Proposition 4.7). But then $A^{-1}x_n \to A^{-1}x$ in $\underline{X}$ and $\underline{T}x_n \to \underline{T}_{-1}x$ in $\underline{X}_{-1}$ by part (a). These imply $\underline{T}A^{-1}x_n = A^{-1}\underline{T}x_n \to \underline{A}_{-1}^{-1}\underline{T}_{-1}x$. Hence we conclude $\underline{T}A^{-1}x = \underline{A}_{-1}^{-1}\underline{T}_{-1}x$ and $A\underline{T}A^{-1}x = \underline{T}_{-1}x$ for $x \in X$. $\qquad\square$

Haase in [126] and Wegner in [238] have constructed the so-called *universal extrapolation space* $X_{-\infty}$ as follows: Suppose $A$ is densely defined (this assumption is *not* made by Haase), then $X_n = \underline{X}_n$ for each $n \in \mathbb{Z}$ and let $X_{-\infty}$ to be the inductive limit of the sequence of Banach spaces $(X_{-n})_{n\in\mathbb{N}}$ (algebraic inductive limit in [126]). One can extend the operator $A$ to an operator $A_{-\infty} : X_{-\infty} \to X_{-\infty}$ such that

$$A_{-\infty}|_{X_n} = A_n, \quad n \in \mathbb{Z}.$$

We now look at a converse situation, and our starting point is the following: Let $\mathscr{E}$ be a locally convex space such that we can embed the Banach space $X$ continuously in $\mathscr{E}$, i.e., there is a continuous injective map $i : X \to \mathscr{E}$, and so we can identify $X$ with a subspace of $\mathscr{E}$. We also assume that we have a continuous operator $\mathcal{A} : \mathscr{E} \to \mathscr{E}$ such that $\lambda - \mathcal{A} : i(X) \to \mathscr{E}$ is injective and that

$$D(A) = \{x \in X : \mathcal{A} \circ i(x) \in i(X)\},$$

and

$$i \circ A = \mathcal{A} \circ i|_{D(A)}.$$

In the next theorem we use this setting to describe the extrapolation spaces $\underline{X}_{-n}$, $X_{-n}$. Notice that we do not assume that $A$ is a Hille–Yosida operator or densely defined.

**Theorem 4.15** *Let $X$ be a Banach space with a continuous embedding $i : X \to \mathscr{E}$ into a locally convex space $\mathscr{E}$, let $A : D(A) \to X$ be a linear operator with $\lambda \in \rho(A)$ such that $A = \mathcal{A}|_X$ (after identifying $X$ with a subspace of $\mathscr{E}$ as described above). We suppose furthermore* *that $\lambda - \mathcal{A}$ is injective on $X$. Then there is a continuous embedding $i_{-1} : X_{-1} \to \mathcal{E}$ which extends $i$. After identifying $X_{-1}$ with a subspace of $\mathcal{E}$ (under $i_{-1}$) we have*

$$X_{-1} = \{(\lambda - \mathcal{A})x : x \in X\}, \quad \underline{X}_{-1} = \{(\lambda - \mathcal{A})x : x \in \underline{X}_0\} \quad and \quad A_{-1} = \mathcal{A}|_{X_{-1}}.$$

***Proof*** Without loss of generality we may assume that $\lambda = 0$. Recall that $A_{-1}|_X = A$ and $A_{-1}$ is an isometric isomorphism $A_{-1} : X \to X_{-1}$. We now define the embedding $i_{-1} : X_{-1} \to \mathcal{E}$ by

$$i_{-1} := \mathcal{A} \circ i \circ A_{-1}^{-1},$$

which is indeed injective and continuous by assumption. Of course, $i_{-1}$ extends $i$ since we have $i = \mathcal{A} \circ i \circ A^{-1}$. We can write

$$i_{-1} \circ A_{-1} = \mathcal{A} \circ i \circ A_{-1}^{-1} \circ A_{-1} = \mathcal{A} \circ i,$$

which yields the following commutative diagram:

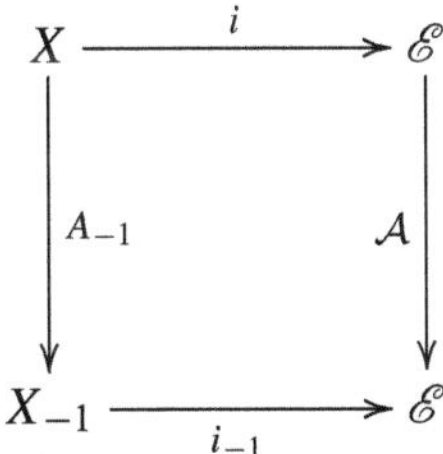

Now all assertions follow easily.  $\square$

The last corollary in this section can be proved by induction based on the previous facts.

**Corollary 4.16** *Let $\mathcal{A}$, $X$, $\mathcal{E}$, and $i$ be as in Theorem 4.15. Then $X_n \subseteq \mathcal{E}$ and $A_n = \mathcal{A}|_{X_n}$ for each $n \in \mathbb{Z}$ (after identifying $X_n$ with a subspace of $\mathcal{E}$ under an embedding $i_n$ compatible with $i$).*

## 4.2   Intermediate Spaces for Operators with Rays of Minimal Growth

The following definition of intermediate, and as a matter of fact interpolation spaces, just as well many results in this section are standard, and we refer, e.g., to the book by Lunardi [175, Chap. 3], and to Engel, Nagel [101, Sect. II.5] for the case of semigroup generators. In this section we suppose the following.

**Assumption C** The operator $A$ on the Banach space $X$ has a *ray of minimal growth*, i.e., $(0, \infty) \subseteq \rho(A)$ and for some $M \geq 0$

$$\|\lambda R(\lambda, A)\| \leq M \quad \text{for all } \lambda > 0. \tag{4.2.1}$$

**Definition 4.17** For $\alpha \in (0, 1]$ and $x \in X$ we define

$$\|x\|_{F_\alpha(A)} := \sup_{\lambda > 0} \|\lambda^\alpha A R(\lambda, A) x\|,$$

and the *abstract Favard space* of order $\alpha$ by

$$F_\alpha(A) := \left\{ x \in X : \|x\|_{F_\alpha(A)} < \infty \right\}.$$

In the literature the notation $D_A(\alpha, \infty)$ is also used, see, e.g., [175]. We further set

$$F_0(A) := F_1(A_{-1}),$$

see [101, Sect. II.5(b)] for the case of semigroup generators.

**Proposition 4.18** (a) *The Favard space $F_\alpha(A)$ becomes a Banach space if endowed with the norm $\| \cdot \|_{F_\alpha(A)}$.*
(b) *The space $X$ is isomorphic to a closed subspace of $F_0(A)$.*

The statement that $X$ is a closed subspace of $F_0(A)$ when $A$ is a Hille–Yosida operator is due to Nagel and Sinestrari [190, Proof of Prop. 2.7].

***Proof*** (a) is trivial.

(b) For $x \in$ we have

$$\|\lambda A_{-1} R(\lambda, A_{-1}) x\|_{X_{-1}(A)} = \|\lambda A R(\lambda, A) x\|_{X_{-1}(A)} = \|\lambda A^{-1} A R(\lambda, A) x\| \leq M \|x\|,$$

yielding

$$\|x\|_{F_0(A)} = \|x\|_{F_1(A_{-1})} \leq M \|x\|.$$

On the other hand, since $A$ and $A_{-1}$ are similar, we have $\sup_{\lambda > 0} \|\lambda R(\lambda, A_{-1})\|_{X_{-1}(A)} \leq M'$ for some $M' \geq 0$. In particular, by Remark 4.5, $\lambda R(\lambda, A_{-1}) x \to x$ for each $x \in \underline{X}_{-1}$. From this we obtain for $x \in X$ that

$$\|x\| = \|A_{-1} x\|_{X_{-1}(A)} = \left\| \lim_{\lambda \to 0} \lambda R(\lambda, A_{-1}) A_{-1} x \right\|_{X_{-1}(A)} \leq \sup_{\lambda > 0} \left\| \lambda A_{-1} R(\lambda, A_{-1}) x \right\|_{X_{-1}(A)}$$

$$= \|x\|_{F_1(A_{-1})} = \|x\|_{F_0(A)},$$

showing the equivalence of the norms $\| \cdot \|$ and $\|x\|_{F_0(A)}$ on $X$. $\qquad\square$

We also need the following well-known result, see, e.g., [175, Chaps. 1 and 3], for which we give a short proof.

**Proposition 4.19** *For $\alpha \in (0, 1]$ we have $F_\alpha(A) \subseteq \overline{D(A)} = \underline{X}$.*

*Proof* We have
$$AR(\lambda, A)x = \lambda R(\lambda, A)x - x,$$

so that
$$\|\lambda R(\lambda, A)x - x\| \leq \frac{\|x\|_{F_\alpha(A)}}{\lambda^\alpha} \to 0 \quad \text{as } \lambda \to \infty.$$

$\square$

**Definition 4.20** Let $A$ be a linear operator on the Banach space $X$ satisfying (4.2.1). For $\alpha \in (0, 1)$ we define the *abstract Hölder space* of order $\alpha$ by
$$\underline{X}_\alpha(A) := \left\{ x \in F_\alpha(A) : \lim_{\lambda \to \infty} \lambda^\alpha AR(\lambda, A)x = 0 \right\},$$

and we recall from Sect. 4.1 that
$$\underline{X}(A) := \overline{D(A)}, \quad \underline{X}_1(A) = D(A|_{\underline{X}(A)}).$$

The proof of the next proposition is straightforward and well known.

**Proposition 4.21** *For $\alpha, \beta \in (0, 1)$ with $\alpha > \beta$ we have*
$$\underline{X}_1(A) \hookrightarrow \underline{X}_\alpha(A) \subseteq F_\alpha(A) \hookrightarrow \underline{X}_\beta(A) \subseteq F_\beta(A) \hookrightarrow \underline{X}(A) \subseteq X(A)$$

*with $\hookrightarrow$ denoting continuous and dense embeddings of Banach spaces, and $\subseteq$ denoting inclusion of closed subspaces.*

*Proof* For $x \in F_\alpha(A)$ we have
$$\|\lambda^\beta AR(\lambda, A)x\| = \lambda^{\beta-\alpha}\|\lambda^\alpha AR(\lambda, A)x\| \leq \lambda^{\beta-\alpha}\|x\|_\alpha \to 0 \quad \text{as } \lambda \to \infty,$$

which also proves the continuity of $F_\alpha(A) \hookrightarrow \underline{X}_\beta(A)$. The other statements can be proved by similar reasonings.

$\square$

**Proposition 4.22** (a) *The spaces $F_\alpha(A)$ and $\underline{X}_\alpha(A)$ are invariant under each $T \in \mathcal{L}(X)$ which commutes with $A^{-1}$.*

(b) *If $T \in \mathcal{L}(X)$ commutes with $A^{-1}$, then the space $F_0(A)$ is invariant under $T_{-1}$.*

**Proof** (a) Suppose that $T \in \mathscr{L}(X)$ commutes with $R(\cdot, A)$ and let $x \in \underline{X}_\alpha(A)$. We have to show that $Tx \in \underline{X}_\alpha(A)$. Since $T$ is assumed to be bounded, we obtain

$$\|\lambda^\alpha A R(\lambda, A) T x\| = \|\lambda^\alpha A T R(\lambda, A) x\| \leq \|T\| \cdot \|\lambda^\alpha A R(\lambda, A) x\|.$$

This implies both assertions.

(b) Follows from part (a) applied to $T_{-1}$ on the space $X_{-1}$.

**Definition 4.23** For $\alpha \in \mathbb{R}$ we write $\alpha = m + \beta$ with $m \in \mathbb{Z}$ and $\beta \in (0, 1]$, and define

$$F_\alpha(A) := F_\beta(A_m),$$

with the corresponding norms. For $\alpha \notin \mathbb{Z}$ we define

$$\underline{X}_\alpha(A) := \underline{X}_\beta(A_m),$$

also with the corresponding norms.

In particular we have for $\alpha \in (0, 1)$ that

$$\underline{X}_{-\alpha}(A) = \underline{X}_{1-\alpha}(A_{-1}) \quad \text{and} \quad F_{-\alpha}(A) = F_{1-\alpha}(A_{-1}).$$

This definition is consistent with Definitions 4.17 and 4.20. The following property of these spaces can be directly deduced from the definitions and the previous assertions (by induction):

**Proposition 4.24** *For any $\alpha, \beta \in \mathbb{R}$ with $\alpha > \beta$ we have*

$$\underline{X}_\alpha(A) \subseteq F_\alpha(A) \hookrightarrow \underline{X}_\beta(A) \subseteq F_\beta(A)$$

*with $\hookrightarrow$ denoting continuous and dense embeddings of Banach spaces, and $\subseteq$ denoting inclusion of closed subspaces.*

Now we put these spaces in the context presented at the end of Sect. 4.1.

**Proposition 4.25** (a) *For $\alpha \in (0, 1]$ we have $A_{-1} F_\alpha = F_{\alpha-1}$ and $A_{-1} \underline{X}_\alpha = \underline{X}_{\alpha-1}$.*
(b) *For $\alpha \in (0, 1]$ and $\mathcal{A}$, $\lambda$ and $\mathscr{E}$ as in Theorem 4.15 we have*

$$F_{-\alpha} = \left\{ (\lambda - \mathcal{A}) y \in X_{-1} : y \in F_{1-\alpha} \right\}.$$

*If $\alpha \in (0, 1)$, then*

$$\underline{X}_{-\alpha} = \left\{ (\lambda - \mathcal{A}) y \in X_{-1} : y \in \underline{X}_{1-\alpha} \right\}.$$

## 4.3   Intermediate and Extrapolation Spaces for Semigroup Generators

In this section, we consider intermediate and extrapolation spaces when the linear operator $A : D(A) \to X$ is the generator of a semigroup $(T(t))_{t \geq 0}$ (meaning that $T : [0, \infty) \to \mathcal{L}(X)$ is a monoid homomorphism) in the sense described in the following.

**Assumption 4.26**  1.  Let $X$ be a Banach space, and let $Y \subseteq X'$ be a norming subspace, i.e.:

$$\|x\| = \sup_{y \in Y, \|y\| \leq 1} |\langle x, y \rangle| \quad \text{for each } x \in X.$$

2. Let $T : [0, \infty) \to \mathcal{L}(X)$ be a semigroup of contractions for which a *generator* $A : D(A) \to X$ exists in the sense that

$$R(\lambda, A)x = \int_0^\infty e^{-\lambda s} T(s)x \, ds \qquad (4.3.1)$$

exists for each $\lambda \geq 0$ as a weak integral with respect to the dual pair $(X, Y)$, i.e., for each $y \in Y$ and $x \in X$

$$\langle R(\lambda, A)x, y \rangle = \int_0^\infty e^{-\lambda s} \langle T(s)x, y \rangle \, ds,$$

and $R(\lambda, A) \in \mathcal{L}(X)$ is the resolvent of a linear operator $A$ (see [164] by Kunze).
3. We also suppose that $T(t)$ commutes with $A^{-1}$ for each $t \geq 0$.

If the semigroup $(T(t))_{t>0}$ is only exponentially bounded of type $\omega$, that is:

$$\|T(t)\| \leq Me^{\omega t} \quad \text{for all } t \geq 0,$$

then one can rescale the semigroup (consider$(e^{-(\omega+1)t} T(t))_{t \geq 0}$), and renorm the Banach space such that the rescaled semigroup becomes a contraction semigroup. Moreover, the new semigroup has negative growth bound, meaning that $T(t) \to 0$ in norm exponentially fast as $t \to \infty$. Then it also has an invertible generator.

**Remark 4.27** (i)  There are several important classes of semigroups, satisfying Assumption 4.26, hence can be treated in a unified manner: $\pi$-semigroups of Priola [202], weakly continuous semigroups of Cerrai [67], bi-continuous semigroups of Kühnemund. We will concentrate on this latter class of semigroups in Sect. 5.1.
(ii)  In this framework Kunze [164] introduced the notion of integrable semigroups, which we briefly describe next. Since we have

$$\|y\| = \sup_{x \in X, \|x\| \leq 1} |\langle x, y \rangle|$$

and, by the norming assumption,

$$\|x\| = \sup_{y \in Y, \|y\| \le 1} |\langle x, y \rangle|,$$

the pair $(X, Y)$ is called a norming dual pair. Kunze has worked out the theory of semigroups on such norming dual pairs in [164]. We recall at least the basic definitions here: assume without loss of generality that $Y$ is a Banach space and consider the weak topology $\sigma = \sigma(X, Y)$ on $X$. An *integrable semigroup* of type $\omega$ on the pair $(X, Y)$ is a semigroup $(T(t))_{t \ge 0}$ of $\sigma$-continuous linear operators satisfying the following.

1. $(T(t))_{t \ge 0}$ is a semigroup, i.e., $T(t + s) = T(t)T(s)$ and $T(0) = I$ for all $t, s \ge 0$.
2. For all $\lambda$ with $\mathrm{Re}(\lambda) > \omega$, there exists an $\sigma$-continuous linear operator $R(\lambda)$ such that for all $x \in X$ and all $y \in Y$

$$\langle R(\lambda)x, y \rangle = \int_0^\infty e^{-\lambda t} \langle T(t)x, y \rangle \, dt.$$

Kunze defines the generator $A$ of the semigroup as the (unique) operator $A : D(A) \to X$ (if it exists at all) with $R(\lambda) = (\lambda - A)^{-1}$, precisely as in Assumption 4.26. Note that $\sigma$-continuity of $T(t)$ can be used to assure that $Y$ is invariant under $T'(t)$, cf. the next remark.

**Remark 4.28** The semigroup $(T(t))_{t \ge 0}$ commutes with the inverse of the generator if $Y$ can be chosen such that it is invariant under $T'(t)$ for each $t \ge 0$:

$$\langle A^{-1}T(t)x, y \rangle = \int_0^\infty \langle T(s)T(t)x, y \rangle \, ds = \int_0^\infty \langle T(s + t)x, y \rangle \, ds$$

$$= \left\langle \int_0^\infty T(s)x \, ds, T'(t)y \right\rangle = \langle T(t)A^{-1}x, y \rangle,$$

for each $x \in X$ and $y \in Y$.

**Remark 4.29**  1. From (4.3.1) it follows that for each $x \in X$

$$T(t)x - x = A \int_0^t T(s)x \, ds. \tag{4.3.2}$$

Indeed, we have by (4.3.1) that

$$x = A \int_0^\infty T(s)x \, ds$$

$$T(t)x = A \int_0^\infty T(s)T(t)x \, ds = \int_t^\infty T(s)x \, ds.$$

Subtracting the first of these equation from the second one we obtain the statement.

2.  If moreover $A$ commutes with $T(t)$ for each $t \geq 0$, then for each $x \in D(A)$ we have

$$T(t)x - x = \int_0^t T(s)Ax \, ds. \tag{4.3.3}$$

Indeed, as in the above, we have by (4.3.1)

$$- x = -A^{-1}Ax = \int_0^\infty T(s)Ax \, ds$$

$$-T(t)x = -A^{-1}T(t)Ax = \int_0^\infty T(s)T(t)Ax \, ds = \int_t^\infty T(s)Ax \, ds.$$

By a simple subtraction we obtain the statement.

The next lemma and its proof are standard for various classes of semigroups.

**Lemma 4.30**  *If $(T(t))_{t \geq 0}$ is (locally) norm bounded, then*

$$X_{\text{cont}} := \{x \in X : t \mapsto T(t)x \text{ is } \| \cdot \|\text{-continuous}\}$$

*is a closed a subspace of $X$ invariant under the semigroup. Under Assumption 4.26 we have*

$$\underline{X} = \overline{D(A)} = X_{\text{cont}}.$$

***Proof*** The closedness and invariance of $X_{\text{cont}}$ are evident. We first show $D(A) \subseteq X_{\text{cont}}$, which implies $\overline{D(A)} \subseteq X_{\text{cont}}$ by closedness of $X_{\text{cont}}$. By (4.3.3) we conclude for $x \in D(A)$ that $T(t)x - x = \int_0^t T(s)Ax \, ds$. Since

$$\|T(t)x - x\| = \sup_{\|y\| \leq 1} |\langle T(t)x - x, y\rangle| \leq \sup_{\|y\| \leq 1} \int_0^t |\langle T(s)Ax, y\rangle| \, ds \leq t\|Ax\| \to 0$$

as $t \to 0$, we obtain $D(A) \subseteq X_{\text{cont}}$ and $\overline{D(A)} \subseteq X_{\text{cont}}$. For the converse inclusion suppose that $x \in X_{\text{cont}}$. Again by (4.3.2) we obtain that the sequence of vectors $x_n := n \int_0^{\frac{1}{n}} T(s)x \, ds \in D(A)$ $(n \in \mathbb{N})$ converges to $x$. Indeed

$$\|x_n - x\| = \sup_{\|y\| \leq 1} |\langle x_n - x, y\rangle| \leq \sup_{\|y\| \leq 1} n \int_0^{\frac{1}{n}} |\langle T(s)x - x, y\rangle| \, ds \leq n \int_0^{\frac{1}{n}} \|T(s)x - x\| \, ds.$$

By the continuity of $s \mapsto T(s)x$ we obtain the inclusion $X_{\mathrm{cont}} \subseteq \overline{D(A)}$. $\qquad \square$

Based on this lemma one can prove the following characterization of the Favard and Hölder spaces:

**Proposition 4.31** *Let* $(T(t))_{t \geq 0}$ *be a semigroup satisfying Assumption 4.26 with negative growth bound and generator A. For* $\alpha \in (0, 1]$ *define*

$$F_\alpha(T) := \left\{ x \in X : \sup_{s>0} \frac{\|T(s)x - x\|}{s^\alpha} < \infty \right\} = \left\{ x \in X : \sup_{s \in (0,1)} \frac{\|T(s)x - x\|}{s^\alpha} < \infty \right\},$$
(4.3.4)

*and for* $\alpha \in (0, 1)$ *define*

$$\underline{X}_\alpha(T) := \left\{ x \in X : \sup_{s>0} \frac{\|T(s)x - x\|}{s^\alpha} < \infty \text{ and } \lim_{s \downarrow 0} \frac{\|T(s)x - x\|}{s^\alpha} = 0 \right\}$$
(4.3.5)
$$= \left\{ x \in X : \lim_{s \downarrow 0} \frac{\|T(s)x - x\|}{s^\alpha} = 0 \right\},$$

*which become Banach spaces if endowed with the norm:*

$$\|x\|_{F_\alpha(T)} := \sup_{s>0} \frac{\|T(s)x - x\|}{s^\alpha}.$$

*The space* $\underline{X}_\alpha(T)$ *is a closed subspace of* $F_\alpha(T)$*. These spaces are invariant under the semigroup* $(T(t))_{t \geq 0}$*, and* $\underline{X}_\alpha(T)$ *is the space of* $\| \cdot \|_{F_\alpha(T)}$*-strong continuity in* $F_\alpha(T)$*. For* $\alpha \in (0, 1]$ *we have* $F_\alpha(A) = F_\alpha(T)$ *and for* $\alpha \in (0, 1)$ *we have* $\underline{X}_\alpha(A) = \underline{X}_\alpha(T)$ *with equivalent norms.*

**Proof** For $x \in F_\alpha(T)$ we have

$$\|T(t)x\|_{F_\alpha(T)} = \sup_{s>0} \frac{\|T(s)T(t)x - T(t)x\|}{s^\alpha} \leq \|T(t)\| \cdot \sup_{s>0} \frac{\|T(s)x - x\|}{s^\alpha} \leq M\|x\|_{F_\alpha(T)},$$

proving the invariance of $F_\alpha(T)$. Similar reasoning proves the invariance of $\underline{X}_\alpha$. Since $F_\alpha(T) \subseteq X_{\mathrm{cont}} = \underline{X} = \overline{D(A)}$ and $F_\alpha(A) \subseteq \underline{X} = \overline{D(A)}$, the rest of the assertions follow from the corresponding results concerning $C_0$-semigroups, see, e.g., [101, Sect. II.5]. $\qquad \square$

To complete the picture we recall a result from [175, Chap. 5]. It is stated there only for $C_0$-semigroup, but Lunardi also remarks that it holds in greater generality. We require here the conditions from Assumption 4.26 and note that the proof is verbatim the same as for the $C_0$-case due to the formulas (4.3.2) and (4.3.3).

**Proposition 4.32** *Let A generate the semigroup* $(T(t))_{t \geq 0}$ *of negative growth bound as described in Assumption 4.26. Then for* $p \in [1, \infty]$ *and* $\alpha \in (0, 1)$ *we have*

$$(X, D(A))_{\alpha,p} = \{x \in X : t \mapsto \psi_x(t) := t^{-\alpha}\|T(t)x - x\| \in \mathrm{L}^p_*(0, \infty)\},$$

*where $\mathrm{L}^p_*(0, \infty)$ denotes the $\mathrm{L}^p$-space with respect to the Haar measure $\frac{dt}{t}$ on the multiplicative group $(0, \infty)$. Moreover, the norms $\|x\|_{\alpha,p}$ and*

$$\|x\|^{**}_{\alpha,p} = \|x\| + \|\psi_x\|_{\mathrm{L}^p_*(0,\infty)}$$

*are equivalent.*

We conclude this section with the construction of the extrapolated semigroup as a direct consequence of Proposition 4.14.

**Proposition 4.33** *Let $A$ generate the semigroup $(T(t))_{t \geq 0}$ of negative growth bound in the sense of Assumption 4.26. Then there is an extension $(T_{-1}(t))_{t \geq 0}$ of the semigroup $(T(t))_{t \geq 0}$ on the extrapolated space $X_{-1}$, whose generator is $A_{-1}$.*

## Notes on This Chapter

Extrapolation spaces for generators of $C_0$-*semigroups* (used here synonymously to *"strongly continuous, one-parameter semigroups of bounded linear operators"*) on Banach spaces, or for more general operators, have been designed to study, e.g., maximal regularity questions by Da Prato and Grisvard [76]; see also Walter [237], Amann [14], van Neerven [226], Nagel, Sinestrari [189], Nagel [187], Sinestrari [217], Magal, Ruan [178, Chap. 3]. These spaces (and the corresponding extrapolated operators) play a central role in recent abstract perturbation results, most prominently in boundary-type or domain perturbations, see, e.g., Desch, Schappacher [83], Greiner [123], Staffans, Weiss [223], Adler, Bombieri, Engel [1], Hadd, Manzo, Rhandi [127]. Extrapolation spaces are also important in the theory of coupled operator matrices, see Engel [96]. Explicit examples of extrapolation spaces can be found in the monograph by Engel and Nagel [101, Chap. II, Sect. 5] or in the work of Nagel, Nickel, and Romanelli [188].

# Intermediate and Extrapolation Spaces for Bi-continuous Semigroups

**5**

In this chapter, we focus on the study of intermediate and extrapolation spaces associated with bi-continuous semigroups. Building upon the foundational results introduced in earlier chapters, we explore how these spaces contribute to a deeper understanding of the generators of bi-continuous semigroups. In particular, we revisit the Hille–Yosida generation theorem for bi-continuous semigroups and present an alternative proof using the structure of extrapolation spaces. This chapter not only extends the classical semigroup theory but also opens up new perspectives for applications in settings where bi-continuity is essential.

## 5.1 A Hille–Yosida Generation-Type Theorem Revised

In this section, we eventually concentrate on extrapolation spaces for generators of bi-continuous semigroups as introduced in Chap. 2. We observe that the class bi-continuous semigroups possess generators as described in Sect. 4.3. We saw in Sect. 3.5 the Hille–Yosida theorem for bi-continuous semigroups, cf. Theorem 3.18, whose proof is based on integrated semigroups. Here, we present a different proof using spaces.

**Theorem 5.1** *Let $(X, \| \cdot \|, \tau)$ be a bi-admissible space, and let $A$ be a linear operator on the Banach space $X$. The following are equivalent:*

(i) *The operator $A$ is the generator of a bi-continuous semigroup $(T(t))_{t \geq 0}$ of type $\omega$, i.e., there exists $M \geq 0$ such that $\|T(t)\| \leq Me^{\omega t}$ for all $t \geq 0$.*

(ii) *The operator $A$ is a Hille–Yosida operator of type $(M, \omega)$, i.e.:*

$$\|R(s, A)^k\| \leq \frac{M}{(s - \omega)^k}$$

© The Author(s), under exclusive license to Springer Nature Switzerland AG 2026
C. Budde, *Bi-Continuous Operator Semigroups*, Frontiers in Mathematics,
https://doi.org/10.1007/978-3-032-12948-2_5

*for all $k \in \mathbb{N}$ and for all $s > \omega$. Moreover, $A$ is bi-densely defined and the family*

$$\left\{ (s - \alpha)^k R(s, A)^k : k \in \mathbb{N}, \ s \geq \alpha \right\} \tag{5.1.1}$$

*is bi-equicontinuous for each $\alpha > \omega$, meaning that for each norm-bounded $\tau$-null sequence $(x_n)$ one has $(s - \alpha)^k R(s, A)^k x_n \to 0$ in $\tau$ uniformly for $k \in \mathbb{N}$ and $s \geq \alpha$ as $n \to \infty$.*

*In this case, we have the Euler formula:*

$$T(t)x := \tau\lim_{m \to \infty} \left( \frac{m}{t} R\left( \frac{m}{t}, A \right) \right)^m x \quad \text{for each } x \in X. \tag{5.1.2}$$

*Moreover, the subspace $\underline{X} := \overline{D(A)} \subseteq X$ is the space of norm strong continuity for $(T(t))_{t \geq 0}$, it is invariant under the semigroup, and $(\underline{T}(t))_{t \geq 0} := (T(t)|_{\underline{X}})_{t \geq 0}$ is the strongly continuous semigroup on $\underline{X}$ generated by the part $\underline{A}$ of $A$ in $\underline{X}$.*

**Proof** It follows from Lemma 4.30 that $\underline{X}$ is the space of norm strong continuity for a bi-continuous semigroup $(T(t))_{t \geq 0}$.

We only prove the implication (ii) $\Rightarrow$ (i) and the Euler formula; the other implication is easy. We may suppose that $\omega < 0$. Since $A$ is a Hille–Yosida operator, the part $\underline{A}$ of $A$ in $\underline{X}$ generates a $C_0$-semigroup $(\underline{T}(t))_{t \geq 0}$ of type $\omega$ on the space $\underline{X} := \overline{D(A)}$. Define the function

$$F(s) := \begin{cases} \frac{1}{s} R(\frac{1}{s}, A) & \text{for } s > 0, \\ I & \text{for } s = 0, \end{cases}$$

which is strongly continuous on $\underline{X}$ by Remark 4.5. Moreover, we have the Euler formula

$$\underline{T}_0(t)x = \lim_{m \to \infty} F\left( \frac{t}{m} \right)^m x$$

for $x \in \underline{X}$ with convergence being uniform for $t$ in compact intervals $[0, t_0]$, see, e.g., [101, Sect. III.5(a)]. Since $R(\lambda, A)|_{\underline{X}} = R(\lambda, \underline{A})$ and since $D(A)$ is bi-dense in $X$, by the local bi-equicontinuity assumption in (5.1.1) we conclude that for $x \in X$ and $t > 0$ the limit

$$S(t)x := \tau\lim_{m \to \infty} F\left( \frac{t}{m} \right)^m x \tag{5.1.3}$$

exists, and the convergence is uniform for $t$ in compact intervals $[0, t_0]$. It follows that $t \mapsto S(t)x$ is $\tau$-strongly continuous for each $x \in X$. The operator family $(S(t))_{t \geq 0}$ is locally bi-equicontinuous because of the bi-equicontinuity assumption in (5.1.1).

Next, we prove that $\underline{T}(t)$ leaves $D(A)$ invariant. Let $x \in D(A)$, so that $x = A^{-1}y$ for some $y \in X$, and insert $x$ in the formula (5.1.3) to obtain

$$\underline{T}(t)x = S(t)A^{-1}y = \tau\!\lim_{m\to\infty} F\!\left(\tfrac{t}{m}\right)^m A^{-1}y = A^{-1}\,\tau\!\lim_{m\to\infty} F\!\left(\tfrac{t}{m}\right)^m y = A^{-1}S(t)y \in D(A),$$

$$(5.1.4)$$

where we have used the bi-continuity of $A^{-1}$ and the boundedness of $\left(\left[\tfrac{m}{t}R\!\left(\tfrac{m}{t}, A\right)\right]^m y\right)_{m\in\mathbb{N}}$. By Proposition 4.14 (b) we can extend $\underline{T}(t)$ to $X$ by setting $T(t) := A\underline{T}(t)A^{-1} \in \mathcal{L}(X)$. It follows that $(T(t))_{t\geq 0}$ is a semigroup. By formula (5.1.4), we have $T(t)y = A\underline{T}(t)A^{-1}y = AA^{-1}S(t)y = S(t)y$ for each $y \in X$. So that $(T(t))_{t\geq 0}$, coinciding with $(S(t))_{t\geq 0}$, is locally bi-equicontinuous, and hence a bi-continuous semigroup.

It remains to show that the generator of $(T(t))_{t\geq 0}$ is $A$. Let $B$ denote the generator of $(T(t))_{t\geq 0}$. Then, for large $\lambda > 0$ and $x \in \underline{X}$, we have

$$R(\lambda, B)x = \int_0^\infty e^{-\lambda s} T(s)x \, \mathrm{d}s = \int_0^\infty e^{-\lambda s} \underline{T}(s)x \, \mathrm{d}s = R(\lambda, \underline{A}_0)x = R(\lambda, A)x.$$

Since $R(\lambda, B)$ and $R(\lambda, A)$ are sequentially $\tau$-continuous on norm-bounded sets and since $D(A)$ is bi-dense in $X$, we obtain $R(\lambda, B) = R(\lambda, A)$. This finishes the proof. $\qquad\square$

The first statement in the next proposition is proved by Nagel and Sinestrari, see [187, 189], while the second one follows directly from the results in Sect. 4.1.

**Proposition 5.2** *Let $A$ be a Hille–Yosida operator on the Banach space $X$ with domain $D(A)$. Denote by $(\underline{T}(t))_{t\geq 0}$ the $C_0$-semigroup on $\underline{X} = \overline{D(A)}$ generated by the part $\underline{A}$ of $A$.*

(a) *There is a one-parameter semigroup $(\overline{T}(t))_{t\geq 0}$ on $F_0(A)$ which extends $(\underline{T}(t))_{t\geq 0}$. This semigroup is strongly continuous for the $\|\cdot\|_{X_{-1}(A)}$ norm.*

(b) *Suppose that for each $t \geq 0$ the operator $\underline{T}(t)$ leaves $D(A)$ invariant. Then the space $X$ is invariant under the semigroup operators $\overline{T}(t)$ for every $t \geq 0$, i.e., for $T(t) := \overline{T}(t)|_X$ we have $T(t) \in \mathcal{L}(X)$.*

### 5.1.1  Extrapolated Semigroups

In this subsection, we extend a bi-continuous semigroup on $X$ to the extrapolation space $X_{-1}$ as a bi-continuous semigroup. We have to handle two topologies, and the next proposition leads to an additional locally convex topology on $X_{-1}$ still satisfying Assumption 2.1.

**Proposition 5.3** *Let $(X, \|\cdot\|, \tau)$ be a bi-admissible space, let $\mathcal{P}$ be as in Remark 2.2, let $E$ be a vector space over $\mathbb{R}$, and let $B : X \to E$ be a bijective linear mapping. We define for $e \in E$ and $p \in \mathcal{P}$*

$$\|e\|_E := \|B^{-1}e\| \quad \text{and} \quad p_E(e) := p(B^{-1}e).$$

*Then the following assertions hold:*

(a) $\|\cdot\|_E$ *is a norm,* $p_E$ *is a seminorm for each* $p \in \mathcal{P}$.
(b) *For the topology* $\tau_E$ *generated by* $\mathcal{P}_E := \{p_E : p \in \mathcal{P}\}$ *the triple* $(E, \|\cdot\|_E, \tau_E)$ *is a bi-admissible space.*
(c) *If* $(T(t))_{t \geq 0}$ *is a bi-continuous semigroup on* $X$ *with respect to the topology* $\tau$, *then* $T_E(t) := BT(t)B^{-1}$ *defines a bi-continuous semigroup on* $E$. *If* $A$ *is the generator of* $(T(t))_{t \geq 0}$, *then* $BAB^{-1}$ *is the generator of* $(T_E(t))_{t \geq 0}$.

***Proof*** Assertion (a) is evident. The conditions (1) and (2) from Assumption 2.1 are satisfied by the definition of $\|\cdot\|_E$ and $p_E$. Since

$$\|e\|_E = \|B^{-1}e\| = \sup_{p \in \mathcal{P}} p(B^{-1}e) = \sup_{p_E \in \mathcal{P}_E} p_E(e),$$

and by Remark 2.2 (3) in Assumption 2.1 is fulfilled. The proof of (b) is complete.

(c) For $e \in E$ we have $\|T_E(t)\|_E = \|B^{-1}BT(t)B^{-1}e\| = \|T(t)B^{-1}e\| \leq \|T(t)\| \cdot \|e\|_E$, which shows that $T_E(t) \in \mathcal{L}(E)$. Clearly, $(T_E(t))_{t \geq 0}$ satisfies the semigroup property. For $e \in E$ and $p_E \in \mathcal{P}_E$ we have

$$p_E(T_E(t)e - e) = p(B^{-1}BT(t)B^{-1}e - B^{-1}e) = p(T(t)B^{-1}e - B^{-1}e) \to 0 \quad \text{for } t \to 0,$$

showing the $\tau_E$-strong continuity of $(T_E(t))_{t \geq 0}$. If $(e_n)$ is a $\|\cdot\|_E$-bounded, $\tau_E$-null sequence, then $(B^{-1}e_n)$ is a $\|\cdot\|$-bounded $\tau$-null sequence, so that by assumption $T_E(t)e_n = T(t)B^{-1}e_n \to 0$ uniformly for $t$ in compact intervals. If $A$ is the generator of $(T(t))_{t \geq 0}$, then by means of (3.3.1) we can conclude that $B^{-1}AB$ is the generator of $(T_E(t))_{t \geq 0}$. $\quad\square$

**Definition 5.4** Let $(T(t))_{t \geq 0}$ be a bi-continuous semigroup on a bi-admissible space $(X, \|\cdot\|, \tau)$ with generator $(A, D(A))$.

(a) For $B = A^{-1} : X \to X_1$ and $E = X_1$ in Proposition 5.3 define $\mathcal{P}_1 := \mathcal{P}_E$, $\tau_1 := \tau_E$, $(T_1(t))_{t \geq 0} := (T_E(t))_{t \geq 0}$.
(b) For $B = A_{-1} : X \to X_{-1}$ and $E = X_{-1}$ in Proposition 5.3 define $\mathcal{P}_{-1} := \mathcal{P}_E$, $\tau_{-1} := \tau_E$, $(T_{-1}(t))_{t \geq 0} := (T_E(t))_{t \geq 0}$.

We obtain immediately the next result.

**Proposition 5.5** *The semigroups* $(T_1(t))_{t \geq 0}$ *and* $(T_{-1}(t))_{t \geq 0}$ *are bi-continuous with generators* $A_1 = A|_{D(A)}$ *and* $A_{-1}$, *respectively.*

Iterating the procedure in Definition 5.4 we obtain the full scale of (extrapolated) semigroups $(T_n(t))_{t \geq 0}$ for $n \in \mathbb{Z}$.

**Definition 5.6**  Let $(T(t))_{t\geq 0}$ be a bi-continuous semigroup on a bi-admissible space $(X, \|\cdot\|, \tau)$ with generator $A$ and suppose that $(T_{\pm n}(t))_{t\geq 0}$ and $\mathcal{P}_{\pm n}$ have been defined for some $n \in \mathbb{N}$ already.

(a)  For $B = A_n^{-1} : X_n \to X_{n+1}$, $E = X_{n+1}$ and the semigroup $(T_n(t))_{t\geq 0}$ in Proposition 5.3 define $\mathcal{P}_{n+1} := \mathcal{P}_E$, $\tau_{n+1} := \tau_E$, $(T_{n+1}(t))_{t\geq 0} := (T_E(t))_{t\geq 0}$.

(b)  For $B = A_{-n-1} : X_{-n} \to X_{-n-1}$, $E = X_{-n-1}$ and the semigroup $(T_{-n}(t))_{t\geq 0}$ in Proposition 5.3 define $\mathcal{P}_{-n-1} := \mathcal{P}_E$, $\tau_{-n-1} := \tau_E$, $(T_{-n-1}(t))_{t\geq 0} := (T_E(t))_{t\geq 0}$.

**Proposition 5.7**  *For each $n \in \mathbb{Z}$ the semigroup $(T_n(t))_{t\geq 0}$ is bi-continuous on $(X_n, \|\cdot\|_n, \tau_n)$ with generator $A_n : X_{n+1} \to X_n$. Its space of norm strong continuity is $\underline{X}_n$.*

**Proof**  The first statement follows directly from Proposition 5.5 by induction. For $n = 0$ the second assertion is the content of Lemma 4.30, for general $n \in \mathbb{Z}$ one can argue inductively.  $\square$

The following diagram summarizes the situation:

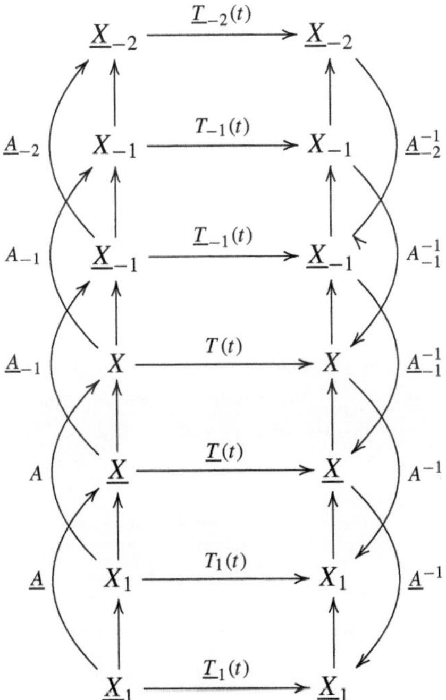

The spaces $\underline{X}_{n+1}$ are bi-dense in $X_n$ for the topology $\tau_n$ and dense in $\underline{X}_n$ for the norm $\|\cdot\|_{X_n}$. The semigroups $(T_n(t))_{t\geq 0}$ are bi-continuous on $X_n$, while $(\underline{T}_n(t))_{t\geq 0}$ are $C_0$-semigroups (strongly continuous for the norm) on $\underline{X}_n$.

### 5.1.2   Hölder Spaces of Bi-continuous Semigroups

Suppose $A$ generates the bi-continuous semigroup $(T(t))_{t\geq 0}$ of negative growth bound on $X$. Recall from Theorem 5.1 that the restricted operators $\underline{T}(t) := T(t)|_{\underline{X}}$ form a $C_0$-semigroup $(\underline{T}(t))_{t\geq 0}$ on $\underline{X}$. Also recall from Proposition 4.31 that for $\alpha \in (0, 1]$

$$F_\alpha(A) = F_\alpha(T) = \left\{ x \in \underline{X} : \sup_{t>0} \frac{\|\underline{T}(t)x - x\|}{t^\alpha} < \infty \right\} = \left\{ x \in X : \sup_{t>0} \frac{\|T(t)x - x\|}{t^\alpha} < \infty \right\}$$

with the norm

$$\|x\|_{F_\alpha} = \sup_{t>0} \frac{\|\underline{T}(t)x - x\|}{t^\alpha},$$

and for $\alpha \in (0, 1)$

$$\underline{X}_\alpha(A) := \left\{ x \in \underline{X} : \lim_{t\to 0} \frac{\|\underline{T}(t)x - x\|}{t^\alpha} = 0 \right\} = \left\{ x \in X : \lim_{t\to 0} \frac{\|T(t)x - x\|}{t^\alpha} = 0 \right\}.$$

We have the (continuous) inclusions

$$\underline{X}_1 \hookrightarrow X_1 \to \underline{X}_\alpha(A) \hookrightarrow F_\alpha(A) \to \underline{X} \hookrightarrow X;$$

all these spaces are invariant under $(T(t))_{t\geq 0}$. We now extend this diagram by a space which lies between $\underline{X}_\alpha$ and $F_\alpha$.

**Definition 5.8**  Let $(T(t))_{t\geq 0}$ be a bi-continuous semigroup of negative growth bound on a Banach space $X$ with respect to a locally convex topology $\tau$ that is generated by a family $\mathcal{P}$ of seminorms satisfying (2.1.2). For $\alpha \in (0, 1)$ we define the space

$$X_\alpha := X_\alpha(T) := \left\{ x \in X : \tau\text{-}\lim_{t\to 0} \frac{T(t)x - x}{t^\alpha} = 0 \text{ and } \sup_{t>0} \frac{\|T(t)x - x\|}{t^\alpha} < \infty \right\},$$

$$(5.1.5)$$

and endow it with the norm $\|\cdot\|_{F_\alpha}$. We further equip $F_\alpha$ and $X_\alpha$ with the locally convex topology $\tau_{F_\alpha}$ generated by the family of seminorms $\mathcal{P}_{F_\alpha} := \{p_{F_\alpha} : p \in \mathcal{P}\}$, where $p_{F\alpha}$ is defined as

$$p_{F_\alpha}(x) := \sup_{t>0} \frac{p(T(t)x - x)}{t^\alpha}.$$

$$(5.1.6)$$

It is easy to see that $X_\alpha$ is a Banach space, i.e., as closed subspace of $F_\alpha$. By construction we have that indeed $\underline{X}_\alpha(A) \subseteq X_\alpha \subseteq F_\alpha(A)$. Next we discuss some properties of this space.

**Lemma 5.9** (a)  *Let $(x_n)_{n\in\mathbb{N}}$ be a $\|\cdot\|_{F_\alpha}$-norm-bounded sequence in $F_\alpha$ with $x_n \to x \in X$ in the topology $\tau$. Then $x \in F_\alpha$.*

(b)  *The triple $(F_\alpha, \|\cdot\|_{F_\alpha}, \tau_{F_\alpha})$ is a bi-admissible space.*

(c)  $X_\alpha$ is bi-closed in $F_\alpha$, i.e., every $\|\cdot\|_{F_\alpha}$-bounded an $\tau_{F_\alpha}$-convergent sequence in $X_\alpha$ has its limit in $X_\alpha$.

**Proof**  (a) The statement follows from the fact that the norm $\|\cdot\|_{F_\alpha}$ is lower semicontinuous for the topology $\tau$. If

$$\frac{\|T(t)x_n - x_n\|}{t^\alpha} \leq \|x_n\|_{F_\alpha} \leq M$$

for each $n \in \mathbb{N}$, $t > 0$ and for some $M \geq 0$ we can estimate

$$\sup_{t>0} \frac{\|T(t)x - x\|}{t^\alpha} = \sup_{t>0} \sup_{p\in\mathcal{P}} p\left(\frac{T(t)x - x}{t^\alpha}\right) = \sup_{t>0} \sup_{p\in\mathcal{P}} \lim_{n\to\infty} p\left(\frac{T(t)x_n - x_n}{t^\alpha}\right)$$

$$\leq \sup_{t>0} \limsup_{n\to\infty} \left\|\frac{T(t)x_n - x_n}{t^\alpha}\right\| \leq \sup_{t>0} \sup_{n\in\mathbb{N}} \left\|\frac{T(t)x_n - x_n}{t^\alpha}\right\| \leq M.$$

(b) We have for $p \in \mathcal{P}$ and $x \in F_\alpha$ that

$$p_{F_\alpha}(x) = \sup_{t>0} \frac{p(T(t)x - x)}{t^\alpha} \leq \sup_{t>0} \frac{\|T(t)x - x\|}{t^\alpha} = \|x\|_{F_\alpha}.$$

This proves that $\tau_{F_\alpha}$ is coarser than the $\|\cdot\|_{F_\alpha}$-topology, but is still Hausdorff by construction. For the second property of Assumption 2.1 let $(x_n)_{n\in\mathbb{N}}$ be a $\tau_{F_\alpha}$-Cauchy sequence in $F_\alpha$ such that there exists $M > 0$ with $\|x_n\|_{F_\alpha} \leq M$ for each $n \in \mathbb{N}$. Since $\tau$ is coarser than $\tau_{F_\alpha}$, we conclude that $(x_n)$ is $\tau$-Cauchy sequence which is also bounded in $\|\cdot\|_{F_\alpha}$, hence in $\|\cdot\|$. By assumption there is $x \in X$ such that $x_n \to x$ in $\tau$. By part (a) we obtain $x \in F_\alpha$. It remains to prove that $x_n \to x$ in $\tau_{F_\alpha}$. Let $\varepsilon > 0$, and take $N \in \mathbb{N}$ such that for each $n, m \in \mathbb{N}$ with $n, m \geq N$ we have $p_{F_\alpha}(x_n - x_m) < \varepsilon$. For $t > 0$

$$p\left(\frac{T(t)(x_n - x) - (x_n - x)}{t^\alpha}\right) = \lim_{m\to\infty} p\left(\frac{T(t)(x_n - x_m) - (x_n - x_m)}{t^\alpha}\right) \leq p_{F_\alpha}(x_n - x_m) < \varepsilon$$

for each $n \geq N$. Taking the supremum in $t > 0$ we obtain $p_{F_\alpha}(x - x_n) \leq \varepsilon$ for each $n \geq N$.

The norming property in (2.1.1) follows again from Remark 2.3 and the fact that the family $\mathcal{P}$ is norming by assumption.

(c) Let $(x_n)_{n\in\mathbb{N}}$ be a $\|\cdot\|_{F_\alpha}$-bounded and $\tau_{F_\alpha}$-convergent sequence in $X_\alpha$ with limit $x \in X$. For $p \in \mathcal{P}$ we then have

$$\sup_{t>0} p\left(\frac{T(t)(x_n - x) - (x_n - x)}{t^\alpha}\right) \to 0.$$

Since $x_n \in X_\alpha$ for each $n \in \mathbb{N}$, we have

$$\lim_{t\to0} p\left(\frac{T(t)x_n - x_n}{t^\alpha}\right) = 0, \quad \text{and} \quad \sup_{t>0} \left\|\frac{T(t)x_n - x_n}{t^\alpha}\right\| < \infty.$$

We now conclude for a fixed $p \in \mathcal{P}$

$$p\left(\frac{T(t)x - x}{t^\alpha}\right) = p\left(\frac{T(t)(x - x_n) - (x - x_n) + T(t)x_n - x_n}{t^\alpha}\right)$$

$$\leq p\left(\frac{T(t)(x - x_n) - (x - x_n)}{t^\alpha}\right) + p\left(\frac{T(t)x_n - x_n}{t^\alpha}\right)$$

$$\leq p_{F_\alpha}(x - x_n) + p\left(\frac{T(t)x_n - x_n}{t^\alpha}\right) < \frac{\varepsilon}{2} + \frac{\varepsilon}{2} = \varepsilon,$$

where we first fix $n \in \mathbb{N}$ such that $p_{F_\alpha}(x - x_n) < \frac{\varepsilon}{2}$, and then we take $\delta > 0$ such that $0 < t < \delta$ implies $p(\frac{T(t)x_n - x_n}{t^\alpha}) < \frac{\varepsilon}{2}$. $\qquad\square$

The next goal is to verify that $(T(t))_{t \geq 0}$ can be restricted to $X_\alpha$ to obtain a bi-continuous semigroup with respect to the topology $\tau_{F_\alpha}$.

**Lemma 5.10**  *If $(T(t))_{t \geq 0}$ is a bi-continuous semigroup, then $X_\alpha$ is invariant under the semigroup.*

***Proof*** We notice that in order to prove

$$\tau\text{-}\lim_{s \to 0} \frac{T(s)x - x}{s^\alpha} = 0$$

we only have to check that

$$\frac{p(T(s_n)x - x)}{s_n^\alpha} \to 0$$

for $n \to \infty$ for every null sequence $(s_n)_{n \in \mathbb{N}}$ in $[0, \infty)$ and for each $p \in \mathcal{P}$. Let $x \in X_\alpha$. Then we have that $y_n := \frac{T(s_n)x - x}{s_n^\alpha}$ converges to 0 with respect to $\tau$ if $(s_n)_{n \in \mathbb{N}}$ is any null sequence and $n \to \infty$. Moreover, this sequence $(y_n)_{n \in \mathbb{N}}$ is $\|\cdot\|$-bounded by the assumption that $x \in X_\alpha$. Whence we conclude

$$\tau\text{-}\lim_{n \to \infty} T(t)y_n = \tau\text{-}\lim_{n \to \infty} \frac{T(s_n)T(t)x - T(t)x}{s_n^\alpha} = 0,$$

so that $T(t)x \in X_\alpha$. $\qquad\square$

We now prove that $(T(t))_{t \geq 0}$ is bi-continuous on $X_\alpha$ and notice first that the local boundedness and the semigroup property are trivial.

**Lemma 5.11**  *If $(T(t))_{t \geq 0}$ is a bi-continuous semigroup on a bi-admissible space $(X, \|\cdot\|, \tau)$ and $\alpha \in (0, 1)$, then $(T(t))_{t \geq 0}$ is strongly $\tau_{F_\alpha}$-continuous on $X_\alpha$.*

***Proof*** We have to show that $p_{F_\alpha}(T(t_n)x - x) \to 0$ for all $p \in \mathcal{P}$ whenever $t_n \downarrow 0$. Let $s_n, t_n > 0$ be with $s_n, t_n \to 0$. Then

$$\frac{p(T(s_n)T(t_n)x - T(s_n)x - T(t_n)x + x)}{s_n^\alpha} \leq \frac{p(T(t_n)T(s_s)x - T(t_n)x)}{s_n^\alpha} + \frac{p(T(s_n)x - x)}{s_n^\alpha}$$

$$= \frac{p(T(t_n)(T(s_n)x - x))}{s_n^\alpha} + \frac{p(T(s_n)x - x)}{s_n^\alpha}.$$

$$(5.1.7)$$

The sequence $(y_n)$ given by $y_n := \frac{T(s_n)x - x}{s_n^\alpha}$ is $\|\cdot\|$-bounded and $\tau$-convergent to $0$, because $x \in X_\alpha$. So that the last term in the previous equation (5.1.7) converges to $0$. But since $\{T(t_n) : n \in \mathbb{N}\}$ is bi-equicontinuous, also the first term in (5.1.7) converges to $0$. This proves strong continuity with respect to $\tau_{F_\alpha}$.   $\square$

**Lemma 5.12** *Let $(T(t))_{t\geq 0}$ be a bi-continuous semigroup on a bi-admissible space $(X, \|\cdot\|, \tau)$. Then $(T(t))_{t\geq 0}$ is locally bi-equicontinuous on $F_\alpha$.*

***Proof*** Let $(x_n)_{n\in\mathbb{N}}$ be a $\|\cdot\|_{F_\alpha}$-bounded sequence which converges to zero with respect to $\tau_{F_\alpha}$ and assume that $(T(t)x_n)_{n\in\mathbb{N}}$ does not converge to zero uniformly for $t \in [0, t_0]$ for some $t_0 > 0$. Hence there exist $p \in \mathcal{P}$, $\delta > 0$ and a sequence $(t_n)_{n\in\mathbb{N}}$ of positive real numbers such that

$$p_{F_\alpha}(T(t_n)x_n) > \delta$$

for all $n \in \mathbb{N}$. As a consequence there exists a null-sequence $(s_n)_{n\in\mathbb{N}}$ in $\mathbb{R}$ such that

$$\frac{p(T(s_n)T(t_n)x_n - T(t_n)x_n)}{s_n^\alpha} > \delta$$

for each $n \in \mathbb{N}$. Now notice that the sequence $(y_n)_{n\in\mathbb{N}}$ defined by $y_n := \frac{T(s_n)x_n - x_n}{s_n^\alpha}$ is a $\tau$-null sequence since

$$\frac{q(T(s_n)x_n - x_n)}{s_n^\alpha} \leq \sup_{s>0} \frac{q(T(s)x_n - x_n)}{s^\alpha}, \quad q \in \mathcal{P},$$

and the term on the right-hand side converges to zero as $n \to \infty$ by assumption. Using the local bi-equicontinuity of the semigroup $(T(t))_{t\geq 0}$ with respect to $\tau$, we conclude that $\frac{T(t)T(s_n)x_n - T(t)x_n}{s_n}$ converges to zero uniformly for $t \in [0, t_0]$, which is a contradiction. Hence $(T(t))_{t\geq 0}$ is locally bi-equicontinuous on $X_\alpha$.   $\square$

**Remark 5.13** Notice that the local bi-equicontinuity with respect to $\tau_{F_\alpha}$ holds on the whole space $F_\alpha$, while strong $\tau_{F_\alpha}$-continuity holds on $X_\alpha$ only. In particular, we will see in Theorem 5.15 that $X_\alpha$ is the space of strong $\tau_{F_\alpha}$-continuity.

We can summarize the previous results in the following theorem.

**Theorem 5.14**  *Let $(T(t))_{t\geq 0}$ be a bi-continuous semigroup on a bi-admissible space $(X, \|\cdot\|, \tau)$. Then the restricted operators $T_\alpha(t) := T(t)|_{X_\alpha}$ on $X_\alpha$ form a bi-continuous semigroup. Moreover, the generator $A_\alpha$ of $(T_\alpha(t))_{t\geq 0}$ is the part of $A$ in $X_\alpha$.*

**Proof**  Because of the previous lemmas it remains to prove that the part of $A$ in $X_\alpha$ generates the restricted semigroup on $X_\alpha$. We can argue as in the proof of the proposition in [101, Chap. II, Par. 2.3]. Since the embedding $X_\alpha \subseteq X$ is continuous for the topologies $\tau_{F_\alpha}$ and $\tau$, we conclude that $A_\alpha \subseteq A|_{X_\alpha}$. For the converse let $C$ denote the generator of $(T_\alpha(t))_{t\geq 0}$ and take $\lambda \in \mathbb{R}$ large enough such that

$$R(\lambda, C)x = \int_0^\infty e^{-\lambda s} T(s)x \, ds = R(\lambda, A)x, \quad x \in X_\alpha.$$

For $x \in D(A|_{X_\alpha})$ we obtain

$$x = R(\lambda, A)(\lambda - A)x = R(\lambda, C)(\lambda - A)x \in D(C)$$

and hence $A|_{X_\alpha} \subseteq A_\alpha$. This proves that the part of $A$ in $X_\alpha$ generates the restricted semigroup. $\qquad\square$

By similar reasoning as in Lemma 4.30 one can prove the following.

**Theorem 5.15**  *Let $\alpha \in (0, 1)$ and let $(T(t))_{t\geq 0}$ be a bi-continuous semigroup on a bi-admissible space $(X, \|\cdot\|, \tau)$. Then $D(A)$ is $\tau_{F_\alpha}$-bi-dense in $X_\alpha$ and*

$$X_\alpha = \left\{ x \in F_\alpha : \tau_{F_\alpha}\lim_{t\to 0} T(t)x = x \right\}, \tag{5.1.8}$$

*i.e., for $x \in F_\alpha$ the mapping $t \mapsto T(t)x$ is $\tau_{F_\alpha}$-continuous if and only if $x \in X_\alpha$.*

**Proof**  Denote by $X_{\alpha,\mathrm{cont}}$ the right-hand side of (5.1.8), i.e., the space of $\tau_{F_\alpha}$-strong continuity. Notice that $D(A) \subseteq \underline{X}_\alpha \subseteq X_\alpha \subseteq X_{\alpha,\mathrm{cont}}$.

Suppose $x \in X_{\alpha,\mathrm{cont}}$. For each $n \in \mathbb{N}$ we have

$$x_n := n \int_0^{\frac{1}{n}} T_\alpha(t)x \, dt = n \int_0^{\frac{1}{n}} T(t)x \, dt \in D(A)$$

as a $\tau$- and $\tau_{F_\alpha}$-convergent Riemann integral. Whence it follows that $x_n \xrightarrow{\tau_{F_\alpha}} x$, whereas the $\|\cdot\|_{F_\alpha}$-boundedness of $(x_n)_{n\in\mathbb{N}}$ clear. We conclude that $x \in X_\alpha$ (because $X_\alpha$ is bi-closed in $F_\alpha$), implying $X_{\alpha,\mathrm{cont}} \subseteq X_\alpha$. As a byproduct we also obtain that $D(A)$ is bi-dense in $X_\alpha$. $\qquad\square$

**Proposition 5.16** *For $0 \leq \alpha < \beta \leq 1$ we have*

$$X_1 = D(A) \hookrightarrow F_\beta \hookrightarrow \underline{X}_\alpha \subseteq X_\alpha,$$

*where the embeddings are continuous for the respective norms and for the respective topologies $\tau_1$, $\tau_{F_\beta}$, $\tau_{F_\alpha}$. The space $D(A)$ is bi-dense in $X_\alpha$, and as a consequence $X_\beta$ is bi-dense in $X_\alpha$.*

### 5.1.3   Characterization of Hölder Spaces by Generators

Analogously to Proposition 4.31 we characterize the Hölder space $X_\alpha$ by means of the semigroup generator.

**Theorem 5.17** *Let $(T(t))_{t\geq 0}$ be a bi-continuous semigroup on a bi-admissible space $(X, \|\cdot\|, \tau)$. Assume that the generator $(A, D(A))$ has negative growth bound. For $\alpha \in (0, 1)$ we have*

$$X_\alpha = \left\{ x \in X : \tau\lim_{\lambda\to\infty} \lambda^\alpha AR(\lambda, A)x = 0 \text{ and } \sup_{\lambda>0} \|\lambda^\alpha AR(\lambda, A)x\| < \infty \right\}. \tag{5.1.9}$$

***Proof*** Suppose $x \in X_\alpha$. From Proposition 4.31 we deduce immediately

$$\sup_{\lambda>0} \|\lambda^\alpha AR(\lambda, A)x\| < \infty.$$

Let now $\varepsilon > 0$ be arbitrary. For $x \in X_\alpha$ and $p \in \mathcal{P}$ we can find $\delta > 0$ such that $0 \leq t < \delta$ implies $\frac{p(T(t)x-x)}{t^\alpha} < \varepsilon$. Recall the following formula:

$$\lambda^\alpha AR(\lambda, A)x = \lambda^{\alpha+1} \int_0^\infty e^{-\lambda s}(T(s)x - x)\,\mathrm{d}s.$$

From this we deduce

$$p(\lambda^\alpha AR(\lambda, A)x) \leq \lambda^{\alpha+1} \int_0^\infty e^{-\lambda s} \cdot \frac{p(T(s)x - x)}{s^\alpha} s^\alpha \, ds$$

$$= \lambda^{\alpha+1} \int_0^\delta e^{-\lambda s} \cdot \frac{p(T(s)x - x)}{s^\alpha} s^\alpha \, ds + \lambda^{\alpha+1} \int_\delta^\infty e^{-\lambda s} \cdot \frac{p(T(s)x - x)}{s^\alpha} s^\alpha \, ds$$

$$< \lambda^{\alpha+1} \varepsilon \int_0^\delta e^{-\lambda s} s^\alpha \, ds + \lambda^{\alpha+1} \int_\delta^\infty e^{-\lambda s} \cdot \frac{\|T(s)x - x\|}{s^\alpha} s^\alpha \, ds$$

$$\leq \lambda^{\alpha+1} \varepsilon \int_0^\delta e^{-\lambda s} s^\alpha \, ds + \|x\|_{F_\alpha} \lambda^{\alpha+1} \int_\delta^\infty e^{-\lambda s} \cdot s^\alpha \, ds$$

$$= \varepsilon \int_0^{\lambda \delta} e^{-t} t^\alpha \, dt + \|x\|_{F_\alpha} \int_{\lambda \delta}^\infty e^{-t} t^\alpha \, dt$$

$$\leq L\varepsilon + \|x\|_{F_\alpha} \int_{\lambda \delta}^\infty e^{-t} t^\alpha \, dt$$

where $L := \int_0^\infty e^{-\lambda s} s^\alpha \, ds < \infty$. Notice that the last part of the sum tends to zero if $\lambda \to \infty$ since $\delta > 0$ is fixed. So we obtain $\tau \lim_{\lambda \to \infty} \lambda^\alpha AR(\lambda, A)x = 0$.

For the converse inclusion suppose that $\tau \lim_{\lambda \to \infty} \lambda^\alpha AR(\lambda, A)x = 0$ and $\sup_{\lambda > 0} \|\lambda^\alpha AR(\lambda, A)x\| < \infty$, the latter immediately implying $\|x\|_{F_\alpha(T)} < \infty$ (see Proposition 4.31). We have to show that $\tau \lim_{t \to 0} \frac{T(t)x - x}{t^\alpha} = 0$. For $\lambda > 0$ define $x_\lambda = \lambda R(\lambda, A)$ and $y_\lambda = AR(\lambda, A)$, then we have

$$x = \lambda R(\lambda, A)x - AR(\lambda, A)x = x_\lambda - y_\lambda.$$

First notice that for $p \in \mathcal{P}$

$$\frac{p(T(t)x_\lambda - x_\lambda)}{t^\alpha} \leq \frac{1}{t^\alpha} p(T(t)\lambda R(\lambda, A)x - \lambda R(\lambda, A)x) \leq \frac{\lambda^{1-\alpha}}{t^\alpha} \int_0^t p(T(s)\lambda^\alpha AR(\lambda, A)x) \, ds.$$

$$(5.1.10)$$

By assumption the term $\lambda^\alpha AR(\lambda, A)x$ is norm-bounded and converges in the topology $\tau$ to zero as $\lambda \to \infty$, hence by the local bi-equicontinuity we conclude that $p(T(s)\lambda^\alpha AR(\lambda, A)x) \to 0$ uniformly for $s \in [0, 1]$. Now let $\varepsilon > 0$ and $\lambda_0 > 1$ so large that for $\lambda > \lambda_0$ and $s \in [0, 1]$ we have $p(T(s)\lambda^\alpha AR(\lambda, A)x) < \varepsilon$. If $t < \frac{1}{\lambda_0}$, then $\lambda := \frac{1}{t} > \lambda_0$ and we obtain that the expression in (5.1.10) becomes smaller than $\varepsilon$.

For the estimate of the second part involving $y_\lambda$ we observe that

$$\frac{p(T(t)y_\lambda - y_\lambda)}{t^\alpha} \leq \frac{1}{(t\lambda)^\alpha} p(T(t)\lambda^\alpha AR(\lambda, A)x) + \frac{1}{(t\lambda)^\alpha} p(\lambda^\alpha AR(\lambda, A)x).$$

By taking $t < \frac{1}{\lambda_0}$ and $\lambda := \frac{1}{t}$ we obtain the estimate

$$\frac{p(T(t)y_\lambda - y_\lambda)}{t^\alpha} \leq p(T(\tfrac{1}{\lambda})\lambda^\alpha AR(\lambda, A)x) + p(\lambda^\alpha AR(\lambda, A)x) < \varepsilon + \varepsilon, \qquad (5.1.11)$$

by the choice of $\lambda_0$. Altogether we obtain for $t < \frac{1}{\lambda_0}$ that $\frac{p(T(t)x - x)}{t^\alpha} < 3\varepsilon$, showing

$$\tau \lim_{t \to 0} \frac{T(t)x - x}{t^\alpha} = 0,$$

i.e., $x \in X_\alpha$ as required. $\qquad\qquad\square$

**Remark 5.18** It is possible to define the space $X_\alpha(A)$ as in (5.1.9) also for operators which are not necessarily generators of bi-continuous semigroups. However, we have to suppose that the resolvent fulfills certain continuity assumptions with respect to a topology satisfying, say, Assumption 2.1.

Again, we put our spaces $X_\alpha$ in the general context of Theorem 4.15.

**Proposition 5.19** *For $\alpha \in (0, 1)$ and $\mathcal{A}$, $\lambda$ and $\mathcal{E}$ as in Theorem 4.15 we have*

$$X_{-\alpha} = \left\{ (\lambda - \mathcal{A})y \in X_{-1} : \sup_{t>0} \frac{\|T(t)y - y\|}{t^{1-\alpha}} < \infty, \ \tau\lim_{t\to 0} \frac{T(t)y - y}{t^{1-\alpha}} = 0 \right\}.$$

Finally, we extend the scale of spaces $X_\alpha$ to the whole range $\alpha \in \mathbb{R}$.

**Definition 5.20** For $\alpha \in \mathbb{R} \setminus \mathbb{Z}$ we write $\alpha = m + \beta$ with $m \in \mathbb{Z}$ and $\beta \in (0, 1]$, and define

$$X_\alpha(A) := X_\beta(A_m),$$

with the corresponding norms. The locally convex topology on $X_\alpha$ comes from $X_\beta$ via the mapping $A_m$.

**Remark 5.21** We summarize all previous results in the following diagram:

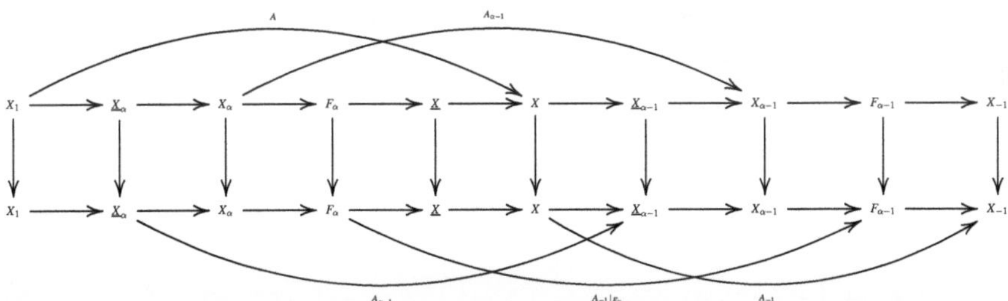

where $\alpha \in (0, 1)$. Here $A_{\alpha-1}$ and $\underline{A}_{\alpha-1}$ are defined to be the part of $A_{-1}$ in $X_{\alpha-1}$ and the part of $\underline{A}_{-1}$ in $\underline{X}_{\alpha-1}$, respectively. They are all continuous with respect to the norms and topologies on these spaces. In addition, we recall that $X_{\alpha-1}$ and $\underline{X}_{\alpha-1}$ are the extrapolation spaces of $X_\alpha(A_{-1})$ and $\underline{X}_\alpha(A_{-1})$, respectively. All horizontal arrows represent continuous inclusions, while the vertical arrows represent the action(s) of the semigroup(s). All the spaces are dense in the underlined ones containing them, while the spaces with underlining are bi-dense in each of the bigger ones.

## 5.2    Examples

In this section, we present examples for extrapolation and intermediate spaces for (generators of) bi-continuous semigroups. We will use Theorem 4.15 and its variants to identify the space $X_\alpha$ for $\alpha < 0$.

### 5.2.1    The Translation Semigroup

From Sect. 2.1.1 we know that $(C_b(\mathbb{R}), \|\cdot\|_\infty, \tau_{co})$ is a bi-admissible space. Moreover, we derived in Sect. 2.2.1 that the left-translation semigroup $(T(t))_{t\geq 0}$ defined by

$$T(t)f(x) = f(x+t), \quad t \geq 0$$

is bi-continuous on $(C_b(\mathbb{R}), \|\cdot\|_\infty, \tau_{co})$. Following Proposition 3.4, the generator $(A, D(A))$ of this semigroup is the first derivative $Af = f'$ on the domain

$$D(A) = \{f \in C_b(\mathbb{R}) : f \text{ is differentiable and } f' \in C_b(\mathbb{R})\}.$$

The space of strong continuity is $\underline{X} = UC_b(\mathbb{R})$, the space of all bounded, uniformly continuous functions. We use Theorem 4.15 to determine the corresponding extrapolation spaces. To this purpose let $\mathscr{E} = \mathscr{D}'(\mathbb{R})$ be the space of all distributions on $\mathbb{R}$, let $\mathcal{A} := D : \mathscr{D}'(\mathbb{R}) \to \mathscr{D}'(\mathbb{R})$ be the distributional derivative, and let $i : C_b(\mathbb{R}) \to \mathscr{D}'(\mathbb{R})$ be the regular embedding. From Theorem 4.15 it then follows that

$$\underline{X}_{-1} = \{F \in \mathscr{D}'(\mathbb{R}) : F = f - Df \text{ for some } f \in UC_b(\mathbb{R})\},$$
$$X_{-1} = \{F \in \mathscr{D}'(\mathbb{R}) : F = f - Df \text{ for some } f \in C_b(\mathbb{R})\}.$$

For the Favard and Hölder spaces we have

$$F_\alpha = \left\{f \in C_b(\mathbb{R}) : \sup_{\substack{x,y\in\mathbb{R} \\ x\neq y}} \frac{|f(x) - f(y)|}{|x-y|^\alpha} < \infty\right\} = C_b^\alpha(\mathbb{R}),$$

$$\underline{X}_\alpha = \left\{f \in UC_b(\mathbb{R}) : \lim_{t\to 0} \sup_{\substack{x,y\in\mathbb{R} \\ 0<|x-y|<t}} \frac{|f(x) - f(y)|}{|x-y|^\alpha} = 0\right\} = h_b^\alpha(\mathbb{R}).$$

Hence $F_\alpha$ is the space of bounded $\alpha$-Hölder-continuous functions and $\underline{X}_\alpha$ is the so-called little Hölder space $h_b^\alpha(\mathbb{R})$ (see also [175]). The abstract Hölder space $X_\alpha$ corresponding to the bi-continuous semigroup yields the local version $h_{b,loc}^\alpha(\mathbb{R})$ of the little Hölder space

$$h_{b,loc}^\alpha = \left\{f \in C_b^\alpha(\mathbb{R}) : \lim_{t\to 0} \sup_{\substack{x,y\in K \\ 0<|x-y|<t}} \frac{|f(x) - f(y)|}{|x-y|^\alpha} = 0 \text{ for each } K \subseteq \mathbb{R} \text{ compact}\right\}.$$

Then $X_\alpha = h^\alpha_{b,loc}(\mathbb{R})$.

It is easy to see $\underline{X}_\alpha \subsetneq X_\alpha \subsetneq F_\alpha$. The extrapolated Favard class $F_0$ can be identified with $L^\infty(\mathbb{R})$. To prove this we argue as follows: We know from the general theory that $F_0(T) = (1 - D)F_1(T)$ where $F_1(T)$ are precisely the bounded Lipschitz functions on $\mathbb{R}$. Now using the fact that $\mathrm{Lip}_b(\mathbb{R}) = W^{1,\infty}(\mathbb{R})$ with equivalent norms we obtain that indeed $F_0 = L^\infty(\mathbb{R})$. For an alternative proof of this fact we refer to [101, Chap. II.5(b)].

Moreover, from Corollary 4.25 we obtain for $\alpha \in (0, 1)$

$$F_{-\alpha} = \left\{ f \in \mathscr{D}'(\mathbb{R}) : \quad F = f - Df \text{ for } f \in C_b^{1-\alpha}(\mathbb{R}) \right\}$$

and

$$X_{-\alpha} = \left\{ f \in \mathscr{D}'(\mathbb{R}) : \quad F = f - Df \text{ for } f \in h^{1-\alpha}_{b,loc}(\mathbb{R}) \right\}.$$

We summarize this example by the diagram:

$$C_b^1(\mathbb{R}) \hookrightarrow \mathrm{Lip}_b(\mathbb{R}) \hookrightarrow h_b^\alpha(\mathbb{R}) \hookrightarrow h^\alpha_{b,loc}(\mathbb{R}) \hookrightarrow C_b^\alpha(\mathbb{R}) \hookrightarrow UC_b(\mathbb{R}) \hookrightarrow C_b(\mathbb{R}) \hookrightarrow L^\infty(\mathbb{R})$$

according to the abstract chain of spaces

$$X_1 \hookrightarrow F_1 \hookrightarrow \underline{X}_\alpha \hookrightarrow X_\alpha \hookrightarrow F_\alpha \hookrightarrow \underline{X} \hookrightarrow X \hookrightarrow F_0$$

for $\alpha \in (0, 1)$. For the higher order spaces, we have

$$X_n := D(A^n) = \left\{ f \in C_b(\mathbb{R}) : \quad f \text{ is } n\text{-times differentiable and } f^{(n)} \in C_b(\mathbb{R}) \right\}$$
$$= \left\{ f \in C_b(\mathbb{R}) : \quad f^{(k)} \in C_b(\mathbb{R}), \ k = 1, \dots, n \right\} = C_b^n(\mathbb{R})$$

for $n \in \mathbb{N}$. For $n \in \mathbb{N}$ and $\alpha \in [0, 1)$

$$F_{n+\alpha} = \left\{ f \in C_b^n(\mathbb{R}) : \quad \sup_{\substack{x,y \in \mathbb{R} \\ x \neq y}} \frac{|f^{(n)}(x) - f^{(n)}(y)|}{|x - y|^\alpha} < \infty \right\} = C_b^{n,\alpha}(\mathbb{R}).$$

This example complements the corresponding one in Nagel, Nickel, Romanelli [188, Sect. 3.2].

### 5.2.2 The Multiplication Semigroup

Let us consider another example that also complements the examples in Sect. 2.2. From Sect. 2.1.1, we know that for $\Omega$ a locally compact space, the space $(C_b(\Omega), \|\cdot\|, \tau_{co})$ is bi-admissible. Now, let $q : \Omega \to \mathbb{C}$ be continuous such that $\sup_{x \in \Omega} \mathrm{Re}(q(x)) < 0$. We define the multiplication operator $M_q : D(M_q) \to C_b(\Omega)$ by $M_q f = qf$ on the maximal domain

$$D(M_q) = \{f \in C_b(\Omega) : qf \in C_b(\Omega)\}.$$

This operator generates the semigroup $(T_q(t))_{t\geq 0}$ defined by

$$(T_q(t)f)(x) = e^{tq(x)}f(x), \quad t \geq 0, x \in \Omega, f \in C_b(\Omega),$$

which is bi-continuous on $C_b(\Omega)$ with respect to the compact-open topology. Now let $\mathscr{E} = C(\Omega)$ the space of all continuous functions on $\Omega$, let $\mathcal{M}_q : C(\Omega) \to C(\Omega)$ be the multiplication operator $\mathcal{M}_q f := qf$ and $i : C_b(\Omega) \to C(\Omega)$ the identical embedding. Then by Theorem 4.15 we obtain

$$X_{-1} = \{g \in C(\Omega) : q^{-1}g \in C_b(\Omega)\}.$$

For $\alpha \in (0, 1)$, the (abstract) Favard space is

$$F_\alpha = \{f \in C_b(\Omega) : |q|^\alpha f \in C_b(\Omega)\}. \tag{5.2.1}$$

To see this suppose first that $f \in F_\alpha$, which means

$$\sup_{t>0} \sup_{x\in\Omega} \frac{|e^{tq(x)}f(x) - f(x)|}{t^\alpha} < \infty.$$

By taking supremum only for $t = \frac{1}{|q(x)|}$ we obtain

$$\sup_{x\in\Omega} \left|e^{\frac{q(x)}{|q(x)|}} - 1\right| \cdot |f(x)| \cdot |q(x)|^\alpha < \infty,$$

since

$$\frac{|e^{tq(x)}f(x) - f(x)|}{t^\alpha} = \frac{|e^{tq(x)} - 1| \cdot |f(x)||q(x)|^\alpha}{|q(x)|^\alpha t^\alpha}. \tag{5.2.2}$$

Hence $|q|^\alpha f \in C_b(\Omega)$, so that the inclusion "$\subseteq$" in (5.2.1) is established. For the converse assume that $|q|^\alpha f \in C_b(\Omega)$. Since the function $g(z) = \frac{|e^z - 1|}{|z|^\alpha}$ is bounded on the left half-plane, we obtain that $f \in F_\alpha$ by (5.2.2). This proves the equality. We also conclude that $F_\alpha = X_\alpha$ since

$$\sup_{x\in K} \left|\frac{e^{tq(x)}f(x) - f(x)}{t^\alpha}\right| = \sup_{x\in K} \left|\frac{e^{tq(x)} - 1}{tq(x)}\right| \cdot |f(x)| \cdot |q(x)|^\alpha t^{1-\alpha}$$

for each compact set $K \subseteq \Omega$. The extrapolated Favard spaces are then given by

$$F_{-\alpha} = \left\{f \in C_b(\Omega) : |q|^{1-\alpha}f \in C_b(\Omega)\right\} = X_{-\alpha}.$$

The spaces $\underline{X}_\alpha$ are more difficult to describe in general since the space of strong continuity $\underline{X}$ depends substantially on the choice of $q$. For example, if $\frac{1}{q} \in C_0(\Omega)$, then $\underline{X} = C_0(\Omega)$.

To see this notice that $C_0(\Omega) \subseteq \underline{X}$ trivially. On the other hand

$$|f| = \left|\frac{1}{q}\right| \cdot |fq|$$

which shows that $D(M_q) \subseteq C_0(\Omega)$ and hence that $\underline{X} \subseteq C_0(\Omega)$. For $\alpha \in [0, 1]$ this yields

$$\underline{X}_\alpha = \{f \in C_0(\Omega) : |q|^\alpha f \in C_0(\Omega)\},$$

and

$$\underline{X}_{-\alpha} = \{qf : f \in C_0(\Omega), \; |q|^{1-\alpha} f \in C_0(\Omega)\} = \{f \in C(\Omega) : |q|^{-\alpha} f \in C_0(\Omega)\}.$$

This example extends Sect. 3.2 in [188] by Nagel, Nickel, and Romanelli.

### 5.2.3 The Diffusion Semigroup

On $(C_b(\mathbb{R}^d), \|\cdot\|_\infty, \tau_{co})$ $(d \geq 1)$ we consider the Gauss–Weierstrass semigroup, see also Sect. 2.2.2 for the one-dimensional case, defined by $T(0) = I$ and

$$T(t)f(x) = \frac{1}{(4\pi t)^{\frac{d}{2}}} \int_{\mathbb{R}^d} e^{-\frac{|x-y|^2}{4t}} f(y) \, \mathrm{d}y, \quad t > 0, \quad x \in \mathbb{R}^d. \tag{5.2.3}$$

Then $(T(t))_{t\geq 0}$ becomes a bi-continuous semigroup, and its space of strong continuity is $UC_b(\mathbb{R}^d)$. From Sect. 3.1.1 as well as [169, Proposition 2.3.6] we know that the generator $A$ of this semigroup is given $Af = \Delta f$ on the maximal domain

$$D(A) = \{f \in C_b(\mathbb{R}^d) : \quad \Delta f \in C_b(\mathbb{R}^d)\},$$

where $\Delta$ is the distributional Laplacian. Now the extrapolation space can again be obtained by Theorem 4.15. If we take $\mathscr{E} = \mathscr{D}'(\mathbb{R}^d)$, $\mathcal{A} = \Delta$ and $i : C_b(\mathbb{R}^d) \to \mathscr{D}'(\mathbb{R}^d)$ the regular embedding, we then have

$$X_{-1} = \{F \in \mathscr{D}'(\mathbb{R}^d) : \quad F = f - \Delta f \text{ for some } f \in C_b(\mathbb{R}^d)\}.$$

The domain of the generator can be given explicitly, see, e.g., [169] or [175]. For $d = 1$ it is

$$D(A) = C_b^2(\mathbb{R}),$$

while for $d \geq 2$

$$D(A) = \left\{f \in C_b(\mathbb{R}^d) \cap W_{loc}^{2,p}(\mathbb{R}^d), \text{ for all } p \in [1, \infty) \text{ and } \Delta f \in C_b(\mathbb{R}^d)\right\}.$$

For $\alpha \in (0, 1) \setminus \{\frac{1}{2}\}$ the Favard spaces are

$$F_\alpha = C_b^{2\alpha}(\mathbb{R}^d),$$

while for $\alpha = \frac{1}{2}$ one obtains

$$F_{\frac{1}{2}} = \left\{ f \in C_b(\mathbb{R}^d) : \sup_{x \neq y} \frac{|f(x) + f(y) - 2f(\frac{x+y}{2})|}{|x - y|} < \infty \right\}.$$

From Corollary 4.25 it follows that for $\alpha \in (0, 1)$, $\alpha \neq \frac{1}{2}$

$$F_{-\alpha} = \left\{ F \in \mathscr{D}'(\mathbb{R}^d) : F = f - \Delta f \text{ for some } f \in C_b^{2(1-\alpha)}(\mathbb{R}^d) \right\},$$

and

$$F_{-\frac{1}{2}} = \left\{ F \in \mathscr{D}'(\mathbb{R}^d) : F = f - \Delta f \text{ for some } f \in F_{\frac{1}{2}} \right\}.$$

### 5.2.4  The Left Implemented Semigroup

Consider the bi-admissible space $(\mathscr{L}(E), \| \cdot \|_{\mathscr{L}(E)}, \tau_{\text{sot}})$, see also Sect. 2.1.3. Let $(S(t))_{t \geq 0}$ be a $C_0$-semigroup with negative growth bound on the Banach space $E$. The semigroup $(\mathcal{U}(t))_{t \geq 0}$ on $X := \mathscr{L}(E)$ defined by

$$\mathcal{U}(t)B = S(t)B, \quad B \in X, \quad t \geq 0$$

is called the semigroup left implemented by $(S(t))_{t \geq 0}$, cf. Sect. 2.2.4. Note that $(\mathcal{U}(t))_{t \geq 0}$ still has negative growth bound and is a bi-continuous semigroup. We determine the intermediate and extrapolation spaces for this semigroup. We can write

$$\|B\|_{F_\alpha(\mathcal{U})} = \sup_{t > 0} \frac{\|\mathcal{U}(t)B - B\|}{t^\alpha} = \sup_{t > 0} \frac{\|S(t)B - B\|}{t^\alpha}$$

$$= \sup_{t > 0} \sup_{\|x\| \leq 1} \frac{\|S(t)Bx - Bx\|}{t^\alpha} = \sup_{\|x\| \leq 1} \sup_{t > 0} \frac{\|S(t)Bx - Bx\|}{t^\alpha} = \sup_{\|x\| \leq 1} \|Bx\|_{F_\alpha(S)}.$$

From this we conclude the following.

**Proposition 5.22** *Let $(\mathcal{U}(t))_{t \geq 0}$ be the semigroup which is left implemented by $(S(t))_{t \geq 0}$. Then*

$$F_\alpha(\mathcal{U}) = \mathscr{L}(E, F_\alpha(S)) \quad \text{for } \alpha \in (0, 1]$$

*with the same norms.*

From the definition we obtain that

$$X_\alpha(\mathcal{U}) = \left\{ B \in \mathscr{L}(E) : \ \tau\lim_{t \to 0} \frac{\mathcal{U}_L(t)B - B}{t^\alpha} = 0, \ \|B\|_{F_\alpha(\mathcal{U})} < \infty \right\}$$

$$= \left\{ B \in \mathscr{L}(E) : \ \lim_{t \to 0} \frac{\|S(t)Bx - Bx\|}{t^\alpha} = 0 \ \text{ for all } x \in E \right\},$$

$$\underline{X}_\alpha(\mathcal{U}) = \left\{ B \in \mathscr{L}(E) : \ \lim_{t \to 0} \frac{S(t)B - B}{t^\alpha} = 0 \right\}.$$

**Proposition 5.23** *Let $(\mathcal{U}(t))_{t \geq 0}$ be the semigroup which is left implemented by $(S(t))_{t \geq 0}$. Then*

$$X_\alpha(\mathcal{U}) = \mathscr{L}(E, X_\alpha(S))$$

*with the same norms.*

We now turn to the extrapolation spaces. For the $C_0$-semigroup $(\underline{\mathcal{U}}(t))_{t \geq 0}$ on the space $\underline{X}$ these have been studied by Alber in [9]. He has shown that the generator $\mathcal{G}$ of $(\mathcal{U}(t))_{t \geq 0}$ is given by

$$\mathcal{G}V = A_{-1}V$$

on

$$D(\mathcal{G}) = \{V \in \mathscr{L}(E) : \quad A_{-1}V \in \mathscr{L}(E)\},$$

where $A_{-1}$ denotes the generator of the extrapolated $C_0$-semigroup $(S_{-1}(t))_{t \geq 0}$ on $E_{-1}$. The extrapolation spaces $X_{-1}$ and $\underline{X}_{-1}$ can now be obtained by Theorem 4.15. For that let

$$\mathscr{E} = \left\{ S : E \to E_{-\infty} : \ \text{linear and continuous} \right\},$$

where $E_{-\infty}$ is the universal extrapolation space of $(S(t))_{t \geq 0}$ (see the paragraph preceding Theorem 4.15), and let $i : \mathscr{L}(E) \to \mathscr{E}$ be the identity. Consider the operator-valued multiplication operator

$$\mathcal{A}V = A_{-\infty}V, \quad V \in \mathscr{E},$$

where $A_{-\infty}x = A_{-(n-1)}x$ for $x \in E_{-n}$. Notice that $\lambda - \mathcal{A} : X \to \mathscr{E}$ is injective for $\lambda > 0$ since $A_{-\infty}$ and $A_{-1}$ coincide on $E$. Hence by applying Theorem 4.15 we obtain

$$X_{-1} = \{A_{-1}V : \quad V \in \mathscr{L}(E)\}$$

and

$$\underline{X}_{-1} = \left\{ A_{-1}V : \quad V \in \underline{X} \right\}.$$

From this we conclude that

$$X_{-1} = \left\{ V \in \mathscr{L}(E, E_{-1}) : \ \exists (V_n)_{n \in \mathbb{N}} \subseteq \mathscr{L}(E) \text{ with } V_n \to V \text{ strongly} \right\} = \overline{\mathscr{L}(E)}^{\mathscr{L}_{\mathrm{sot}}(E, E_{-1})}.$$

Since for any $C \in \mathscr{L}(E, E_{-1})$ we have $nR(n, A_{-1})C \in \mathscr{L}(E)$ and $nR(n, A_{-1})C \to C$ strongly as $n \to \infty$, we obtain

$$X_{-1} = \mathscr{L}(E, E_{-1}).$$

For $\underline{X}_{-1}$ we have

$$\underline{X}_{-1} = \left\{ V \in \mathscr{L}(E, E_{-1}) : \exists (V_n)_{n \in \mathbb{N}} \subseteq \mathscr{L}(E) \text{ with } V_n \to V \text{ in } \mathscr{L}(E, E_{-1}) \right\} = \overline{\mathscr{L}(E)}^{\mathscr{L}(E, E_{-1})}.$$

This last statement is a result of Alber, see [9], which we could recover as a simple consequence of the abstract techniques described in this chapter. Finally, we obtain by Corollary 4.25 and Remark 5.21 that for $\alpha \in [0, 1)$

$$F_{-\alpha}(\mathcal{U}) = A_{-1}\mathscr{L}(E, F_{1-\alpha}(S)) = \mathscr{L}(E, F_{-\alpha}(S))$$

and

$$X_{-\alpha}(\mathcal{U}) = A_{-1}\mathscr{L}(E, X_{1-\alpha}(S)) = \mathscr{L}(E, X_{-\alpha}(S)).$$

## Notes on This Chapter

In this chapter, we discussed the construction as well as explicit examples of extrapolation spaces for bi-continuous semigroups. This extends the theory in [101, Chap. II, Sect. 5] as well as the examples mentioned in [188]. The approach to extrapolation spaces of bi-continuous semigroups has been partially developed in Chap. 4. The results presented in Chap. 5 as well as in this chapter were first presented by Budde and Farkas in [59]. The existence of extrapolation spaces for bi-continuous semigroups is the groundwork for the Desch–Schappacher perturbations that we present in Chap. 8. This current chapter also concludes the first part of the book and particularly the basic theory of bi-continuous semigroups. With the theory developed in this part we can now turn to applications of bi-continuous semigroups.

## Part II

# Approximation and Perturbations

# Approximation of Bi-continuous Semigroups

**6**

This chapter introduces key approximation techniques for bi-continuous semigroups, focusing on the foundational results that allow for effective handling of these semigroups as we will point out in some examples, see Sect. 6.3. The Trotter–Kato approximation theorems form the basis, offering precise conditions under which convergence of generators (in a certain sense) of the convergence of resolvents imply convergence of the associated bi-continuous semigroups, see Theorems 6.3 and 6.6. This framework is further enriched by the Chernoff product formula (Theorem 6.8), which enables the approximation of semigroups via compositions of simpler operators. We also examine the Post–Widder inversion formula, cf. Corollary 6.9, providing a method to reconstruct semigroups from Laplace transforms as well as the Lie–Trotter product formula which gives a first taste for perturbations, cf. Corollary 6.10. Together, these methods form a versatile toolkit for approximating and analyzing bi-continuous semigroups, with broad implications in the study of evolution equations.

## 6.1 The Trotter–Kato Approximation Theorems

The goal is to investigate the relation between the convergence of bi-continuous semigroups, their resolvents and generators. To that purpose, we introduce the notion of uniformly bi-continuous semigroups which will play the role of the stability condition essential for the subsequent approximation theory. As always, we assume that $(X, \| \cdot \|, \tau)$ is a bi-admissible space, see also Assumption 2.1.

**Definition 6.1** For each $k \in \mathbb{N}$ let $(T_k(t))_{t \geq 0}$ be a bi-continuous semigroup on a bi-admissible space $(X, \| \cdot \|, \tau)$. These semigroups are called *uniformly bi-continuous* (of type $\omega$) if the following conditions are satisfied:

© The Author(s), under exclusive license to Springer Nature Switzerland AG 2026
C. Budde, *Bi-Continuous Operator Semigroups*, Frontiers in Mathematics,
https://doi.org/10.1007/978-3-032-12948-2_6

(a) There exists $M \geq 1$ and $\omega \in \mathbb{R}$ such that $\|T_k(t)\| \leq Me^{\omega t}$ for all $t \geq 0$ and all $k \in \mathbb{N}$.

(b) $(T_k(t))_{t \geq 0}$ are locally bi-equicontinuous uniformly for $k \in \mathbb{N}$, i.e., for every $t_0 \geq 0$ and for every $\|\cdot\|$-bounded $\tau$-null sequence $(x_n)_{n \in \mathbb{N}}$ one has $\tau\text{lim}_{n \to \infty} T_k(t)x_n = 0$ uniformly for $t \in [0, t_0]$ and $k \in \mathbb{N}$.

Before we state and prove the first Trotter–Kato approximation theorem for bi-continuous semigroups, we need the following result.

**Lemma 6.2** *Let $(T_k(t))_{t \geq 0}$, $k \in \mathbb{N}$, be uniformly bi-continuous semigroups (of type $\omega$) on a bi-admissible space $(X, \|\cdot\|, \tau)$. The corresponding generator will be denoted by $(A_k, D(A_k))$, $k \in \mathbb{N}$. For every $\|\cdot\|$-bounded $\tau$-null sequence $(x_n)_{n \in \mathbb{N}}$ and every $\alpha > \omega$ one has*

(a) $\displaystyle \tau\lim_{n \to \infty} e^{-\alpha t} T_k(t)x_n = 0$ *uniformly for all $k \in \mathbb{N}$ and $t \geq 0$,*

(b) $\displaystyle \tau\lim_{n \to \infty} (\lambda - \alpha)^l R(\lambda, A_k)^l x_n = 0$ *uniformly for all $k, l \in \mathbb{N}$ and $\lambda > \alpha$.*

***Proof*** Notice that assertion (a) is a consequence of the fact that the rescaled semigroup $(e^{-\alpha t} T(t))_{t \geq 0}$ is globally bi-equicontinuous, cf. Theorem 3.15(c), and Definition 6.1. In order to prove (b), let $x \in X, \varepsilon > 0, p \in \mathcal{P}$ and $\alpha > \omega$. For a $\|\cdot\|$-bounded $\tau$-null sequence $(x_n)_{n \in \mathbb{N}}$ in $X$ we obtain as a result of a combination of assertion (a), formula (3.4.2) as well as Theorem 3.15(c) that there exists $N \in \mathbb{N}$ such that

$$p((\lambda - \alpha)^l R(\lambda, A_k)^l x_n) \leq \frac{(\lambda - \alpha)^l}{(l-1)!} \int_0^\infty t^{l-1} e^{-(\lambda - \alpha)t} p(e^{-\alpha t} T_k(t)x_n) \, dt,$$

for all $n \geq N$ and uniformly for $k, l \in \mathbb{N}$ and $\lambda > \alpha$, which proves (b). $\qquad \square$

We are now able to state and prove the first Trotter–Kato approximation theorem which tells us about the link between the convergence of resolvents and semigroups.

**Theorem 6.3** *Let $(T_k(t))_{t \geq 0}$, $k \in \mathbb{N}$, and $(T(t))_{t \geq 0}$ be uniformly bi-continuous semigroups of type $\omega$ on a bi-admissible space $(X, \|\cdot\|, \tau)$. By $(A_k, D(A_k))$, $k \in \mathbb{N}$, and $(A, D(A))$ we denote the corresponding semigroup generators. Moreover, let $D \subseteq \overline{D(A)}^{\|\cdot\|}$ be a $\|\cdot\|$-dense subset. If there exists $\lambda > \alpha > \omega$ such that*

$$R(\lambda, A_k)x \xrightarrow{\|\cdot\|} R(\lambda, A)x, \quad k \to \infty,$$

*for all $x \in D$, then*

$$T_k(t)x \xrightarrow{\tau} T(t)x, \quad k \to \infty,$$

*for all $x \in X$ uniformly for $t$ in compact intervals of $\mathbb{R}_+$.*

***Proof*** Let us fix $t_0 \geq 0$. Firstly, we show that

$$\tau\text{-}\lim_{k \to \infty} (T_k(t) - T(t))R(\lambda, A)y = 0, \tag{6.1.1}$$

uniformly for all $y \in D$ and $t \in [0, t_0]$. Let $\mathcal{P}$ be the family of seminorms generating the locally convex topology $\tau$ on $X$. For $p \in \mathcal{P}$ we then have

$$p((T_k(t) - T(t))R(\lambda, A)y)$$
$$\leq p(T_k(t)(R(\lambda, A) - R(\lambda, A_k))y) + p(R(\lambda, A_k)(T_k(t) - T(t))y) + p((R(\lambda, A_k) - R(\lambda, A))T(t)y)$$
$$\leq \underbrace{\|T_k(t)(R(\lambda, A) - R(\lambda, A_k))y\|}_{\alpha_k(t)} + \underbrace{p(R(\lambda, A_k)(T_k(t) - T(t))y)}_{\beta_k(t)} + \underbrace{\|(R(\lambda, A_k) - R(\lambda, A))T(t)y\|}_{\gamma_k(t)}.$$
$$\tag{6.1.2}$$

We will show that each summand converges to zero uniformly on $[0, t_0]$ for $k \to \infty$. Let us start with $\alpha_k(t)$. Since there exists $C > 0$ such that $\|T(t)\| \leq C$ for all $t \in [0, t_0]$ and by assumption $R(\lambda, A_k)x \xrightarrow{\|\cdot\|} R(\lambda, A)x$ for $k \to \infty$ and for each $x \in D$, we conclude that $\alpha_k(t) \to 0$ uniformly on $[0, t_0]$ for $k \to \infty$. Let us continue with $\gamma_k(t)$, since we will treat $\beta_k(t)$ separately afterwards. Observe, that by Proposition 4.30 the map $t \mapsto T(t)y$ is $\|\cdot\|$-continuous for each $y \in D$ implying that the set $\{T(t)y : t \in [0, t_0]\} \subseteq D(A)$ is $\|\cdot\|$-compact. By the assumption that $D$ is $\|\cdot\|$-dense in $\overline{D(A)}^{\|\cdot\|}$ we obtain that $\gamma_k(t) \to 0$ uniformly on $[0, t_0]$ for $k \to \infty$, cf. [208, Chap. III, Thm. 4.5]. The challenging part is indeed to show that $\beta_k(t) \to 0$ uniformly on $[0, t_0]$ for $k \to \infty$. For that purpose, we consider for each $t \in [0, t_0]$, $k \in \mathbb{N}$ and $y \in D$ the map

$$[0, t] \ni s \mapsto T_k(t - s)R(\lambda, A_k)T(s)R(\lambda, A)y \in D(A_k),$$

which actually is $\tau$-differentiable on $[0, t]$ with derivative

$$[0, t] \ni s \mapsto T_k(t - s)R(\lambda, A_k)T(s)AR(\lambda, A)y - T_k(t - s)A_k R(\lambda, A_k)T(s)R(\lambda, A)y \in X.$$

Therefore, for all $t \in [0, t_0]$ and $y \in D$ we obtain

$$p(R(\lambda, A_k)(T_k(t) - T(t))R(\lambda, A)y)$$
$$\leq \int_0^t p(T_k(t - s)\,(R(\lambda, A_k)T(s)A - A_k R(\lambda, A_k)T(s))\,R(\lambda, A)y)\,\mathrm{d}s$$
$$= \int_0^t p(T_k(t - s)\,((I - \lambda R(\lambda, A_k))\,R(\lambda, A) + R(\lambda, A_k)\,(\lambda R(\lambda, A) - I))\,T(s)y)\,\mathrm{d}s$$
$$\leq C \int_0^{t_0} \|(R(\lambda, A) - R(\lambda, A_k))T(s)y\|\,\mathrm{d}s.$$

By the compactness argument we used previously, as well as the assumption, we conclude that indeed $\beta_k(t) \to 0$ uniformly on $[0, t_0]$ for $k \to \infty$ and therefore

$$R(\lambda, A)(T_k(t) - T(t))y \xrightarrow{\tau} 0,$$

for all $y \in R(\lambda, A)D$ uniformly on $[0, t_0]$ as $k \to \infty$. Now, by (6.1.2) and the fact that $R(\lambda, A)D$ is bi-dense in $D(A)$ as well, we obtain

$$\tau\text{-}\lim_{k \to \infty} (T_k(t) - T(t))R(\lambda, A)y = 0,$$

uniformly for all $y \in D$ and $t \in [0, t_0]$. It follows that

$$T_k(t)x \xrightarrow{\tau} T(t)x,$$

for all $x \in R(\lambda, A)D$ and $t \in [0, t_0]$. Finally, we show that the convergence holds for each $x \in X$. To do so, let $x \in X$, $\varepsilon > 0$ and $p \in \mathcal{P}$. We know that $D(A)$ is bi-dense in $X$ and together with the assumption that $D$ is $\|\cdot\|$-dense in $D(A)$, there exists a $\|\cdot\|$-bounded sequence $(x_n)_{n \in \mathbb{N}}$ in $D$ such that $x_n \xrightarrow{\tau} x$. By the assumption that the semigroups are uniformly bi-continuous, cf. Definition 6.1, we find $N \in \mathbb{N}$ such that

$$p((T_k(t) - T(t))(x_n - x)) < \frac{\varepsilon}{3},$$

for all $n \geq N$ uniformly for $k \in \mathbb{N}$ and $t \in [0, t_0]$. Again by the fact that $R(\lambda, A)D$ is bi-dense in $D(A)$, there exists a sequence $(y_n)_{n \in \mathbb{N}}$ in $D$ such that $(R(\lambda, A)y_n)_{n \in \mathbb{N}}$ is a $\|\cdot\|$-bounded sequence in $D(A)$ $\tau$-converging to $x_N$. In particular, there exists $\widetilde{N} \geq N$ such that

$$p((T_k(t) - T(t))(x - x_N)) + p((T_k(t) - T(t))(x_N - R(\lambda, A)y_n)) < \frac{2\varepsilon}{3},$$

for all $n \geq \widetilde{N}$ uniformly for $k \in \mathbb{N}$ and $t \in [0, t_0]$. By (6.1.1), there exists $\widehat{K} \in \mathbb{N}$ such that

$$p((T_k(t) - T(t))R(\lambda, A)x_{\widetilde{N}}) < \frac{\varepsilon}{3},$$

for all $k \geq \widehat{K}$ and uniformly for $t \in [0, t_0]$ and hence one obtains

$$\begin{aligned}
p(T_k(t)x - T(t)x) &\leq p((T_k(t) - T(t))(x - x_N)) \\
&\quad + p((T_k(t) - T(t))(x_N - R(\lambda, A)y_n)) \\
&\quad + p((T_k(t) - T(t))R(\lambda, A)x_{\widetilde{N}}) < \varepsilon,
\end{aligned}$$

for all $k \geq \widehat{K}$ and uniformly for $t \in [0, t_0]$. $\qquad\qquad\qquad\qquad\qquad\qquad \square$

Before we tend to the second Trotter–Kato approximation theorem, we need the following result, which in fact is a bi-continuous version of [101, Chap. III, Prop. 4.4]. For a better understanding, recall the definition of a pseudoresolvent from Sect. 3.3.

**Lemma 6.4** *Let $(T_k(t))_{t \geq 0}$, $k \in \mathbb{N}$, be uniformly bi-continuous semigroups (of type $\omega$) on a bi-admissible space $(X, \|\cdot\|, \tau)$ with generators $(A_k, D(A_k))$. If*

$$\tau\lim_{k\to\infty} R(\lambda_0, A_k)x,$$

*exists for all $x \in X$ and some $\lambda_0 > \omega$, then*

$$R(\lambda)x := \tau\lim_{k\to\infty} R(\lambda, A_k)x,$$

*exists for all $x \in X$ and $\mathrm{Re}(\lambda) > \omega$ and defines a pseudoresolvent.*

**Proof** Firstly, we notice that $R(\lambda, A_k)x$ can be expressed as a power series around $\lambda_0$, i.e., one has

$$R(\lambda, A_k)x = \sum_{j\in\mathbb{N}} (\lambda - \lambda_0)^j R(\lambda_0, A_k)^{j+1}x,$$

where the series converges in $X$ uniformly for $k \in \mathbb{N}$ if $|\lambda - \lambda_0| \leq \varepsilon(\lambda_0 - \omega)$ for each $\varepsilon \in (0, 1)$. Let us prove that $\tau\lim_{k\to\infty} R(\lambda, A_k)x$ exists for all $\lambda$ such that $|\lambda - \lambda_0| \leq \varepsilon(\lambda_0 - \omega)$. Since $X$ is sequentially complete on $\|\cdot\|$-bounded sets, it suffices to show that the sequence $(R(\lambda, A_k)x)_{k\in\mathbb{N}}$ is a Cauchy sequence in $X$ with respect to the locally convex topology $\tau$. To do so, let $p \in \mathcal{P}$ and $\varepsilon > 0$. There exists $N \in \mathbb{N}$ such that

$$p\left( \sum_{j>N} (\lambda - \lambda_0)^j (R(\lambda_0, A_k)^{j+1} - R(\lambda_0, A_l)^{j+1})x \right) < \frac{\varepsilon}{2},$$

for all $k, l \geq N$. Further, by Lemma 6.2 the sequence $(R(\lambda_0, A_k)^j x)_{k\in\mathbb{N}}$ is a $\tau$-Cauchy sequence in $X$ for each $x \in X$ and all $j \in \mathbb{N}$. Therefore, there exists $N' \geq N$ such that

$$p(R(\lambda, A_k)x - R(\lambda, A_l)x)$$

$$= p\left( \sum_{j\in\mathbb{N}} (\lambda - \lambda_0)^j (R(\lambda_0, A_k)^{j+1} - R(\lambda_0, A_l)^{j+1})x \right)$$

$$\leq \sum_{j=0}^{N} (\lambda - \lambda_0)^j p(R(\lambda_0, A_k)^{j+1}x - R(\lambda_0, A_l)^{j+1}x) + \frac{\varepsilon}{2} < \varepsilon,$$

for all $k, l \geq N'$. The above also implies that the set of all $\lambda \in \mathbb{C}$ with $\mathrm{Re}(\lambda) > \omega$ such that $\tau\lim_{k\to\infty} R(\lambda, A_k)x$ exists for all $x \in X$ is open and relatively closed. Therefore, it coincides with the set $\{\lambda \in \mathbb{C} : \mathrm{Re}(\lambda) > \omega\}$. Clearly, the resolvent equation also holds for the operators $R(\lambda), \mathrm{Re}(\lambda) > \omega$, which concludes the proof. $\qquad\square$

The following useful lemma will be left as an exercise.

**Lemma 6.5** *Let $(C, D(C))$ be a bi-continuous operator on a bi-admissible space $(X, \|\cdot\|, \tau)$, i.e., for all sequence $(x_n)_{n\in\mathbb{N}}$ in $D(C)$ such that $(x_n)_{n\in\mathbb{N}}$ and $(Cx_n)_{n\in\mathbb{N}}$ are $\|\cdot\|$-*

*bounded and $x_n \xrightarrow{\tau} x$ one has $Cx_n \xrightarrow{\tau} Cx$. If $(C, D(C))$ is bi-closed, then its domain $D(C)$ is bi-closed.*

Let us now state and prove the second Trotter–Kato approximation theorem, which gives conditions such that the strong convergence of the resolvent to some limit implies that this limit is the resolvent of a certain generator. Moreover, we obtain that the convergence of a sequence of generators to an operator already implies that it is a generator.

**Theorem 6.6**  *Let $(T_k(t))_{t \geq 0}$, $k \in \mathbb{N}$, be uniformly bi-continuous semigroups (of type $\omega$) on a bi-admissible space $(X, \|\cdot\|, \tau)$ with generators $(A_k, D(A_k))$. Consider for $\lambda > \alpha > \omega$ the following assertions:*

(a)  *There exists a bi-densely defined operator $(A, D(A))$ such that $A_k x \xrightarrow{\|\cdot\|} Ax$ for all $x$ in a bi-core $D$ for $A$ and $\mathrm{Ran}(\lambda - A)$ is bi-dense in $X$.*

(b)  *There exists an operator $R \in \mathscr{L}(X)$ such that $R(\lambda, A_k)x \xrightarrow{\|\cdot\|} Rx$ for all $x$ in a bi-dense subset of $\mathrm{Ran}(R)$.*

(c)  *There exists a bi-continuous semigroup $(T(t))_{t \geq 0}$ with generator $(B, D(B))$ such that $T_k(t)x \xrightarrow{\tau} T(t)x$ for all $x \in X$ and uniformly for $t$ in compact intervals.*

*Then the implications (a) $\Rightarrow$ (b) $\Rightarrow$ (c) hold. In particular, if (a) holds, then the bi-closure of $A$ is equal to $B$.*

***Proof***  Let us start with the implication (a)$\Rightarrow$(b). To do so, take $y := (\lambda - A)x$ for $x \in D$ and observe that

$$R(\lambda, A_k)y = R(\lambda, A_k)\left((\lambda - A_k)x - (\lambda - A_k)x + (\lambda - A)x\right)$$
$$= x + R(\lambda, A_k)(A_k x - Ax),$$

which converges to $x =: Ry$ as $k \to \infty$ by using the assumption of assertion (a). Moreover, $\mathrm{Ran}(R)$ contains $D$. Since $D$ is a bi-core for the operator $(A, D(A))$ and $D(A)$ is bi-dense in $X$ by assumption, we have that $D$ is bi-dense in $X$, showing that $\mathrm{Ran}(A)$ is bi-dense in $X$. It remains to show that $R$ can be $\|\cdot\|$-continuously be extended to $X$. For that purpose, we show that $\|\cdot\|$-bounded sequence $(R(\lambda, A_k)x)_{k \in \mathbb{N}}$ is a $\tau$-Cauchy sequence. To do so, let $x \in X$, $p \in \mathcal{P}$, and $\varepsilon > 0$. By assumption, $\mathrm{Ran}(\lambda - A)$ is bi-dense in $X$, i.e., there exists a sequence $(z_n)_{n \in \mathbb{N}}$ in $D(A)$ and $N \in \mathbb{N}$ such that for all $n \geq N$ one has that

$$p(x - (\lambda - A)z_n) < \frac{\varepsilon}{2}.$$

Making use of the assumption that $D$ is a bi-core for $(A, D(A))$, there exists a $\|\cdot\|$-bounded sequence $y_n := (\lambda - A)x_n$, $x_n \in D$, $n \in \mathbb{N}$, and $\widetilde{N} \geq N$ such that

$$p(x - y_n) \leq p(x - (\lambda - A)z_N) + p((\lambda - A)z_N - y_n) < \varepsilon,$$

for all $n \geq \widetilde{N}$. By Lemma 6.2, there exists $\widehat{N} \in \mathbb{N}$ such that

$$p(R(\lambda, A_k)(x - y_n)) < \frac{\varepsilon}{3}, \tag{6.1.3}$$

for all $n \geq \widehat{N}$ and uniformly in $k \in \mathbb{N}$. Taking all together, there exists $K \in \mathbb{N}$ such that

$$p(R(\lambda, A_k)x - R(\lambda, A_l)x)$$
$$\leq p(R(\lambda, A_k)(x - y_N)) + \|R(\lambda, A_k)y_N - R(\lambda, A_l)y_N\| + p(R(\lambda, A_l)(y_N - x)) < \varepsilon \tag{6.1.4}$$

for all $k, l \geq K$. Since by our general assumptions $\tau$ is sequentially complete on $\|\cdot\|$-bounded sets, we may conclude that

$$Rx := \tau\lim_{k \to \infty} R(\lambda, A_k)x,$$

exists for all $x \in X$. Using the norming property of $(X, \tau)'$ we observe that for all $x \in X$ holds that

$$\|Rx\| = \left\|\tau\lim_{k \to \infty} R(\lambda, A_k)x\right\| = \sup_{\substack{\varphi \in (X, \tau)' \\ \|\varphi\| \leq 1}} \left|\lim_{k \to \infty} \varphi(R(\lambda, A_k)x)\right| \leq \frac{\|x\|}{\lambda - \omega},$$

which shows that $R \in \mathscr{L}(X)$.

Let us continue with the implication $(b) \Rightarrow (c)$. For that we observe that by the same arguments as used before $\tau\lim_{k \to \infty} R(\lambda, A_k)x$ exists for all $x \in X$ and some $\lambda > \alpha > \omega$, i.e., make use of (6.1.3) and (6.1.4). Without loss of generality, we assume that $\alpha = 0$. By Lemma 6.4 we conclude that

$$R(\lambda)x := \tau\lim_{k \to \infty} R(\lambda, A_k)x$$

exists for all $x \in X$ and $\lambda > 0$ and defines a pseudoresolvent which in fact has bi-dense range $\mathrm{Ran}(R(\lambda)) = \mathrm{Ran}(R)$ in $X$. Moreover, we have that $\|\lambda R(\lambda)\| \leq M$ for all $\lambda > 0$ and some $M \geq 1$. Moreover, since

$$R(\lambda)^l x := \tau\lim_{k \to \infty} R(\lambda, A_k)^l x,$$

for all $x \in X$ and $l \in \mathbb{N}$, we also have that $\left\|\lambda^l R(\lambda)^l\right\| \leq M$ for all $x \in X$ and $l \in \mathbb{N}$. Let us show that the family of operators $(R(\lambda))_{\lambda > 0}$ is indeed a resolvent. To do so, let $(x_n)_{n \in \mathbb{N}}$ be a $\|\cdot\|$-bounded sequence in $X$ which is $\tau$-convergent to $x \in X$. Let $p \in \mathcal{P}$ and $\varepsilon > 0$ and determine by using Lemma 6.2 a natural number $N \in \mathbb{N}$ such that

$$p(\lambda^l R(\lambda, A_k)^l (x - x_n)) < \varepsilon,$$

for all $n \geq N$ uniformly in $k, l \in \mathbb{N}$ and $\lambda > 0$ and hence

$$p(\lambda^l R(\lambda)^l (x - x_n)) < \varepsilon,$$

for all $n \geq N$ uniformly for $\lambda > 0$ and $l \in \mathbb{N}$. Therefore, there exists a bi-densely defined operator $(B, D(B))$ such that $R(\lambda) = R(\lambda, B)$ for all $\lambda > 0$. It satisfies $\left\| \lambda^l R(\lambda, B)^l \right\| < \varepsilon$ for all $\lambda > 0$ and uniformly in $l \in \mathbb{N}$ and the family of operators given by $\left\{ \lambda^l R(\lambda, B)^l : \lambda > 0, l \in \mathbb{N} \right\}$ is bi-equicontinuous. Hence, by the Hille–Yosida theorem, cf. Theorem 3.17 or Theorem 5.1, we conclude that $(B, D(B))$ generates a bi-continuous semigroup $(T(t))_{t \geq 0}$. By an application of the first Trotter–Kato approximation theorem, cf. Theorem 6.3, we finally conclude that $T_k(t)x \xrightarrow{\tau} T(t)x$ for all $x \in X$ and $k \to \infty$ uniformly for $t$ in compact intervals.

In the final step, we show that condition (a) implies that the bi-closure of $A$ is equal to $B$. Since $R(\lambda, B) = R$ we have that $R(\lambda, B)(\lambda - A)x = x$ for all $x \in D$ showing that $D \subseteq D(B)$. Since $D$ is a bi-core for $A$, we can find for a given $x \in D(A)$ a sequence $(x_n)_{n \in \mathbb{N}}$ in $D \subseteq D(B)$ such that both sequences $(x_n)_{n \in \mathbb{N}}$ and $(Ax_n)_{n \in \mathbb{N}}$ are $\|\cdot\|$-bounded and $x_n \xrightarrow{\tau} x$ and $Ax_n \xrightarrow{\tau} Ax$. Since $Bx_n = Ax_n$, we obtain that $Bx_n \xrightarrow{\tau} Ax$. By Theorem 3.3(c) the operator $B$ is bi-closed showing that $x \in D(B)$ and $Bx = Ax$. Therefore, we have $D(A) \subseteq D(B)$ and $Bx = Ax$ for all $x \in D(A)$. Moreover, $R(\lambda, A)$ exists and its bi-closure $R(\lambda, \overline{A})$ is contained in $R(\lambda, B)$. Since $R(\lambda, B)$ is bi-continuous, one obtains that $R(\lambda, \overline{A})$ is bi-continuous. Moreover, the domain $D(R(\lambda, \overline{A}))$ contains $\mathrm{Ran}(\lambda - A)$ which is bi-dense in $X$ by assumption. Applying Lemma 6.5 we obtain $D(R(\lambda, \overline{A})) = X$. Consequently, $R(\lambda, \overline{A}) = R(\lambda, B)$ showing that the bi-closure of $A$ is equal to $B$. $\qquad\square$

**Remark 6.7** In the proof of Theorem 6.6 one observes that operator $R$ in assertion (b) gives rise to a pseudoresolvent that is used to define operator $(B, D(B))$ in assertion (c).

## 6.2    Approximation Formulas

In this section, we are going to apply the approximation results from Sect. 6.1 to obtain a generalization of the Chernoff product formula for bi-continuous semigroups. From this we obtain the Post–Widder formula as a direct consequence for bi-continuous semigroups.

**Theorem 6.8** *Let $(X, \|\cdot\|, \tau)$ be a bi-admissible space and $V : \mathbb{R}_+ \to \mathscr{L}(X)$ be a function satisfying the following properties:*

*(i)*   $V(0) = I$.

*(ii)*   *There exists $M \geq 1$ and $\omega \in \mathbb{R}$ such that $\| V(t)^m \| \leq M e^{m\omega t}$ for all $t \geq 0$ and $m \in \mathbb{N}$.*

*(iii)*   *The family of operators $\{ V(t)^m : t \geq 0 \}$ is locally bi-equicontinuous for each $m \in \mathbb{N}$.*

*(iv)*   *The limit $Ax := \lim_{s \to 0} \frac{V(s)x - x}{s}$ exists with respect to the norm for all $x \in D \subseteq X$, where $D$ and $(\lambda - A)D$ are bi-dense subsets of $X$ for some $\lambda > \alpha > \omega$.*

*Then the bi-closure of A generates a bi-continuous semigroup $(T(t))_{t \geq 0}$ which is given by the* Chernoff product formula, *i.e.:*

$$T(t)x = \tau\text{-}\lim_{k \to \infty} \left( V\left(\tfrac{t}{k}\right)^k \right) x,$$

*for all $x \in X$ and uniformly for $t \geq 0$ on compact intervals.*

**Proof** Without loss of generality we assume that $\omega = 0$. For $s > 0$ we define for each $k \in \mathbb{N}$ a bounded operator by

$$A_k := \frac{V\left(\tfrac{s}{k}\right) - I}{s},$$

and observe that $A_k x \xrightarrow{\|\cdot\|} Ax$ for all $x \in D$ for $k \to \infty$. Due to the boundedness of these operators, they yield bounded uniformly continuous semigroups $(e^{tA_k})_{t \geq 0}$, i.e., by assumption there exists $M \geq 1$ such that

$$\left\| e^{tA_k} \right\| \leq e^{-\frac{tk}{s}} \sum_{m \in \mathbb{N}_0} \frac{\left(\tfrac{tk}{s}\right)^m}{m!} \left\| V\left(\tfrac{s}{k}\right)^m \right\| \leq M,$$

for all $t \geq 0$. By the third assumption, we also get that for a $\|\cdot\|$-bounded sequence $(x_n)_{n \in \mathbb{N}}$ which $\tau$-converges to some $x \in X$ that for each $\varepsilon > 0$ and each $p \in \mathcal{P}$ there exists $N \in \mathbb{N}$ such that $p\left(\left(V(\tfrac{s}{k}\right)^m (x_n - x)\right) < \varepsilon$ for all $n \geq N$ uniformly for $k, m \in \mathbb{N}$. Hence

$$p\left(e^{tA_k}(x_n - x)\right) \leq e^{-\frac{tk}{s}} \sum_{m \in \mathbb{N}_0} \frac{\left(\tfrac{tk}{s}\right)^m}{m!} p\left(\left(V(\tfrac{s}{k}\right)^m (x_n - x)\right) < \varepsilon,$$

for all $n \geq N$ and uniformly for $k \in \mathbb{N}$ and $t \geq 0$. Therefore, the semigroups $(e^{tA_k})_{t \geq 0}$ are bi-equicontinuous uniformly for $k \in \mathbb{N}$. By Definition 6.1 this means that we deal with uniformly bi-continuous semigroups. Hence, we can apply the second Trotter–Kato approximation theorem since Theorem 6.6(a) is satisfied. This also yields that the bi-closure of $A$ generates a bi-continuous semigroup $(T(t))_{t \geq 0}$ such that

$$e^{tA_k}x \to T(t)x, \tag{6.2.1}$$

for all $x \in X$ for $k \to \infty$ uniformly for $t$ in compact intervals. Furthermore, we get that

$$\left\| e^{tA_k}y - V(\tfrac{t}{k})^k y \right\| \leq \frac{tM}{\sqrt{n}} \left\| A_k y \right\|, \tag{6.2.2}$$

for all $y \in D$. As $k \to \infty$ we obtain convergence to zero uniformly for $t$ in compact intervals. Finally, let $x \in X$, $t \in [0, t_0]$, $p \in \mathcal{P}$, and $\varepsilon > 0$. By (iii) and (iv) there exists a sequence $(y_n)_{n \in \mathbb{N}}$ in $D$ which is $\|\cdot\|$-bounded and $y_n \xrightarrow{\tau} x$. Moreover, there exists $N \in \mathbb{N}$ such that

$$p\left(T(t)(x - y_n)\right) + p\left(V(\tfrac{t}{k})^k(y_n - x)\right) < \frac{\varepsilon}{2}, \tag{6.2.3}$$

for all $n \geq N$, uniformly for $k \in \mathbb{N}$ and $t \in [0, t_0]$. Combining (6.2.1)–(6.2.3), we conclude that there exists $k_0 \in \mathbb{N}$ such that

$$p(T(t)x - V(\tfrac{t}{k})^k x) \leq p(T(t)(x - y_N)) + p(T(t)y_N - V(\tfrac{t}{k})^k y_N) + p(V(\tfrac{t}{k})^k(y_N - x))$$

$$\leq \frac{\varepsilon}{2} + \left\| e^{tA_k} y_N - V(\tfrac{t}{k})^k y_N \right\| < \varepsilon,$$

for all $k \geq k_0$ and uniformly for $t \in [0, t_0]$. Therefore, $(T(t))_{t \geq 0}$ is given by the Chernoff product formula and the proof is complete. $\qquad\square$

**Corollary 6.9** *For every bi-continuous semigroup $(T(t))_{t \geq 0}$ on a bi-admissible space $(X, \| \cdot \|, \tau)$ with generator $(A, D(A))$ the* Post–Widder Inversion Formula *holds, i.e.:*

$$T(t)x = \tau\lim_{k \to \infty} \left[ \frac{k}{t} R\left( \frac{k}{t}, A \right) \right]^k x,$$

*for all $x \in X$ and uniformly for $t$ in compact intervals.*

The next corollary states that under stability and consistency conditions on two bi-continuous semigroups generated by $(A, D(A))$ and $(B, D(B))$, respectively, the closure of the sum of $(A, D(A))$ and $(B, D(B))$ is a generator and the perturbed semigroup can be represented by the so-called Lie–Trotter product formula.

**Corollary 6.10** *Let $(T(t))_{t \geq 0}$ and $(S(t))_{t \geq 0}$ be two bi-continuous semigroups on a bi-admissible space $(X, \| \cdot \|, \tau)$ with generators $(A, D(A))$ and $(B, D(B))$, respectively. Assume that the following conditions hold:*

*(i)*  $\left\| T(\tfrac{t}{m})^m S(\tfrac{t}{m})^m \right\| \leq M e^{\omega t}$ *for all $t \geq 0$ and some constants $M \geq 1$ and $\omega \in \mathbb{R}$.*
*(ii) The operator family $\left\{ \left[ T(\tfrac{t}{m})S(\tfrac{t}{m}) \right]^m : t \geq 0 \right\}$ is locally bi-equicontinuous uniformly for $m \in \mathbb{N}$.*

*Consider the sum $A + B$ on a subspace $D \subseteq D(A_0) \cap D(B_0)$ and assume that $D$ and $(\lambda_0 - A - B)D$ are bi-dense in $X$ for some $\lambda_0 > \alpha > \omega$. Then the bi-closure of $A + B$ exists and generates a bi-continuous semigroup $(U(t))_{t \geq 0}$ given by the Lie–Trotter product formula, i.e.:*

$$U(t)x = \tau\lim_{k \to \infty} \left[ T\left( \frac{t}{m} \right) S\left( \frac{t}{m} \right) \right]^m x, \tag{6.2.4}$$

*where the limit exists for all $x \in X$ and uniformly for $t$ in compact intervals of $\mathbb{R}_+$.*

**Proof** Define $V(t) := T(t)S(t)$ for $t \geq 0$. Then

$$\lim_{t \to 0} \frac{V(t)x - x}{t} = Ax + Bx,$$

for all $x \in D$ and where the limit is taken with respect to the norm on $X$. The assertion is now a consequence of Theorem 6.8. $\qquad\square$

## 6.3  Examples

Before we start discussing specific examples, it is worth to recall the notion of locally equicontinuous operator semigroups. In order to deal with phenomena in which the underlying space is not a Banach space, one can use operator semigroups on locally convex spaces. In 1957, L. Schwartz generalized the classical Hille–Yosida theorem to those semigroups, cf. [209]. Thereafter, many authors studied operator semigroups on locally convex spaces and tried to develop a systematic theory parallel to the one in Banach spaces, see, for example, [146, 147, 196, 241]. The theory divides basically into two classes of semigroups: equicontinuous and locally equicontinuous semigroups, respectively, and it suffices to suppose that the underlying vector space is sequentially complete.

**Definition 6.11** Let $(X, \tau)$ be a sequentially complete locally convex space and assume that $\tau$ is induced by a family of seminorms $\mathcal{P}_\tau$. A family $(T(t))_{t \geq 0}$ of continuous linear operators on $X$ is called an *equicontinuous semigroup* if

(i)  $T(t + s) = T(t)T(s)$ and $T(0) = I$ for all $t, s \geq 0$.
(ii)  $(T(t))_{t \geq 0}$ is strongly $\tau$-continuous.
(iii)  For each $p \in \mathcal{P}_\tau$ there exists $q \in \mathcal{P}_\tau$ such that $p(T(t)x) \leq q(x)$ for all $t \geq 0$ and $x \in X$.

The family of operators $(T(t))_{t \geq 0}$ is called *quasi-equicontinuous* if there exists $\alpha > 0$ such that $(e^{-\alpha t} T(t))_{t \geq 0}$ is equicontinuous.

**Definition 6.12** Let $(X, \tau)$ be a sequentially complete locally convex space and assume that $\tau$ is induced by a family of seminorms $\mathcal{P}_\tau$. A family $(T(t))_{t \geq 0}$ of continuous linear operators on $X$ is called a *locally equicontinuous semigroup* if

(i)  $T(t + s) = T(t)T(s)$ and $T(0) = I$ for all $t, s \geq 0$.
(ii)  $(T(t))_{t \geq 0}$ is strongly $\tau$-continuous.
(iii)  For each $t_0 > 0$ the subset $\{T(t) : t \in [0, t_0]\}$ is equicontinuous.

As in the case of $C_0$-semigroups and bi-continuous semigroups, one has the notion of an (infinitesimal) generator.

**Definition 6.13** Let $(X, \tau)$ be a sequentially complete locally convex space and $(T(t))_{t \geq 0}$ either a equicontinuous or locally equicontinuous operator semigroup on $X$. The generator $(A, D(A))$ is defined by

$$Ax := \tau\lim_{t \to 0} \frac{T(t)x - x}{t}, \quad D(A) := \left\{ x \in X : \tau\lim_{t \to 0} \frac{T(t)x - x}{t} \text{ exists} \right\}.$$

Although the theory of operator semigroups on locally convex spaces is well established, it seems that there are only very few serious applications of this theory. In addition, no systematic qualitative theory has been developed, certainly due to the lack of an appropriate spectral theory. However, in many concrete situations the underlying space is indeed a Banach space $X$, but has an additional locally convex topology $\tau$ satisfying Assumption 2.1. Moreover, the equicontinuous (resp. locally equicontinuous) semigroups to be considered on $(X, \tau)$ are exponentially norm bounded. This leads us to bi-continuous semigroups as discussed in Chaps. 2 and 3. In fact, we have the next result which follows immediately from Definitions 2.4 and 6.12.

**Proposition 6.14** *Let $(X, \| \cdot \|, \tau)$ be a bi-admissible space. If the family of bounded linear operators $(T(t))_{t \geq 0}$ is a locally equicontinuous operator semigroup on $(X, \tau)$ such that $\|T(t)\| \leq M e^{t\omega}$ for all $t \geq 0$ and some $M \geq 0$ and $\omega \in \mathbb{R}$, then $(T(t))_{t \geq 0}$ is a bi-continuous semigroup on $X$ with respect to $\tau$.*

### 6.3.1   Semigroups Induced by Flows

We already discussed semigroups induced by jointly continuous flows in Sect. 3.5.1. Especially, we saw how generators of semigroups induced by flows and the corresponding Lie generator are related to each other. Now, we want to extend what we saw in Sect. 3.5.1 by approximation theory as developed in this chapter. Using the Lie–Trotter product formula stated in Corollary 6.10, we obtain the following product formula for two semigroups induced by flows.

**Proposition 6.15** *Let $(T(t))_{t \geq 0}$ and $(S(t))_{t \geq 0}$ be two bi-continuous semigroups on $C_b(\Omega)$ induced by jointly continuous flows $\varphi$ and $\psi$ on $\Omega$ and generators $(A, D(A))$ and $(B, D(B))$, respectively. Suppose that for every $t_0 > 0$ and every compact set $K \subseteq \Omega$ there exists a compact set $\widetilde{K} \subseteq \Omega$ such that $\left[ \psi_{\frac{t}{m}} \varphi_{\frac{t}{m}} \right]^m K \subseteq \widetilde{K}$ for all $t \in [0, t_0]$ and $m \in \mathbb{N}$. If there exists a subset $D \subseteq D(A_0) \cap D(B_0)$ such that $D$ and $(\lambda_0 - A - B)D$ are bi-dense in $C_b(\Omega)$, then the bi-closure of $A + B$ generates a bi-continuous semigroup $(U(t))_{t \geq 0}$ on $C_b(\Omega)$ which is induced by a jointly continuous flow $\xi$ on $\Omega$ such that*

$$U(t)f = f \circ \xi_t = \tau\lim_{m \to \infty} f \circ \left[ \psi_{\frac{t}{m}} \varphi_{\frac{t}{m}} \right]^m,$$

*for all $f \in C_b(\Omega)$ and uniformly for $t$ in compact intervals of $\mathbb{R}_+$.*

**Proof** Since both $(T(t))_{t\geq 0}$ and $(S(t))_{t\geq 0}$ are contraction semigroups, we directly obtain that $\left\| \left[ T(\frac{t}{m}) S(\frac{t}{m}) \right]^m \right\| \leq 1$ for all $t \geq 0$ and $m \in \mathbb{N}$. In order to apply Corollary 6.10, we need to show that $\left\{ \left[ T(\frac{t}{m}) S(\frac{t}{m}) \right]^m : t \geq 0 \right\}$ is locally bi-equicontinuous uniformly for $m \in \mathbb{N}$. Let $t_0 > 0$ and $(f_n)_{n\in\mathbb{N}}$ be a sequence in $C_b(\Omega)$ which is $\|\cdot\|$-bounded and $\tau_{co}$-convergent to some $f \in C_b(\Omega)$. In particular, for each compact subset $K \subseteq \Omega$ there exists another compact subset $\widetilde{K} \subseteq \Omega$ such that

$$\sup_{x\in K} \left| \left[ T(\tfrac{t}{m}) S(\tfrac{t}{m}) \right]^m (f_n(x) - f(x)) \right| = \sup_{x\in K} \left| f_n\left( \left[ \psi_{\frac{t}{m}} \varphi_{\frac{t}{m}} \right]^m (x) \right) - f\left( \left[ \psi_{\frac{t}{m}} \varphi_{\frac{t}{m}} \right]^m (x) \right) \right|$$
$$\leq \sup_{x\in\widetilde{K}} |f_n(x) - f(x)|$$

which converges to zero uniformly for $t \in [0, t_0]$ and $m \in \mathbb{N}$ as $n \to \infty$. Together with the fact that $\left\| \left[ T(\frac{t}{m}) S(\frac{t}{m}) \right]^m \right\| \leq 1$ for all $t \geq 0$ and $m \in \mathbb{N}$ we can conclude the local bi-equicontinuity uniformly for $m \in \mathbb{N}$. Thus, we are able to apply Corollary 6.10 and obtain a bi-continuous semigroup $(U(t))_{t\geq 0}$ on $C_b(\Omega)$ generated by the bi-closure of $A + B$ such that the Lie–Trotter product formula holds. Furthermore, using the same arguments as in the proof of Theorem 3.22, we obtain the existence of a jointly continuous flow $\xi$ on $\Omega$ such that $U(t)f = f \circ \xi_t$ for all $f \in C_b(\Omega)$. $\qquad\square$

**Example 6.16** We will illustrate how Proposition 6.15 works. On the space $C_b(\mathbb{R})$ we consider the operators $(A, D(A))$ and $(B, D(B))$ defined by

$$(Af)(x) := x^{\frac{2}{3}} f'(x), \quad D(A) := \{f \in C_b(\mathbb{R}) : Af \in C_b(\mathbb{R})\}$$

and

$$(Bf)(x) := xf'(x), \quad D(B) := \{f \in C_b(\mathbb{R}) : Bf \in C_b(\mathbb{R})\}.$$

Both operators generate operator semigroups induced by jointly continuous flows on $C_b(\mathbb{R})$ which are bi-continuous with respect to the $\tau_{co}$-topology and which are given by

$$(T(t)f)(x) = (f \circ \varphi_t)(x) \text{ where } \varphi_t(x) = (x^{\frac{1}{3}} + \frac{t}{3})^3,$$

and

$$(S(t)f)(x) = (f \circ \psi_t)(x) \text{ where } \psi_t(x) = e^t x,$$

for all $f \in C_b(\mathbb{R})$, $x \in \mathbb{R}$ and $t \geq 0$, respectively. By induction we obtain that

$$\left[ \psi_{\frac{t}{m}} \varphi_{\frac{t}{m}} \right]^m (x) = \left[ e^{\frac{t}{3}} x^{\frac{1}{3}} + \frac{t}{3m} \sum_{k=1}^{m} e^{\frac{kt}{3m}} \right]^3, \quad x \in \mathbb{R}, \ t \geq 0, \ m \in \mathbb{N}. \qquad (6.3.1)$$

We observe that $\frac{t}{3m} \sum_{k=1}^{m} e^{\frac{kt}{3m}} \leq \frac{t}{3} e^{\frac{t}{3}}$ for all $m \in \mathbb{N}$ and $t \geq 0$, showing that the products $\left[\psi_{\frac{t}{m}} \varphi_{\frac{t}{m}}\right]^m$ map a fixed compact set $K \subseteq \mathbb{R}$ into some fixed compact set $\widetilde{K} \subseteq \mathbb{R}$. Furthermore, the space of Schwartz functions $\mathcal{S}(\mathbb{R})$ is contained in $C_0^1(\mathbb{R})$ and therefore $\mathcal{S}(\mathbb{R}) \subseteq D(A_0) \cap D(B_0)$. By solving an ordinary differential equation, there exists $\lambda_0 > 0$ such that $(\lambda_0 - A - B)\mathcal{S}(\mathbb{R})$ is bi-dense in $C_b(\mathbb{R})$. Thus, the assumptions of Proposition 6.15 are fulfilled and we obtain that the bi-closure of $A + B$ generates a semigroup $(U(t))_{t \geq 0}$ on $C_b(\mathbb{R})$ which is induced by a jointly continuous flow $\xi$ on $\mathbb{R}$ such that

$$U(t)f = f \circ \xi_t = \tau\!\lim_{m \to \infty} f \circ \left[\psi_{\frac{t}{m}} \varphi_{\frac{t}{m}}\right]^m,$$

for all $f \in C_b(\Omega)$ and uniformly for $t$ in compact intervals of $\mathbb{R}_+$. In our case, the flow $\xi$ can explicitly be calculated. In fact, by using (6.3.1) we observe that

$$\frac{t}{3m} \sum_{k=1}^{m} e^{\frac{kt}{3m}} = \frac{t \left(e^{\frac{t(m+1)}{3m}} - e^{\frac{t}{3m}}\right)}{3m \left(e^{\frac{t}{3m}} - 1\right)}$$

for all $t \geq 0$ and $m \in \mathbb{N}$. By taking the limit for $m \to \infty$ we see that the right-hand side tends to $e^{\frac{t}{3}} - 1$ for $t \geq 0$ showing that the jointly continuous flow $\xi$ is given by

$$\xi_t(x) = e^t \left(x^{\frac{1}{3}} + 1 - e^{-\frac{t}{3}}\right)^3, \quad x \in \mathbb{R}, \ t \geq 0.$$

### 6.3.2  The Ornstein–Uhlenbeck Semigroup

In Sect. 2.2.5, we considered the Ornstein–Uhlenbeck semigroup on $C_b(\mathcal{H})$ for an infinite-dimensional, separable Hilbert space $\mathcal{H}$. We will show that the Lie–Trotter product formula can be applied to the Ornstein–Uhlenbeck semigroup on $C_b(\mathbb{R}^n)$. Let $A := (a_{ij})$ be a symmetric, positive definite matrix and $B = (b_{ij}) \in \mathscr{L}(\mathbb{R}^n)$. In this case, the generator of the Ornstein–Uhlenbeck semigroup can be written as

$$
\begin{aligned}
(\mathcal{O}f)(x) &= \sum_{i,j=1}^{n} a_{ij} \frac{\partial^2}{\partial x_i \partial x_j} f(x) + \sum_{i,j=1}^{n} b_{ij} x_j \frac{\partial}{\partial x_i} f(x) \\
&= \langle \nabla, A\nabla f(x)\rangle + \langle Bx, \nabla f(x)\rangle \\
&= (\mathcal{A}f)(x) + (\mathcal{B}f)(x),
\end{aligned}
\tag{6.3.2}
$$

for all $f \in \mathcal{S}(\mathbb{R}^n)$ and $x \in \mathbb{R}^n$, cf. [77]. Again by [77] the Ornstein–Uhlenbeck semigroup $(P(t))_{t \geq 0}$ generated by the operator defined by (6.3.2) has the following representation:

$$(P(t)f)(x) = \begin{cases} \dfrac{1}{(2\pi)^{\frac{n}{2}}\sqrt{\det(Q_t)}} \int_{\mathbb{R}^n} e^{-\frac{\langle Q_t y, y\rangle}{2}} f(e^{tB}x - y)\, \mathrm{d}y, & t > 0 \\ f(x), & t = 0, \end{cases} \tag{6.3.3}$$

for all $f \in C_b(\mathbb{R}^n)$ and $x \in \mathbb{R}^n$ where $Q_t := \int_0^t e^{sB} A e^{sB'}\, \mathrm{d}s$. From (6.3.2) we see that the operator $\mathcal{O}$ is the sum of two operators $\mathcal{A}$ and $\mathcal{B}$. Since $(P(t))_{t\geq 0}$ is not strongly continuous on $C_b(\mathbb{R}^n)$ the classical Lie–Trotter product formula cannot be applied. However, we will show that the Lie–Trotter product formula for bi-continuous semigroups can be used.

First of all, we construct a locally convex Hausdorff topology $\tau$ on $C_b(\mathbb{R}^n)$ which is weaker than $\tau_{co}$ and becomes sequentially complete on $\|\cdot\|$-bounded sets. Then $C_b(\mathbb{R}^n)$ equipped with $\tau$ satisfies Assumption 2.1 and $(P(t))_{t\geq 0}$ is bi-continuous on it. To do so, let us denote by $C_0(\mathbb{R}^n)$ the space of continuous functions vanishing at infinity and let

$$\Gamma := \left\{ \gamma \in C_0(\mathbb{R}^n) : \gamma > 0 \text{ and } \lim_{\|x\|\to\infty} \|x\|^2 \gamma(x) \text{ exists in } \mathbb{R} \right\}.$$

We observe that $\Gamma \neq \varnothing$ since for arbitrary $\alpha, r > 0$ the function defined by

$$\gamma(x) := \begin{cases} \alpha, & \|x\| \leq r, \\ \dfrac{\alpha r^2}{\|x\|^2}, & \|x\| > r \end{cases} \tag{6.3.4}$$

satisfies $\gamma \in \Gamma$. Moreover, let $(\Omega_m)_{m\in\mathbb{N}}$ be an exhaustion of $\mathbb{R}^n$ and $(\alpha_m)_{m\in\mathbb{N}}$ and increasing sequence of integers satisfying $\alpha_m \geq \max\{m, d(0, \Omega_m)\}$. If $(\gamma_m)_{m\in\mathbb{N}}$ is a sequence in $C_0(\mathbb{R}^n)$ satisfying

- $0 \leq \gamma_m \leq 1$ on $\mathbb{R}^n$,
- $\gamma_m = 1$ on $\overline{\Omega}_{m-1}$, and
- $\gamma_m = 0$ on $\mathbb{R}^n \setminus \Omega_m$,

for all $m \in \mathbb{N}$, then the function defined by

$$\gamma_0(x) := \sum_{m=1}^{\infty} \frac{\gamma_m(x)}{2^{\alpha_m}}, \quad x \in \mathbb{R}^n \tag{6.3.5}$$

also belongs to $\Gamma$. Now, we are able to define a family $\mathcal{P}_\tau$ of seminorms by $\mathcal{P}_\tau := \{p_\gamma : \gamma \in \Gamma\}$, where

$$p_\gamma(f) := \sup_{x\in\mathbb{R}^n} \gamma(x)\,|f(x)|, \quad f \in C_b(\mathbb{R}^n).$$

By construction, $\mathcal{P}_\tau$ generates a locally convex Hausdorff topology $\tau$ which is coarser than $\tau_{co}$. Obviously, the inclusion $(C_b(\mathbb{R}^n), \|\cdot\|_\infty) \hookrightarrow (C_b(\mathbb{R}^n), \tau)$ is continuous. Indeed, if $\gamma \in \Gamma$, there exists $M > 0$ such that $\sup_{x\in\mathbb{R}^n} \gamma(x) > 0$ and hence $p_\gamma(f) \leq M \|f\|_\infty$ for all

$f \in C_b(\mathbb{R}^n)$. Moreover, the inclusion map $(C_b(\mathbb{R}^n), \tau) \hookrightarrow (C_b(\mathbb{R}^n), \tau_{co})$ is continuous as well. In fact, for each $m \in \mathbb{N}$ there exists $\gamma \in \Gamma$ defined by (6.3.4) with $\alpha = 1$ and $r = m$ such that

$$\sup_{\|x\| \leq m} |f(x)| \leq \sup_{x \in \mathbb{R}^n} \gamma(x) |f(x)| = p_\gamma(f),$$

for all $f \in C_b(\mathbb{R}^n)$.

**Remark 6.17**  We can make the following further observations:

(i)  If $A$ is a real and non-zero matrix and $\gamma \in \Gamma$ and $s > 0$, one can define for a given $\gamma \in \Gamma$ the function

$$\widetilde{\gamma}_s(x) := \sup_{0 \leq t \leq s} \gamma\left(e^{-tA}x\right), \quad x \in \mathbb{R}^n.$$

We observe that $\widetilde{\gamma}_s \in \Gamma$ and $\widetilde{\gamma}_s \leq \gamma$.

(ii)  By following [116, Prop. 2.3 & Prop. 2.4] with minor changes and by choosing the function as defined by (6.3.5) we see that $C_0(\mathbb{R}^n)$ is bi-dense in $(C_b(\mathbb{R}^n), \tau)$.

(iii)  Furthermore, $(C_b(\mathbb{R}^n), \tau)$ is sequentially complete on $\|\cdot\|$-bounded sets.

By Remark 6.17(ii)&(iii) we can conclude that $C_b(\mathbb{R}^n)$ equipped with $\tau$ satisfies Assumption 2.1, as promised above. Now, we are able to show that the closure of the operators $(\mathcal{A}, \mathcal{S}(\mathbb{R}^n))$ and $(\mathcal{B}, \mathcal{S}(\mathbb{R}^n))$ as they appear in (6.3.2) are generators of bi-continuous semigroups on $C_b(\mathbb{R}^n)$ with respect to $\tau$.

**Proposition 6.18**  *The operator semigroup on $C_b(\mathbb{R}^n)$ defined by*

$$(S(t)f)(x) := f(e^{tB}x), \quad t \geq 0, \ f \in C_b(\mathbb{R}^n), \ x \in \mathbb{R}^n$$

*is bi-continuous with respect to the locally convex topology $\tau$ and its generator coincides with the bi-closure of the operator given by*

$$(\mathcal{B}f)(x) = \sum_{i,j=1}^{n} b_{ij} x_j \frac{\partial}{\partial x_i} f(x) = \langle Bx, \nabla f(x) \rangle, \quad f \in \mathcal{S}(\mathbb{R}^n).$$

***Proof***  First of all, we observe that $S(t) \in \mathcal{L}(C_b(\mathbb{R}^n))$ and $\|S(t)\|_\infty \leq 1$ for all $t \geq 0$. Let us show that $(S(t))_{t \geq 0}$ is $\tau$-strongly continuous. Indeed, we first observe that for each compact set $K \subseteq \mathbb{R}^n$ one has

$$\lim_{t \to 0} \sup_{x \in K} \left\| e^{tB}x - x \right\| = 0. \tag{6.3.6}$$

Let $f \in C_b(\mathbb{R}^n)$ and denote $M := \sup_{x \in \mathbb{R}^n} |f(x)|$. For arbitrary $\varepsilon > 0$ and $\gamma \in \Gamma$ there exists $r > 0$ such that $0 < \gamma(x) < \frac{\varepsilon}{4M}$ whenever $|x| > r$. Therefore,

$$\sup_{\|x\|>r} \gamma(x) \left| f(e^{tB}x) - f(x) \right| \le \frac{\varepsilon}{4M} \sup_{\|x\|>r} \left| f(e^{tB}x) - f(x) \right| \le \frac{\varepsilon}{4M} 2M = \frac{\varepsilon}{2}, \quad (6.3.7)$$

for all $t \ge 0$. Now, let $K \subseteq \mathbb{R}^n$ be the compact set $K := \{x \in \mathbb{R}^n : \|x\| \le r\}$ and $0 < d := \max_{x \in K} \gamma(x) < \infty$. By (6.3.6), there exists $\delta > 0$ such that for all $t \in (0, \delta)$ one has

$$\sup_{\|x\|\le r} \left| f(e^{tB}x) - f(x) \right| < \frac{\varepsilon}{2d}.$$

Therefore, we have for all $t \in (0, \delta)$ that

$$\sup_{\|x\|\le r} \gamma(x) \left| f(e^{tB}x) - f(x) \right| \le d \sup_{\|x\|\le r} \left| f(e^{tB}x) - f(x) \right| < d\frac{\varepsilon}{2d} = \frac{\varepsilon}{2}. \quad (6.3.8)$$

By summarizing (6.3.7) and (6.3.8) we obtain that for all $f \in C_b(\mathbb{R}^n)$ holds that

$$\tau\lim_{t \to 0} S(t)f = f. \quad (6.3.9)$$

In what follows, we also show that $(S(t))_{t\ge 0}$ is locally bi-equicontinuous. Let $s > 0$ and $\gamma \in \Gamma$ and let $\widetilde{\gamma}_s$ be the element in $\Gamma$ as defined in Remark 6.17(i) (with $B$ instead of $A$). Then

$$p_\gamma(S(t)f) = \sup_{x\in\mathbb{R}^n} \gamma(x) \left| f(e^{tB}x) \right| \le \sup_{x\in\mathbb{R}^n} \widetilde{\gamma}_s(x) \, |f(x)| = p_{\widetilde{\gamma}_s}(f), \quad (6.3.10)$$

for all $f \in C_b(\mathbb{R}^n)$ and $t \in [0, s]$. In particular, $(S(t))_{t\ge 0}$ is a locally equicontinuous semigroup with respect to the locally convex topology $\tau$. In fact, by combining (6.3.9) and (6.3.10) as well as Proposition 6.14 we see that $(S(t))_{t\ge 0}$ yields a bi-continuous semigroup on $C_b(\mathbb{R}^n)$ with respect to $\tau$. Finally, assume $(\mathcal{C}, D(\mathcal{C}))$ is the generator of $(S(t))_{t\ge 0}$. For $f \in \mathcal{S}(\mathbb{R}^n)$ consider $g := f - \mathcal{B}f \in \mathcal{S}(\mathbb{R}^n)$. By Lemma 3.10 and integration by parts we obtain

$$(R(1, \mathcal{C})g)(x) = \int_0^\infty e^{-t} f(e^{tB}x) \, dt - \int_0^\infty e^{-t} \left\langle Be^{tB}x, \nabla f(e^{tB}x) \right\rangle dt = f(x).$$

Therefore, $\mathcal{S}(\mathbb{R}^n) \subseteq D(\mathcal{C})$. On the other hand, $\mathcal{S}(\mathbb{R}^n)$ is invariant under $(S(t))_{t\ge 0}$ and bi-dense in $C_b(\mathbb{R}^n)$ with respect to $\tau$. Hence, it is a bi-core by Proposition 3.9. This completes the proof. $\qquad\square$

**Proposition 6.19** *The semigroup $(T(t))_{t\ge 0}$ defined*

$$(T(t)f)(x) := \begin{cases} \dfrac{1}{(2\pi t)^{\frac{n}{2}}\sqrt{\det(A)}} \displaystyle\int_{\mathbb{R}^n} e^{-\frac{1}{2t}\langle A^{-1}(x-y),(x-y)\rangle} f(y) \, dy, & t > 0, \\[2mm] f(x), & t = 0, \end{cases}$$

*for $t \geq 0$, $f \in C_b(\mathbb{R}^n)$ and $x \in \mathbb{R}^n$ is bi-continuous on $C_b(\mathbb{R}^n)$ with respect to $\tau$ and its generator coincides with the bi-closure of the operator*

$$(\mathcal{A}f)(x) = \frac{1}{2} \sum_{i,j=1}^{n} a_{ij} \frac{\partial^2}{\partial x_i \partial x_j} f(x) = \langle \nabla, A\nabla f(x) \rangle, \quad f \in \mathcal{S}(\mathbb{R}^n).$$

***Proof*** Of course, one has $T(t) \in \mathcal{L}(C_b(\mathbb{R}^n))$ and $\|T(t)\| \leq 1$ for all $t \geq 0$. As next step we show that $(T(t))_{t\geq 0}$ is locally equicontinuous with respect to $\tau$. For this, let $\gamma \in \Gamma$, $f \in C_b(\mathbb{R}^n)$, and $x \in \mathbb{R}^n$ and observe the following:

$$
\begin{aligned}
\gamma(x)\,|(T(t)f)(x)| &\leq \frac{\gamma(x)}{(2\pi t)^{\frac{n}{2}}\sqrt{\det(A)}} \int_{\mathbb{R}^n} e^{-\frac{1}{2t}\langle A^{-1}(x-y),(x-y)\rangle} \cdot \left(1 + \|y\|^2\right) \frac{|f(y)|}{1+\|y\|^2}\, dy \\
&\leq \frac{\gamma(x)}{(2\pi t)^{\frac{n}{2}}\sqrt{\det(A)}} \int_{\mathbb{R}^n} e^{-\frac{1}{2t}\langle A^{-1}(x-y),(x-y)\rangle} \cdot \left(1 + \|y\|^2\right) dy \cdot \sup_{z\in\mathbb{R}^n} \frac{|f(z)|}{1+\|z\|^2} \\
&\leq \gamma(x)\left(1 + \|x\|^2 + nt\,\|A\|^2\right) \cdot \sup_{z\in\mathbb{R}^n} \frac{|f(z)|}{1+\|z\|^2} \\
&\leq M(1+t) \cdot \sup_{z\in\mathbb{R}^n} \frac{|f(z)|}{1+\|z\|^2},
\end{aligned}
$$

where $M := 2\max\left\{M_\gamma, n\,\|A\|^2\right\}$ with $M_\gamma := \sup_{x\in\mathbb{R}^n}\left(1 + \|x\|^2\right)\gamma(x) < \infty$. With $\widetilde{\gamma}(x) := \frac{M}{1+\|x\|^2}$ for $x \in \mathbb{R}^n$ one has $\widetilde{\gamma} \in \Gamma$. It follows that

$$p_\gamma(T(t)f) \leq (1+t)p_{\widetilde{\gamma}}(f), \tag{6.3.11}$$

for all $f \in C_b(\mathbb{R}^n)$ showing that $(T(t))_{t\geq 0}$ is locally equicontinuous on $C_b(\mathbb{R}^n)$ with respect to $\tau$. By taking for granted that $(T(t))_{t\geq 0}$ is strongly continuous on $C_0(\mathbb{R}^n)$ with respect to the $\|\cdot\|_\infty$-norm, cf. [64], we conclude that

$$\tau\lim_{t\to 0} T(t)f = f, \tag{6.3.12}$$

for all $f \in C_0(\mathbb{R}^n)$ as $\tau$ is coarse then the $\|\cdot\|_\infty$-topology. We want to show that (6.3.12) indeed holds for all $f \in C_b(\mathbb{R}^n)$. By Remark 6.17(ii), we know that $C_0(\mathbb{R}^n)$ is dense in $C_b(\mathbb{R}^n)$ with respect to $\tau$. Hence, for given $f \in C_b(\mathbb{R}^n)$, $\gamma \in \Gamma$, and $\varepsilon > 0$ there exists $f_0 \in C_0(\mathbb{R}^n)$ such that

$$p_{\widetilde{\gamma}}(f - f_0) < \frac{\varepsilon}{4},$$

where $\widetilde{\gamma} \in \Gamma$ is defined as above. By (6.3.12) there exists $\delta \in (0, 1)$ such that for all $t \in (0, \delta)$ one has that

$$p_\gamma(T(t)f_0 - f_0) < \frac{\varepsilon}{4}.$$

Together with (6.3.11) this yields

$$p_\gamma(T(t)f - f) \le p_\gamma(T(t)(f - f_0)) + p_\gamma(T(t)f_0 - f_0) + p_\gamma(f - f_0)$$
$$\le (1 + t)p_{\tilde\gamma}(f - f_0) + p_\gamma(T(t)f_0 - f_0) + p_\gamma(f - f_0) < \varepsilon,$$

for all $t \in (0, \delta)$. This shows that

$$\tau\lim_{t \to 0} T(t)f = f,$$

for all $f \in C_b(\mathbb{R}^n)$. By Proposition 6.14 we conclude that $(T(t))_{t \ge 0}$ is a bi-continuous semigroup on $C_b(\mathbb{R}^n)$ with respect to $\tau$. Let $(\mathcal{C}, D(\mathcal{C}))$ denote the generator of $(T(t))_{t \ge 0}$. As mentioned previously, $(T(t))_{t \ge 0}$ is a $C_0$-semigroup on $C_0(\mathbb{R}^n)$ and its generator is given by the $\|\cdot\|$-closure of $(\mathcal{A}, \mathcal{S}(\mathbb{R}^n))$. Furthermore, the space $\mathcal{S}(\mathbb{R}^n)$ is left invariant under $(T(t))_{t \ge 0}$ so that $\mathcal{S}(\mathbb{R}^n) \subseteq D(\mathcal{C})$. On the other hand, $\mathcal{S}(\mathbb{R}^n)$ is bi-dense in $C_b(\mathbb{R}^n)$ with respect to the locally convex topology $\tau$. Hence, it is a bi-core by Proposition 3.9. This completes the proof. $\qquad\square$

With the previous propositions we are now able to approximate the operator semigroup $(P(t))_{t \ge 0}$ by the Lie–Trotter products of $(T(t))_{t \ge 0}$ and $(S(t))_{t \ge 0}$.

**Theorem 6.20**  *Let $(T(t))_{t \ge 0}$ and $(S(t))_{t \ge 0}$ be the two bi-continuous semigroups on $C_b(\mathbb{R}^n)$ as defined in Propositions 6.19 and 6.18, respectively. Furthermore, we denote their generators by $(\mathcal{A}, D(\mathcal{A}))$ and $(\mathcal{B}, D(\mathcal{B}))$, respectively. Then the Ornstein–Uhlenbeck semigroup on $C_b(\mathbb{R}^n)$ given by (6.3.3) is bi-continuous with respect to $\tau$, generated by the bi-closure of $\mathcal{A} + \mathcal{B}$ and represented by the Lie–Trotter product formula, i.e.:*

$$P(t)f = \tau\lim_{n \to \infty} \left[T(\tfrac{t}{n})S(\tfrac{t}{n})\right]^n f,$$

*for all $f \in C_b(\mathbb{R}^n)$ and uniformly for $t$ in compact intervals of $\mathbb{R}_+$.*

***Proof***  Clearly $\left\|[T(t)S(t)]^m\right\| \le 1$ for all $t \ge 0$ and $m \in \mathbb{N}$. Let $m \in \mathbb{N}, t \ge 0, f \in C_b(\mathbb{R}^n)$, and $x \in \mathbb{R}^n$. Then

$$[T(t)S(t)]^m f(x)$$
$$= \frac{1}{(2\pi t)^{\frac{nm}{2}} \sqrt{\det(A)^m}} \int_{\mathbb{R}^n} e^{-\frac{1}{2t}\langle A^{-1}(x-y_1),(x-y_1)\rangle} \, dy_1 \cdot \prod_{i=2}^{m-1} \int_{\mathbb{R}^n} e^{-\frac{1}{2t}\langle A^{-1}(e^{tB}y_{i-1}-y_i),(e^{tB}y_{i-1}-y_i)\rangle} \, dy_i$$
$$\cdot \int_{\mathbb{R}^n} e^{-\frac{1}{2t}\langle A^{-1}(e^{tB}y_{m-1}-y_m),(e^{tB}y_{m-1}-y_m)\rangle} f(e^{tB}y_m) \, dy_m.$$

Now, put $\alpha_t := (2\pi t)^{\frac{n}{2}} \sqrt{\det(A)}$ and take $\gamma \in \Gamma$. It follows that

$$\gamma(x)\left|\left[T(t)S(t)\right]^m f(x)\right|$$

$$\leq \frac{\gamma(x)}{\alpha_t^m} \int_{\mathbb{R}^n} e^{-\frac{1}{2t}\langle A^{-1}(x-y_1),(x-y_1)\rangle}\, dy_1 \cdot \prod_{i=2}^{m-1} \int_{\mathbb{R}^n} e^{-\frac{1}{2t}\langle A^{-1}(e^{tB}y_{i-1}-y_i),(e^{tB}y_{i-1}-y_i)\rangle}\, dy_i$$

$$\cdot \int_{\mathbb{R}^n} e^{-\frac{1}{2t}\langle A^{-1}(e^{tB}y_{m-1}-y_m),(e^{tB}y_{m-1}-y_m)\rangle} \cdot \left(1+\|y_m\|^2\right) \frac{\left|f(e^{tB}y_m)\right|^2}{1+\|y_m\|}\, dy_m$$

$$\leq \frac{\gamma(x)}{\alpha_t^m} \cdot \sup_{z\in\mathbb{R}^n} \frac{|f(z)|}{1+\|z\|^2} \int_{\mathbb{R}^n} e^{-\frac{1}{2t}\langle A^{-1}(x-y_1),(x-y_1)\rangle}\, dy_1 \cdot \prod_{i=2}^{m-1} \int_{\mathbb{R}^n} e^{-\frac{1}{2t}\langle A^{-1}(e^{tB}y_{i-1}-y_i),(e^{tB}y_{i-1}-y_i)\rangle}\, dy_i$$

$$\cdot \int_{\mathbb{R}^n} e^{-\frac{1}{2t}\langle A^{-1}(e^{tB}y_{m-1}-y_m),(e^{tB}y_{m-1}-y_m)\rangle} \cdot \left(1+\|y_m\|^2\right)\, dy_m.$$

Now fix $s > 0$ and put $\widetilde{\gamma}_0(z) := \sup_{t\in[0,s]} \frac{1}{1+\|e^{-tB}z\|^2}$, $z \in \mathbb{R}^n$, so that $\widetilde{\gamma}_0 \in \Gamma$ by Remark 6.17(i). By the previous estimate we obtain for $t \in [0, s]$ the following:

$$\gamma(x)\left|\left[T(t)S(t)\right]^m f(x)\right|$$

$$\leq \frac{\gamma(x)}{\alpha_t^m} \cdot \sup_{z\in\mathbb{R}^n} \widetilde{\gamma}_0(z)\,|f(z)| \int_{\mathbb{R}^n} e^{-\frac{1}{2t}\langle A^{-1}(x-y_1),(x-y_1)\rangle}\, dy_1 \cdot \prod_{i=2}^{m-1} \int_{\mathbb{R}^n} e^{-\frac{1}{2t}\langle A^{-1}(e^{tB}y_{i-1}-y_i),(e^{tB}y_{i-1}-y_i)\rangle}\, dy_i$$

$$\cdot \int_{\mathbb{R}^n} e^{-\frac{1}{2t}\langle A^{-1}(e^{tB}y_{m-1}-y_m),(e^{tB}y_{m-1}-y_m)\rangle} \cdot \left(1+\|y_m\|^2\right)\, dy_m$$

$$\leq \sup_{z\in\mathbb{R}^n} \widetilde{\gamma}_0(z)\,|f(z)| \cdot \left(\gamma(x) + \frac{\gamma(x)}{\alpha_t^m} \int_{\mathbb{R}^n} e^{-\frac{1}{2t}\langle A^{-1}(x-y_1),(x-y_1)\rangle}\, dy_1\right)$$

$$\cdot \prod_{i=2}^{m-2} \int_{\mathbb{R}^n} e^{-\frac{1}{2t}\langle A^{-1}(e^{tB}y_{i-1}-y_i),(e^{tB}y_{i-1}-y_i)\rangle}\, dy_i$$

$$\cdot \int_{\mathbb{R}^n} e^{-\frac{1}{2t}\langle A^{-1}(e^{tB}y_{m-2}-y_{m-1}),(e^{tB}y_{m-2}-y_{m-1})\rangle} \cdot \left(\left\|e^{tB}y_{m-1}\right\|^2 + nt\,\|A\|^2\right) \alpha_t^m\, dy_{m-1}$$

$$\leq \sup_{z\in\mathbb{R}^n} \widetilde{\gamma}_0(z)\,|f(z)| \cdot \left(\gamma(x)\left(1+nt\,\|A\|^2\right) + \frac{\gamma(x)e^{2t\|B\|}}{\alpha_t^{m-1}} \int_{\mathbb{R}^n} e^{-\frac{1}{2t}\langle A^{-1}(x-y_1),(x-y_1)\rangle}\, dy_1\right)$$

$$\cdot \prod_{i=2}^{m-2} \int_{\mathbb{R}^n} e^{-\frac{1}{2t}\langle A^{-1}(e^{tB}y_{i-1}-y_i),(e^{tB}y_{i-1}-y_i)\rangle}\, dy_i$$

$$\cdot \int_{\mathbb{R}^n} e^{-\frac{1}{2t}\langle A^{-1}(e^{tB}y_{m-2}-y_{m-1}),(e^{tB}y_{m-2}-y_{m-1})\rangle} \|y_{m-1}\|^2\, dy_{m-1}$$

$$\leq \sup_{z\in\mathbb{R}^n} \widetilde{\gamma}_0(z)\,|f(z)| \cdot \gamma(x)\left(1+mnt\,\|A\|^2 + e^{2mt\|B\|^2}\,\|x\|^2\right)$$

showing that

$$\gamma(x)\left|\left[T(t)S(t)\right]^m f(x)\right| \leq \sup_{z\in\mathbb{R}^n} \widetilde{\gamma}_0(z)\,|f(z)| \cdot \gamma(x)\left(1+mnt\,\|A\|^2 + e^{2mt\|B\|^2}\,\|x\|^2\right).$$

By taking $\omega := \max\{2\,\|B\|, 1\}$, which is independent of the choice of $\gamma \in \Gamma$, $s > 0$ and $f \in C_b(\mathbb{R})$, and $M := 2\max\{M_\gamma, n\,\|A\|^2\}$ with $M_\gamma := \sup_{x\in\mathbb{R}^n} \left(1+\|x\|^2\right)\gamma(x) < \infty$, it follows by the previous estimate that there exists $\widetilde{\gamma} := M\widetilde{\gamma}_0 \in \Gamma$ such that for all $f \in C_b(\mathbb{R}^n)$

$$p_\gamma\left([T(t)S(t)]^m\, f\right) \le \mathrm{e}^{m\omega t}\, p_{\widetilde\gamma}(f),$$

whenever $t \in [0, s]$ and $m \in \mathbb{N}$. Since $\gamma \in \Gamma$ and $s > 0$ were arbitrary, we have that there exists $\omega_0 \in \mathbb{R}_+$ such that for all $\gamma \in \Gamma$ and $s > 0$ there exists $\widetilde\gamma \in \Gamma$ such that

$$p_\gamma\left(\left[T(\tfrac{t}{m})S(\tfrac{t}{m})\right]^m f\right) \le \mathrm{e}^{\omega_0}\, p_{\widetilde\gamma}(f),$$

for all $f \in C_b(\mathbb{R}^n)$, $t \in [0, s]$, and $m \in \mathbb{N}$. Hence, the family $\left\{[T(\tfrac{t}{m})S(\tfrac{t}{m})]^m : t \ge 0\right\}$ is locally bi-equicontinuous uniformly for $m \in \mathbb{N}$. For the last part, we recall that by Remark 6.17(ii), the space $\mathcal{S}(\mathbb{R}^n)$ is bi-dense in $C_b(\mathbb{R}^n)$ with respect to $\tau$. Furthermore, $\mathcal{S}(\mathbb{R}^n) \subseteq D(\mathcal{A}_0) \cap D(\mathcal{B}_0)$, where $D(\mathcal{A}_0)$ and $D(\mathcal{B}_0)$ denote the domains of the parts of $\mathcal{A}$ and $\mathcal{B}$ in $C_0(\mathbb{R}^n)$, respectively. As mentioned previously, by [64], the Ornstein–Uhlenbeck semigroup $(P(t))_{t\ge 0}$ is strongly continuous on $C_0(\mathbb{R}^n)$. Moreover, by [162, Prop. 12] the semigroup $(P(t))_{t\ge 0}$ can be represented on $C_0(\mathbb{R}^n)$ by means of the Lie–Trotter product formula. In particular, its generator coincides with $\mathcal{A} + \mathcal{B}$ restricted to $\mathcal{S}(\mathbb{R}^n)$. Hence, by the invariance of $\mathcal{S}(\mathbb{R}^n)$ under the Ornstein–Uhlenbeck semigroup we obtain that $(\lambda - \mathcal{A} - \mathcal{B})\mathcal{S}(\mathbb{R}^n)$ is bi-dense in $C_b(\mathbb{R}^n)$ with respect to $\tau$ for $\lambda > 0$. By Proposition 6.10 we obtain the desired result. $\qquad\square$

## 6.4   A Lumer–Phillips Generation-Type Theorem

In order to present the Lumer–Phillips-type generation theorem for bi-continuous semigroups, we first need to discuss the mixed topology. This topology has been mentioned already in Chap. 2 and in particular in Remark 2.23. For more detailed information, we also refer again to Appendix A. Let us start with the following definition, see also [157, Def. 2.2 and 5.4].

**Definition 6.21**  Let $(X, \|\cdot\|, \tau)$ be a bi-admissible space and let $\gamma$ denote the corresponding mixed topology.

(i)  We call $(X, \|\cdot\|, \tau)$ *(sequentially) complete* if $(X, \gamma)$ is (sequentially) complete.

(ii)  We call $(X, \|\cdot\|, \tau)$ *C-sequential* if $(X, \gamma)$ is C-sequential, i.e., every convex sequentially open subset of $(X, \gamma)$ is already open.

**Remark 6.22** (a)  The property of $C$-sequential locally convex spaces will appear later again in Chaps. 7 and 10. The definition originates from [220, p. 273]. More on $C$-sequential locally convex space, their properties as well as connection to bi-continuous semigroups can be found here [154].

(b)  Let $(X, \|\cdot\|, \tau)$ be sequentially complete according to Definition 6.21. A semigroup of linear operators $(T(t))_{t\ge 0}$ from $X$ to $X$ is $\gamma$-strongly continuous and *locally sequentially*

$\gamma$-*equicontinuous* (i.e., for all $\gamma$-null sequences $(x_n)_{n\in\mathbb{N}}$ in $X$, $t_0 > 0$ and $p \in \Gamma_\gamma$ we have $\lim_{n\to\infty} \sup_{t\in[0,t_0]} p(T(t)x_n) = 0$) if and only if it is a bi-continuous semigroup with respect to $\tau$ on $X$. This is a consequence of Proposition B.4 and [154, Rem. 2.6(b)].

**Definition 6.23** Let $(X, \upsilon)$ be a Hausdorff locally convex space and $\Gamma_\upsilon$ a directed system of continuous seminorms that generates $\upsilon$. A family $(T(t))_{t\in I}$ of linear maps from $X$ to $X$ is called $\upsilon$-*equicontinuous* if

$$\forall\, p \in \Gamma_\upsilon \ \exists\, \widetilde{p} \in \Gamma_\upsilon,\ C \geq 0\,\forall\, t \in I,\ x \in X :\ p(T(t)x) \leq C\widetilde{p}(x).$$

The family $(T(t))_{t\geq 0}$ is called *locally $\upsilon$-equicontinuous* if $(T(t))_{t\in[0,t_0]}$ is $\upsilon$-equicontinuous for all $t_0 \geq 0$. The family $(T(t))_{t\geq 0}$ is called *quasi-$\upsilon$-equicontinuous* if there is $\alpha \in \mathcal{R}$ such that $(e^{-\alpha t}T(t))_{t\geq 0}$ is $\upsilon$-equicontinuous. Note that one often drops the $\upsilon$ if the topology is clear.

The following definition is inspired by the work of Albanese and Jornet [6, Def. 3.9].

**Definition 6.24** Let $(X, \upsilon)$ be a Hausdorff locally convex space and $\Gamma_\upsilon$ a directed system of continuous seminorms that generates $\upsilon$. A linear operator $(A, D(A))$ on $X$ is called $\Gamma_\upsilon$-*dissipative* if

$$\forall\, \lambda > 0,\ x \in D(A),\ p \in \Gamma_\upsilon :\ p((\lambda - A)x) \geq \lambda p(x).$$

It is important to note that the notion of dissipativity depends on the selection of the directed system of continuous seminorms that generates the topology $\upsilon$, see also [6, Rem. 3.10].

**Remark 6.25** Let $(X, \|\cdot\|, \tau)$ be a bi-admissible space, $\upsilon$ a Hausdorff locally convex topology on $X$, and $(A, D(A))$ a $\Gamma_\upsilon$-dissipative operator on $X$. If $\Gamma_\upsilon$ is norming in the sense of Assumption 2.1, then it follows that

$$\forall\, \lambda > 0,\ x \in D(A) :\ \|(\lambda - A)x\| \geq \lambda\|x\|.$$

Thus $(A, D(A))$ is also a dissipative operator on the Banach space $(X, \|\cdot\|)$ in the sense of [101, Chap. II, Def. 3.13]. This is also related to the observation made by Budde and Wegner for the case $\upsilon = \tau$, cf. [62, Rem. 3.3(i)]. We also denote such kind of dissipativity on a Banach space $(X, \|\cdot\|)$ by $\|\cdot\|$-*dissipativity*.

**Proposition 6.26** *Let $(X, \|\cdot\|, \tau)$ be a sequentially complete Saks space, $\upsilon$ a Hausdorff locally convex topology on $X$, and $(A, D(A))$ a $\Gamma_\upsilon$-dissipative operator on $X$. Then the following assertions hold:*

*(a)* $\lambda - A$ *is injective for all* $\lambda > 0$. *Moreover, we have*

$$\forall \, \lambda > 0, \; x \in \mathrm{Ran}(\lambda - A), \; p \in \Gamma_v : \; p((\lambda - A)^{-1}x) \le \frac{1}{\lambda} p(x). \qquad (6.4.1)$$

*(b)  If* $\mathrm{Ran}(\lambda - A)$ *is (sequentially)* $v$-*closed for some* $\lambda > 0$, *then* $(A, D(A))$ *is (sequentially)* $v$-*closed. If* $v = \gamma$, *then the converse even holds for all* $\lambda > 0$.

*(c)  Let* $\Gamma_v$ *be norming. Then* $\lambda - A$ *is surjective for some* $\lambda > 0$ *if and only if it is surjective for all* $\lambda > 0$. *In such a case,* $(0, \infty) \subseteq \rho(A)$.

*(d)  Let* $v = \gamma$. *Then* $\lambda - A$ *is surjective for some* $\lambda > 0$ *if and only if it is surjective for all* $\lambda > 0$. *In such a case,* $(0, \infty) \subseteq \rho_\gamma(A)$.

**Proof** Parts (a), (b), and (d) are just [6, Prop 3.11] in combination with the sequential completeness of $(X, \gamma)$. Part (c) is a consequence of [101, Chap. II, Propo. 3.14] and Remark 6.25. $\qquad \square$

**Theorem 6.27** *Let* $(X, \| \cdot \|, \tau)$ *be a bi-admissible space,* $v$ *a Hausdorff locally convex topology on* $X$ *with* $\tau \subseteq v \subseteq \| \cdot \|$ *such that* $\gamma$-*convergent sequences are* $v$-*convergent,* $(A, D(A))$ *is a bi-densely defined,* $\Gamma_v$-*dissipative operator on* $X$ *and* $\Gamma_v$ *norming. Then the following assertions are equivalent:*

*(a)  $(A, D(A))$ generates a bi-continuous contraction semigroup with respect to* $\tau$ *on* $X$.

*(b)  $\lambda - A$ is surjective for some* $\lambda > 0$.

**Proof** In order to prove this result, we make use of the Hille–Yosida-type generation theorem for bi-continuous semigroups, cf. Theorem 3.18.

(a)$\Rightarrow$(b): Let $(A, D(A))$ generate a bi-continuous contraction semigroup with respect to $\tau$ on $X$. Due to Theorem 3.18 with $\omega = 0$ we obtain that $(0, \infty) \subseteq \rho(A)$, in particular, that $\lambda - A$ is surjective for all $\lambda > 0$.

(b)$\Rightarrow$(a): Let $\lambda - A$ be surjective for some $\lambda > 0$. In order to conclude that $(A, D(A))$ generates a bi-continuous contraction semigroup with respect to $\tau$ on $X$, we need to check the conditions of Theorem 3.18(b). Since $(A, D(A))$ is supposed to be $\Gamma_v$-dissipative, we obtain that $\lambda - A$ is bijective for all $\lambda > 0$ and $\rho(A) \subseteq (0, \infty)$ by Proposition 6.26(a)&(c). From Remark 6.25 and the assumption that $\Gamma_v$ is norming we conclude that $\|R(\lambda, A)x\| \le \frac{1}{\lambda}\|x\|$ for all $\lambda > 0$ and $x \in \mathrm{Ran}(\lambda - A) = X$, yielding $\|R(\lambda, A)^n\|_{\mathcal{L}(X)} \le \frac{1}{\lambda^n}$ for all $n \in \mathbb{N}$ and $\lambda > 0$. Now, let $\Gamma_\tau$ be a directed system of continuous seminorms that generates the topology $\tau$ and $q \in \Gamma_\tau$. Thanks to (6.4.1) we know that $p((\lambda - A)^{-1}x) \le \frac{1}{\lambda}p(x)$ for all $p \in \Gamma_v$, $\lambda > 0$, and $x \in \mathrm{Ran}(\lambda - A) = X$. As $\tau \subseteq v$, there are $p \in \Gamma_v$ and $C \ge 0$ such that for each $\alpha > 0$ we have

$$q((\lambda - \alpha)^n R(\lambda, A)^n x) \leq C(\lambda - \alpha)^n p((\lambda - A)^{-n} x) \leq C \frac{(\lambda - \alpha)^n}{\lambda^n} p(x)$$

$$\leq C\left(1 - \frac{\alpha}{\lambda}\right)^n p(x) \leq C p(x)$$

for all $x \in X, n \in \mathbb{N}$ and $\lambda \geq \alpha$. Since $\|\cdot\|$-bounded $\tau$-null sequences are exactly the $\gamma$-null sequences by Proposition B.4 and $\gamma$-convergent sequences are assumed to be $\upsilon$-convergent, this inequality implies that $\{(\lambda - \alpha)^n R(\lambda, A)^n \mid n \in \mathbb{N}, \lambda \geq \alpha\}$ is bi-equicontinuous for all $\alpha > 0$. This concludes the proof. $\qquad\square$

In the case $\upsilon = \tau$ we know that $\gamma$-convergent sequences are $\tau$-convergent and thus Theorem 6.27 implies [62, Thm. 3.6]. The most interesting choice is $\upsilon = \gamma$. Our second generation result involves complete spaces.

**Theorem 6.28** *Let $(X, \|\cdot\|, \tau)$ be a complete bi-admissible space (according to Definition 6.21) and $(A, D(A))$ a $\gamma$-densely defined, $\Gamma_\gamma$-dissipative operator. Assume that $\mathrm{Ran}(\lambda - A)$ is $\gamma$-dense in $X$ for some $\lambda > 0$. Then the following assertions hold:*

*(a)  The $\gamma$-closure $(\overline{A}, D(\overline{A}))$ generates a $\gamma$-strongly continuous, $\gamma$-equicontinuous semigroup $(T(t))_{t \geq 0}$ on $X$.*
*(b)  If $\Gamma_\gamma$ is norming, then $(T(t))_{t \geq 0}$ is a contraction semigroup.*

**_Proof_** (a) Due to [6, Thm. 3.14], the operator $(\overline{A}, D(\overline{A}))$ generates a $\gamma$-equicontinuous, $\gamma$-strongly continuous semigroup $(T(t))_{t \geq 0}$ on $X$.

(b) By [6, Prop. 3.13,] the operator $(\overline{A}, D(\overline{A}))$ is also $\Gamma_\gamma$-dissipative and $\lambda - \overline{A}$ is surjective for all $\lambda > 0$. As a consequence of part (a) and Remark 6.22, the semigroup $(T(t))_{t \geq 0}$ is also a bi-continuous semigroup with respect to $\tau$ on $X$. By [155, p. 5] the generator $(\overline{A}, D(\overline{A}))$ of $(T(t))_{t \geq 0}$ as a $\gamma$-strongly continuous, $\gamma$-equicontinuous semigroup (see [6, p. 922]) and the generator of $(T(t))_{t \geq 0}$ as a $\tau$-bi-continuous semigroup coincide. Thus $(\overline{A}, D(\overline{A}))$ is bi-densely defined by Theorem 3.15(a). Hence we get that $(T(t))_{t \geq 0}$ is a contraction semigroup by Theorem 6.27 with $\upsilon = \gamma$ and the norming property of $\Gamma_\gamma$. $\qquad\square$

**Remark 6.29** So far there is no nice characterization of the completeness of $(X, \gamma_s)$ that is assumed in [62, Thm. 3.15]. However, there is a nice characterization of the completeness of the bi-admissible space $(X, \|\cdot\|, \tau)$. By definition the Saks space is complete if and only if $(X, \gamma)$ is complete. The space $(X, \gamma)$ is complete if and only if $B_{\|\cdot\|} = \{x \in X \mid \|x\| \leq 1\}$ is $\tau$-complete by [73, Chap. I, Prop. 1.14 Proposition, p. 11]. But, since $\gamma_s$ is in general a weaker topology than $\gamma$ (see Chap. A), the completeness of $(X, \gamma)$ does in general not imply the completeness of $(X, \gamma_s)$.

Let us take a closer look at the completeness assumption on the Saks space $(X, \|\cdot\|, \tau)$ in Theorem 6.28, which is actually fulfilled for many important examples, and its characterization in Remark 6.29. Especially, $(X, \gamma)$ is complete, thus $(X, \| \cdot \|, \tau)$ as well, if $B_{\|\cdot\|}$ is $\tau$-compact, which is Proposition B.6(i) and a sufficient condition for $\gamma = \gamma_s$. We consider and list some important examples as they have also been investigated by Kruse and Schwenninger [154]. We encountered some of those spaces already in Sect. 2.1.

**Example 6.30** (a)  Let $\Omega$ be a Hausdorff $k_{\mathbb{R}}$-space and recall that a completely regular space $\Omega$ is called $k_{\mathbb{R}}$-*space* if any map $f : \Omega \to \mathbb{R}$, whose restriction to each compact $K \subset \Omega$ is continuous, is already continuous on $\Omega$ (see [182, p. 487]). Further, let $C_b(\Omega)$ be the space of bounded continuous functions on $\Omega$, and $\| \cdot \|_\infty$ the sup-norm as well as $\tau_{co}$ the compact-open topology, i.e., the topology of uniform convergence on compact subsets of $\Omega$. Then $(C_b(\Omega), \| \cdot \|_\infty, \tau_{co})$ is a complete bi-admissible space and $\gamma(\| \cdot \|_\infty, \tau_{co}) = \gamma_s(\| \cdot \|_\infty, \tau_{co})$.

Let $\mathcal{V}$ denote the set of all non-negative bounded functions $\nu$ on $\Omega$ that vanish at infinity, i.e., for every $\varepsilon > 0$ the set $\{x \in \Omega \mid \nu(x) \geq \varepsilon\}$ is compact. Let $\beta_0$ be the Hausdorff locally convex topology on $C_b(\Omega)$ that is induced by the seminorms

$$|f|_\nu := \sup_{x \in \Omega} |f(x)| \nu(x), \quad f \in C_b(\Omega),$$

for $\nu \in \mathcal{V}$. Then we have $\gamma(\| \cdot \|_\infty, \tau_{co}) = \beta_0$. If $\Omega$ is locally compact, then $\mathcal{V}$ may be replaced by the functions in $C_0(\Omega)$ that are non-negative where $C_0(\Omega)$ is the space of real-valued continuous functions on $\Omega$ that vanish at infinity.

If $\Omega$ is a hemicompact Hausdorff $k_{\mathbb{R}}$-space or a Polish space, then we even have

$$\gamma(\| \cdot \|_\infty, \tau_{co}) = \beta_0 = \mu(C_b(\Omega), M_t(\Omega))$$

where $M_t(\Omega) = (C_b(\Omega), \beta_0)'$ is the space of bounded Radon measures and $\mu(C_b(\Omega), M_t(\Omega))$ the Mackey topology of the dual pair $(C_b(\Omega), M_t(\Omega))$.

(b)  Let $(X, \| \cdot \|)$ be a Banach space and $\sigma^* := \sigma(X', X)$ the weak*-topology. Then condition Proposition B.6(i) is fulfilled, $(X', \| \cdot \|_{X'}, \sigma^*)$ is a complete bi-admissible space and $\gamma(\| \cdot \|_{X'}, \sigma^*) = \gamma_s(\| \cdot \|_\infty, \sigma^*) = \tau_c(X', X)$ where $\tau_c(X', X)$ is the topology of uniform convergence on compact subsets of $X$.

(c)  Let $(X, \| \cdot \|)$ be a Banach space and $\mu^* := \mu(X', X)$ the dual Mackey topology. Then $(X', \| \cdot \|_{X'}, \mu^*)$ is a complete Saks space, where the completeness follows from [142, p. 74], and $\gamma(\| \cdot \|_{X'}, \mu^*) = \mu^*$. If $X$ is a *Schur space*, i.e., every $\sigma(X, X')$-convergent sequence is $\| \cdot \|$-convergent (see [105, p. 253]), then condition Proposition B.6(i) is fulfilled and $\gamma(\| \cdot \|_{X'}, \mu^*) = \gamma_s(\| \cdot \|_{X'}, \mu^*)$.

(d)  Let $(X, \| \cdot \|_X)$ and $(Y, \| \cdot \|_Y)$ be Banach spaces, and $\tau_{sot}$ the strong operator topology on $\mathcal{L}(X, Y)$. Then $(\mathcal{L}(X, Y), \| \cdot \|_{\mathcal{L}(X,Y)}, \tau_{sot})$ is a Sbi-admissible space. Let $(T_i)_{i \in I}$ be a $\tau_{sot}$-Cauchy net in $B_{\|\cdot\|_{\mathcal{L}(X,Y)}} = \{T \in \mathcal{L}(X, Y) \mid \|T\|_{\mathcal{L}(X,Y)} \leq 1\}$. Then for each $x \in X$ the net $(T_i x)_{i \in I}$ is $\| \cdot \|_Y$-convergent to some $Tx \in Y$ with $\|Tx\|_Y \leq \|x\|_X$ in the Banach space $(Y, \| \cdot \|_Y)$. Thus the map $T : x \mapsto Tx$ belongs to $\mathcal{L}(X, Y)$ with

$\|T\|_{\mathscr{L}(X,Y)} \leq 1$ and $(T_i)_{i \in I}$ is $\tau_{\text{sot}}$-convergent to $T$. Hence $B_{\|\cdot\|_{\mathscr{L}(X,Y)}}$ is $\tau_{\text{sot}}$-complete and so $(\mathscr{L}(X,Y), \|\cdot\|_{\mathscr{L}(X,Y)}, \tau_{\text{sot}})$ is complete. If $Y$ is in addition finite-dimensional, then Proposition B.6(i) is fulfilled and $\gamma(\|\cdot\|_{\mathcal{L}(X;Y)}, \tau_{\text{sot}}) = \gamma_s(\|\cdot\|_{\mathcal{L}(X;Y)}, \tau_{\text{sot}})$.

(e) Let $H$ be a separable Hilbert space and $\mathscr{N}(H)$ the space of trace class operators in $\mathscr{L}(H)$ and note that $\mathscr{L}(H) = \mathscr{N}(H)'$. Let $\tau_{\text{sot}*}$ be the symmetric strong operator topology, i.e., the Hausdorff locally convex topology on $\mathscr{L}(H)$ generated by the directed system of seminorms

$$p_N(R) := \max\Big(\sup_{x \in N} \|Rx\|_H, \sup_{x \in N} \|R^*x\|_H\Big), \quad R \in \mathcal{L}(H),$$

for finite $N \subset H$ where $R^*$ is the adjoint of $R$. We denote by $\beta_{\text{sot}*}$ the mixed topology $\gamma(\|\cdot\|_{\mathscr{L}(H)}, \tau_{\text{sot}*})$. Then the triple $(\mathscr{L}(H), \|\cdot\|_{\mathscr{L}(H)}, \tau_{\text{sot}*})$ is a complete bi-admissible space and $\beta_{\text{sot}*} = \mu(\mathscr{L}(H), \mathscr{N}(H))$.

(f) Let $\Omega$ be a completely regular Hausdorff space, $M_t(\Omega)$ the space of bounded Radon measures on $\Omega$, and $\|\cdot\|_{M_t(\Omega)}$ the total variation norm on $M_t(\Omega)$. Then $(M_t(\Omega), \|\cdot\|_{M_t(\Omega)}, \sigma(M_t(\Omega), C_b(\Omega)))$ is a complete Saks space where the completeness follows from $B_{\|\cdot\|_{M_t(\Omega)}}$ being $\sigma(M_t(\Omega), C_b(\Omega))$-compact by [154, Cor. 3.23(a)]. In fact, Proposition B.6(i) is fulfilled. Furthermore, we have

$$\begin{aligned} \beta_0' :=& \gamma(\|\cdot\|_{M_t(\Omega)}, \sigma(M_t(\Omega), C_b(\Omega))) = \gamma_s(\|\cdot\|_{M_t(\Omega)}, \sigma(M_t(\Omega), C_b(\Omega))) \\ =& \tau_c(M_t(\Omega), (C_b(\Omega), \|\cdot\|_\infty)). \end{aligned}$$

**Remark 6.31** Having the notion of $C$-sequential spaces as well as Example 6.30, we can now also mention that Theorem 2.16 can be extended in a more general framework. Firstly, by [106, Thm. 2.2.6], for a $\sigma$-compact, locally compact space (or a Polish space) every bi-continuous semigroup $(T(t))_{t \geq 0}$ on $C_b(\Omega)$ is both local and locally equicontinuous for the mixed topology $\gamma$. Moreover, every operator semigroup on $C_b(\Omega)$ which is strongly continuous and locally equicontinuous with respect to $\gamma$ is bi-continuous.

The above-mentioned connection can even be more generalized to spaces which are not necessarily $C_b(\Omega)$. In fact, by [150, Thm. 7.4] and [154, Prop. 3.16 and Thm. 3.17] one obtains that on a bi-admissible space $(X, \|\cdot\|, \tau)$ such that $(X, \gamma)$ and $C$-sequential and $\gamma = \gamma_s$ the following statements are equivalent:

(i)   $(T(t))_{t \geq 0}$ is bi-continuous on $X$ with respect to $\tau$.
(ii)  $(T(t))_{t \geq 0}$ is strongly continuous and locally equicontinuous with respect to $\gamma$.
(iii) $(T(t))_{t \geq 0}$ is local and strongly continuous with respect to $\gamma$.

Let us consider an example for an application of Theorem 6.28, namely, the multiplication operator on $C_b(\Omega)$, which we will revisit for other generation results and which we will also consider later in Sect. 5.2.2.

**Example 6.32** Let $\Omega$ be a Hausdorff $k_{\mathbb{R}}$-space and $q : \Omega \to C$ be continuous with $C := \sup_{x \in \Omega} \mathrm{Re}(q(x)) < \infty$. We define the multiplication operator $(M_q, D(M_q))$ by setting

$$D(M_q) := \{f \in C_b(\Omega) \mid qf \in C_b(\Omega)\}$$

and $M_q := qf$ for $f \in D(M_q)$. By solving the equation $(\lambda - q)f = g$ we can compute the resolvent $R(\lambda, M_q)$ of $M_q$ explicitly by

$$R(\lambda, M_q)f = \frac{1}{\lambda - q}f, \quad f \in C_b(\Omega),$$

for all $\lambda \in (C \setminus \overline{q(\Omega)}) = \rho(M_q)$, which shows that $\lambda - M_q$ is surjective, i.e., $\mathrm{Ran}(\lambda - M_q) = C_b(\Omega)$, for all $\lambda \in C \setminus \overline{q(\Omega)}$. Suppose that $C \leq 0$. Then $(0, \infty) \in \rho(M_q)$ and $\mathrm{Ran}(\lambda - M_q) = C_b(\Omega)$ for all $\lambda > 0$. Furthermore, we have for all $\lambda > 0$, $f \in C_b(\Omega)$ and $\nu \in \mathcal{V}$ from Example 6.30(a) that

$$|R(\lambda, M_q)f|_\nu = \sup_{x \in \Omega} \frac{1}{|\lambda - q(x)|}|f(x)|\nu(x) \underset{C \leq 0}{\leq} \sup_{x \in \Omega} \frac{1}{\lambda - \mathrm{Re}(q(x))}|f(x)|\nu(x)$$

$$\leq \frac{1}{\lambda} \sup_{x \in \Omega} |f(x)|\nu(x) = \frac{1}{\lambda}|f|_\nu.$$

Therefore $(M_q, D(M_q))$ is $\Gamma_{\beta_0}$-dissipative for the directed system of seminorms $\Gamma_{\beta_0} := (|\cdot|_\nu)_{\nu \in \mathcal{V}}$ that generates the mixed topology $\beta_0 = \gamma(\|\cdot\|_\infty, \tau_{\mathrm{co}})$. Moreover, due to Proposition 6.26(b) and $\mathrm{Ran}(\lambda - M_q) = C_b(\Omega)$ for all $\lambda > 0$ the operator $(M_q, D(M_q))$ is $\beta_0$-closed and thus generates a $\beta_0$-strongly continuous, $\beta_0$-equicontinuous semigroup $(T(t))_{t \geq 0}$ on $C_b(\Omega)$ by Theorem 6.28(a) and Example 6.30(a). Choosing $\mathcal{V}_1 := \{\nu \in \mathcal{V} \mid \forall x \in \Omega : \nu(x) \leq 1\}$ instead of $\mathcal{V}$, we get a norming directed system of continuous seminorms that generates $\beta_0$ for which $(M_q, D(M_q))$ is dissipative, too. Hence $(T(t))_{t \geq 0}$ is also a contraction semigroup by Theorem 6.28(b) $\beta_0 = \gamma(\|\cdot\|_\infty, \tau_{\mathrm{co}}) = \gamma_s(\|\cdot\|_\infty, \tau_{\mathrm{co}})$ by Example 6.30(a).

We close this section with formulating the general version of the Lumer–Phillips-type generation theorem. This result is a refinement of Theorem 6.27 and also covers [62, Prop. 3.11].

**Proposition 6.33** *Let $(X, \|\cdot\|, \tau)$ be a sequentially complete bi-admissible space, $\upsilon$ a Hausdorff locally convex topology on $X$ with $\tau \subseteq \upsilon \subseteq \tau_{\|\cdot\|}$ such that $\gamma$-convergent sequences are $\upsilon$-convergent, and $(A, D(A))$ bi-densely defined. Then the following assertions are equivalent:*

(a) $(A, D(A))$ generates a bi-continuous contraction semigroup $(T(t))_{t\geq 0}$ with respect to $\tau$ on $X$ and there exists a norming directed system of continuous seminorms $\Gamma_v$ that generates $v$ such that $p(T(t)x) \leq p(x)$ for all $t \geq 0$, $p \in \Gamma_v$ and $x \in X$.

(b) $\lambda - A$ is surjective for some $\lambda > 0$ and $(A, D(A))$ is a $\Gamma_v$-dissipative operator on $X$ for some norming directed system of continuous seminorms $\Gamma_v$ that generates $v$.

**Proof** (a)$\Rightarrow$(b): First, we show that $(A, D(A))$ is $\Gamma_v$-dissipative. We note that $(0, \infty) \subseteq \rho(A)$ and

$$R(\lambda, A)x = \int_0^\infty e^{-\lambda t} T(t)x\,dt$$

for all $\lambda > 0$ and $x \in X$ by Definition 3.13 and Theorem 3.14 where the integral is an improper $\tau$-Riemann integral. The sequence of Riemann sums that approximate the integral on the right-hand side with respect to $\tau$ are $\|\cdot\|$-bounded for each $\lambda > 0$ and $x \in X$. Due to Proposition B.4 this means that this sequence of Riemann sums is actually $\gamma$-convergent and thus $v$-convergent by assumption. Therefore we have for all $\lambda > 0$, $p \in \Gamma_v$, and $x \in X$ that

$$p\left(R(\lambda, A)x\right) \leq \int_0^\infty e^{-\lambda t} p(T(t)x)\,dt \leq \int_0^\infty e^{-\lambda t} p(x)\,dt = \frac{1}{\lambda}p(x)$$

where we used that $p$ is $v$-continuous for the first inequality. Hence $(A, D(A))$ is $\Gamma_v$-dissipative. In combination with Theorem 6.27 this yields that $\lambda - A$ is surjective for some $\lambda > 0$.

(b)$\Rightarrow$(a): Due to Theorem 6.27, $(A, D(A))$ generates a bi-continuous contraction semigroup with respect to $\tau$ on $X$, and $(0, \infty) \subseteq \rho(A)$ by Proposition 6.26(c). Furthermore, we have by the Post–Widder inversion formula, cf. Corollary 6.9 that

$$T(t)x = \tau\lim_{n\to\infty} \left(\frac{n}{t} R\left(\frac{n}{t}, A\right)\right)^n x$$

for all $t > 0$ and $x \in X$. As a consequence of Remark 6.25 the $\tau$-convergent sequence $((\frac{n}{t}R(\frac{n}{t}, A))^n x)_{n\in\mathbb{N}}$ is $\|\cdot\|$-bounded for each $t > 0$ and $x \in X$, thus $\gamma$-convergent by Proposition B.4 and so $v$-convergent by assumption. We deduce from (6.4.1) that for all $t > 0$, $p \in \Gamma_v$ and $x \in X$ it holds that

$$p(T(t)x) = \lim_{n\to\infty} \left(\frac{n}{t}\right)^n p\left(R\left(\frac{n}{t}, A\right)^n x\right) \underset{(6.4.1)}{\leq} p(x)$$

where we used that $p$ is $v$-continuous for the first equality. Further, for $t = 0$ we have $p(T(t)x) = p(x)$. We conclude that statement (a) holds. $\qquad\square$

## Notes on This Chapter

The original work on the approximation theorem was done by H.F. Trotter [225] and T. Kato [143, Chap. IX, Thm. 3.6]. These results can also be found in the monograph by Engel and Nagel; see [101, Chap. III, Thm. 4.8 and Thm. 4.9]. Another version of the Trotter–Kato approximation theorems, which is applicable in the numerical approximation of solutions to partial differential equations, can be found in [135]. Applications, as well as extensions of the results from the previously mentioned works, can also be found in [54].

It is important to mention that the results on approximations, presented in Sects. 6.1–6.3 including the examples in Sects. 6.3.1 and 6.3.2, are taken from the thesis by F. Kühnemund [159]. In fact, the above-mentioned sections from the book correspond to [159, Chap. 2] as well as [159, Sects. 3.2 and 3.3.2]. After that A. Albanese and E. Mangino [8] contributed with a more general approach to the Trotter–Kato approximation theorems for bi-continuous semigroups. An application of the Lie–Trotter approximation formula is given in [161]. Furthermore, in the context of locally convex spaces, Trotter–Kato approximation theorems were established here [2].

The original Lumer–Phillips generation theorem for strongly continuous semigroups on Banach spaces goes back to the work of Lumer and Phillips [173]. The first attempt to develop a Lumer–Phillips-type generation theorem was by Budde and Wegner [62]. However, the more general version has been proven by Kruse and Seifert [156]. In this book, for the readers sake and in order to present the state of the art we stick to the latter one. Moreover, in [156] the authors do even more than presented here. In fact, they also deal with semi-reflexive spaces and $\gamma$-dual operators in order to cover some more examples. The interested reader is of course more than welcome to read through [156] in addition.

# Bounded and Miyadera–Voigt Perturbations

**7**

In this chapter, we will cover two standard types of perturbations, namely, bounded perturbations and Miyadera–Voigt perturbations. As a first step towards the perturbation theory for bi-continuous semigroups, we consider bounded perturbations $B \in \mathscr{L}(X)$. It will turn out that an additional property of the perturbing operator $B$ is needed. We will establish a positive result on bounded perturbations and investigate their properties. We also give a characterization of semigroups that are "linearly close" to a preliminary given bi-continuous semigroup.

## 7.1 Admissibility Space

Let $(T(t))_{t \geq 0}$ be a bi-continuous semigroup on a bi-admissible space $(X, \| \cdot \|, \tau)$. Moreover, let $B \in \mathscr{L}(X)$ and suppose that it is in addition $\tau$-sequentially continuous on $\|\cdot\|$-bounded sets. For $t_0 > 0$ we define a new space $\mathfrak{X}_{t_0}$ by

$$
\mathfrak{X}_{t_0} := \left\{ F : [0, t_0] \to \mathscr{L}(X) \,\middle|\, \begin{array}{l} F \text{ is } \tau\text{-strongly continuous, norm bounded} \\ \text{and } \{F(t) : t \in [0, t_0]\} \text{ is bi-equicontinuous} \end{array} \right\}. \tag{7.1.1}
$$

The following results tell us more about the geometric structure of $\mathfrak{X}_{t_0}$.

**Lemma 7.1** *Let $t_0 > 0$. Then $\mathfrak{X}_{t_0}$ is a Banach space if equipped with the norm given by*

$$
\|F\| := \sup_{t \in [0, t_0]} \|F(t)\|, \quad F \in \mathfrak{X}_{t_0}.
$$

© The Author(s), under exclusive license to Springer Nature Switzerland AG 2026
C. Budde, *Bi-Continuous Operator Semigroups*, Frontiers in Mathematics,
https://doi.org/10.1007/978-3-032-12948-2_7

***Proof*** We show that $\mathfrak{X}_{t_0}$ is a closed subspace of $\mathrm{B}\left([0, t_0], \mathcal{L}(X)\right)$, the space of bounded functions from $[0, t_0]$ to $\mathcal{L}(X)$ endowed with the supremum norm. Let $(F_n)_{n \in \mathbb{N}}$ be a sequence in $\mathfrak{X}_{t_0}$ converging to $F \in \mathrm{B}\left([0, t_0], \mathcal{L}(X)\right)$. Let $x \in X$ be arbitrary, then $F_n(t)x \to F(t)x$ with respect to $\tau$ uniformly for $t \in [0, t_0]$. To see this, let $\mathcal{P}$ denote the family of seminorms generating $\tau$. Then for $p \in \mathcal{P}$

$$p(F_n(t)x - F(t)x) \leq \|F_n(t)x - F(t)x\| \leq \|F_n - F\| \cdot \|x\| \to 0.$$

Hence, $F : [0, t_0] \to \mathcal{L}(X)$ is strongly $\tau$-continuous. Since the norm boundedness is obvious, it is left to show that the set $\{F(t) : t \in [0, t_0]\}$ is bi-equicontinuous. To do so, let $(x_n)_{n \in \mathbb{N}}$ be any norm-bounded $\tau$-null sequence and choose $\varepsilon > 0$ arbitrary. Choose $m \in \mathbb{N}$ such that $\|F_m - F\| < \frac{\varepsilon}{2}$ and observe that for $p \in \mathcal{P}$

$$p(F(t)x_n) \leq p(F(t)x_n - F_m(t)x_n) + p(F_m(t)x_n) < \frac{\varepsilon}{2} + \frac{\varepsilon}{2} = \varepsilon,$$

for $n$ sufficiently large and by the bi-equicontinuity of $\{F_m(t) : t \in [0, t_0]\}$.  $\square$

The space $\mathfrak{X}_{t_0}$ is not only a vector space, but also admits an algebra structure, which makcs $\mathfrak{X}_{t_0}$ a Banach algebra.

**Lemma 7.2** *Let $t_0 > 0$. Then $\mathfrak{X}_{t_0}$ is a Banach algebra.*

***Proof*** Firstly, we have to show that if $F, G \in \mathfrak{X}_{t_0}$, then $F \cdot G \in \mathfrak{X}_{t_0}$. Let $x \in X$ and take $0 < t < t_0$ and a sequence $(h_n)_{n \in \mathbb{N}}$ in $\mathbb{R}$ such that $h_n \to 0$. Then

$$\begin{aligned} &F(t + h_n)G(t + h_n)x - F(t)G(t)x \\ =\,&F(t + h_n)G(t + h_n)x - F(t + h_n)G(t)x + F(t + h_n)G(t)x - F(t)G(t)x. \end{aligned}$$

Clearly, one has that

$$F(t + h_n)G(t)x - F(t)G(t)x \to 0,$$

with respect to the locally convex topology $\tau$. Since $\{F(t) : t \in [0, t_0]\}$ is bi-equicontinuous by assumption, one obtains

$$F(t + h_n)G(t + h_n)x - F(t + h_n)G(t)x = F(t + h_n)\left(G(t + h_n) - G(t)\right)x \to 0$$

with respect to $\tau$. The left (respectively, the right) $\tau$-strong continuity of $F \cdot G$ in the endpoints of $[0, t_0]$ can be proved analogously. The $\|\cdot\|$-boundedness of the map $t \mapsto T(t)S(t)$ is obvious. It remains to show that the family $\{F(t)G(t) : t \in [0, t_0]\}$ is bi-equicontinuous. We prove this by contradiction: assume that there exists a $\|\cdot\|$-bounded $\tau$-null sequence $(x_n)_{n \in \mathbb{N}}$ in $X$, an $\varepsilon > 0$ and $p \in \mathcal{P}$ such that for all $n \in \mathbb{N}$ there exists $t_n \in [0, t_0]$ such that $p(F(t_n)G(t_n)x_n) > \varepsilon$. However, since $(G(t_n)x_n)_{n \in \mathbb{N}}$ is $\|\cdot\|$-bounded and a $\tau$-null sequence by the bi-equicontinuity of $G$, we conclude by the bi-equicontinuity of $F$ that $p(F(t_n)G(t_n)x_n) \to 0$. Contradiction.                                                          $\square$

The bi-equicontinuity of the operator family $\{T(t) : t \in [0, t_0]\}$ appearing in Definition 2.4 is an important assumption, for example, to establish the results in Chap. 3. For unbounded perturbations, we shall require more as introduced below. Clearly, if $(T(t))_{t \geq 0}$ is a locally bi-equicontinuous function, then each of the operators $T(t)$, $t \geq 0$, has to be sequentially $\tau$-continuous on $\|\cdot\|$-bounded sets. The following definition adjusts this notion of equicontinuity to our needs.

**Definition 7.3** A $\|\cdot\|$-bounded linear operator $B$ is called *local* if for all $p \in \mathcal{P}$ and $\varepsilon > 0$ there exists $K > 0$ and $q \in \mathcal{P}$ such that for all $x \in X$ one has that

$$p(Bx) \leq Kq(x) + \varepsilon \|x\| . \tag{7.1.2}$$

**Definition 7.4** A family $\{F(t) : t \in [0, t_0]\}$ of bounded linear operators is called *local* if for all $p \in \mathcal{P}$ and $\varepsilon > 0$ there exists $K > 0$ and $q \in \mathcal{P}$ such that for all $t \in [0, t_0]$ and all $x \in X$ one has that

$$p(F(t)x) \leq Kq(x) + \varepsilon \|x\| . \tag{7.1.3}$$

We denote the set of all local functions in $\mathfrak{X}_{t_0}$, as defined by (7.1.1), by $\mathfrak{X}_{t_0}^{\mathrm{loc}}$.

**Lemma 7.5** $\mathfrak{X}_{t_0}^{\mathrm{loc}}$ *is a closed subspace of* $\mathfrak{X}_{t_0}$.

**Proof** Let $(F_n)_{n \in \mathbb{N}}$ be a sequence in $\mathfrak{X}_{t_0}^{\mathrm{loc}}$ such that $F_n \to F$ with respect to the norm of $\mathfrak{X}_{t_0}$ as defined in Lemma 7.1 as $n \to \infty$. Let $p \in \mathcal{P}$ and $\varepsilon > 0$ be arbitrary. Fix $x \in X$ and find $N \in \mathbb{N}$ such that $\|F - F_n\| < \frac{\varepsilon}{2}$ for all $n \geq N$. Fix such a $n \geq N$ and find $K > 0$ and $q \in \mathcal{P}$ such that $p(F_n(t)x) \leq Kq(x) + \frac{\varepsilon}{2} \|x\|$ for all $t \in [0, t_0]$. Then

$$\begin{aligned}
p(F(t)x) &\leq p(F(t)x - F_n(t)x) + p(F_n(t)x) \\
&\leq \|F(t)x - F_n(t)x\| + p(F_n(t)x) \\
&\leq \|F - F_n\| \cdot \|x\| + p(F_n(t)x) \\
&\leq \left( \|F - F_n\| + \frac{\varepsilon}{2} \right) \|x\| + Kq(x),
\end{aligned}$$

showing that $F$ is local.                                                                    $\square$

The following result yields an important example of a local family of operators.

**Lemma 7.6** *Let $(T(t))_{t\geq 0}$ be a local bi-continuous semigroup on a bi-admissible space $(X, \|\cdot\|, \tau)$ with generator $(A, D(A))$. Moreover, let $\omega < \alpha < \beta$, then $\{R(\lambda, A) : \lambda \in [\alpha, \beta]\}$ is local.*

***Proof*** By Lemma 3.10 and Theorem 3.14 we know that the resolvent can be expressed by means of the Laplace transform (with respect to the topology $\tau$). Let $\lambda \in [\alpha, \beta]$, $\varepsilon > 0$, and $p \in \mathcal{P}$ be arbitrary. There exists $q \in \mathcal{P}$ such that $p(T(t)x) \leq Mq(x) + \varepsilon \|x\|$ for $t \in [0, t_0]$ and $x \in X$. Hence

$$
\begin{aligned}
p(R(\lambda, A)x) &\leq \int_0^\infty e^{-\lambda s} p(T(s)x)\, ds \\
&= \int_0^t e^{-\lambda s} p(T(s)x)\, ds + \int_t^\infty e^{-\lambda s} p(T(s)x)\, ds \\
&\leq \int_0^t e^{-\lambda s} Mq(x)\, ds + \int_0^t e^{-\lambda s} \varepsilon \|x\|\, ds + \int_t^\infty e^{-\lambda s} Me^{\omega s} \|x\|\, ds \\
&= \frac{e^{-\lambda t} - 1}{\lambda}(Mq(x) + \varepsilon \|x\|) + M \frac{e^{(\omega - \lambda)t}}{(\omega - \lambda)} \|x\| \\
&\leq Kq(x) + \varepsilon \|x\|,
\end{aligned}
$$

by taking $t > 0$ large enough and where $K > 0$ is an appropriate constant. $\qquad\square$

An improved version of Lemma 7.6 can be found [154, Prop. 3.6].

## 7.2   Bounded Perturbations

For a given bi-continuous semigroup $(T(t))_{t\geq 0}$ on a bi-admissible space $(X, \|\cdot\|, \tau)$ we introduce on $\mathfrak{X}_{t_0}$ the associated *Volterra operator* $V$ as follows:

$$
(VF)(t)x := \int_0^t T(t - s)BF(s)x\, ds, \quad t \in [0, t_0], \ F \in \mathfrak{X}_{t_0}, \ x \in X. \tag{7.2.1}
$$

This integral exists within the topology $\tau$ since $t \mapsto T(t - s)BF(s)$ is $\tau$-continuous by assumption and by an application of Lemma 7.2. The following result shows that $V$ yields a bounded operator on $\mathfrak{X}_{t_0}$.

**Lemma 7.7** *The Volterra operator defined by (7.2.1) is a bounded linear operator on $\mathfrak{X}_{t_0}$ and has spectral radius $r(V) = 0$.*

***Proof*** We have to show that $V$ maps $\mathfrak{X}_{t_0}$ into $\mathfrak{X}_{t_0}$. To do so, let $F \in \mathfrak{X}_{t_0}$, $t \in [0, t_0]$, and $x \in X$. Since $(T(t))_{t\geq 0}$ is a bi-continuous semigroup there exists $M > 0$ such that $\sup_{t\in[0,t_0]} \|T(t)\| \leq M$. Using that one observes

$$\|(VF)(t)x\| = \sup_{\substack{\varphi \in (X,\tau)' \\ \|\varphi\| \le 1}} \left| \varphi \left( \int_0^t T(t-s)BF(s)x \, ds \right) \right|$$

$$\le \sup_{\substack{\varphi \in (X,\tau)' \\ \|\varphi\| \le 1}} \int_0^t |\varphi(T(t-s)BF(s)x)| \, ds$$

$$\le M \, \|B\| \cdot \|F\| \cdot \|x\| ,$$

showing that $(VF)(t) \in \mathcal{L}(X)$. Next, we show that $\{(VF)(t) : t \in [0, t_0]\}$ is bi-equicontinuous. Therefore, let $(x_n)_{n \in \mathbb{N}}$ be a $\|\cdot\|$-bounded $\tau$-null sequence. Take $p \in \mathcal{P}$ and $\varepsilon > 0$ and observe that by Lemma 7.2 one obtains bi-equicontinuity of the set $\{T(t-s)BF(s) : s \in [0, t_0]\}$ and hence for sufficient large $n$ one gets

$$p((VF)(t)x_n) = p\left( \int_0^t T(t-s)BF(s)x_n \, ds \right) \le \int_0^t p(T(t-s)BF(s)x_n) \, ds \le t_0 \varepsilon.$$

Obviously, $(VF)(\cdot)$ is $\|\cdot\|$-bounded and $\tau$-strongly continuous, showing that $VF \in \mathfrak{X}_{t_0}$.

The spectral radius $r(V)$ can be computed by following [101, Chapter III, Lemma 1.9]. In fact, for $F \in \mathfrak{X}_{t_0}, x \in X$, and $t \in [0, t_0]$ we obtain with $M := \sup_{t \in [0, t_0]} \|T(t)\|$ the following estimate:

$$\|(VF)(t)x\| \le \int_0^t \|T(t-s)BF(s)x\| \, ds \le M \cdot \|B\| \cdot \|F\| \cdot tx.$$

By an induction argument, it follows that

$$\|V^n F\| \le \frac{(t_0 M \|B\|)^n}{n!} \|F\| , \tag{7.2.2}$$

for every $n \in \mathbb{N}$, showing that $r(V) = 0$.  $\square$

**Remark 7.8**  From Lemma 7.7 we conclude that $1 \in \rho(V)$ and that

$$R(1, V) = \sum_{n=0}^{\infty} V^n,$$

where the series converges in the operator norm of $\mathcal{L}(\mathfrak{X}_{t_0})$.

**Theorem 7.9**  *Let $(T(t))_{t \ge 0}$ be a bi-continuous semigroup on a bi-admissible space $(X, \|\cdot\|, \tau)$ with generator $(A, D(A))$ and suppose that $B \in \mathcal{L}(X)$ is $\tau$-sequentially continuous on $\|\cdot\|$-bounded sets. Then $(A + B, D(A))$ is the generator of a bi-continuous semigroups $(S(t))_{t \ge 0}$. Moreover, the semigroup $(S(t))_{t \ge 0}$ is given by the Dyson–Phillips series*

$$S(t) = \sum_{n=0}^{\infty} T_n(t), \quad t \geq 0, \tag{7.2.3}$$

*with*

$$T_0(t) = T(t), \quad T_n(t) = \int_0^t T(t-s) B T_{n-1}(s) \, ds, \quad n > 0,$$

*where the integral is understood in the $\tau$-strong topology and the series is uniformly $\|\cdot\|$-convergent on compact intervals.*

**Proof** Let $t > 0$ be arbitrary and let $V_B$ the abstract Volterra operator on $\mathfrak{X}_{t_0}$ as defined by (7.2.1). We set

$$S_{t_0}(t) := \delta_t(R(1, V_B) T_{|[0,t_0]}), \quad t \in [0, t_0],$$

and observe that by Lemma 7.7 and Remark 7.8 one has that

$$S_{t_0}(t) = \sum_{n=0}^{\infty} \left[ V_B^n T \right](t), \quad t \in [0, t_0]. \tag{7.2.4}$$

It is obvious that $S_{t_0}(t)x = S_{t_0'}(t)x$ whenever $x \in X$ and $t \leq t_0' \leq t_0$. Hence, we are able to define

$$S(t) := \begin{cases} I, & t = 0, \\ S_t(t), & t > 0. \end{cases}$$

We first show that $(S(t))_{t \geq 0}$ is a semigroup. In what follows we abbreviate $T_{|[0,t_0]}$ by $T$. Let $0 \leq s, t \leq s + t \leq t_0$ and observe that by (7.2.4) one has

$$S_{t_0}(t) S_{t_0}(s) = \sum_{n=0}^{\infty} \left[ V_B^n T \right](t) \cdot \sum_{n=0}^{\infty} \left[ V_B^n T \right](s) = \sum_{n=0}^{\infty} \sum_{k=0}^{n} \left[ V_B^k T \right](t) \left[ V_B^{n-k} T \right](s),$$

whereas the second equality holds since the series converges in the operator norm. In order to show that $S_{t_0}(t) S_{t_0}(s) = S_{t_0}(t + s)$ we observe that it remains to show that

$$\sum_{k=0}^{n} \left[ V_B^k T \right](t) \left[ V_B^{n-k} T \right](s) = \left[ V_B^n T \right](t + s). \tag{7.2.5}$$

To show that (7.2.5) holds, we use induction on $n \in \mathbb{N}$. Obviously, (7.2.5) holds for $n = 0$. Now, assume that (7.2.5) holds for some $n \in \mathbb{N}$, then

$$\left[V_B^{n+1}T\right](t+s) = \int_0^{t+s} T(t+s-r)B\left[V_B^nT\right](r)x\,\mathrm{d}r$$

$$= \int_0^s T(s-r)B\left[V_B^nT\right](t+r)x\,\mathrm{d}r + \int_s^{s+t} T(r)B\left[V_B^nT\right](t+s-r)BT(r)x\,\mathrm{d}r$$

$$= \int_0^s T(s-r)B\left[V_B^nT\right](t+r)x\,\mathrm{d}r + \int_0^t T(s+r)B\left[V_B^nT\right](t-r)x\,\mathrm{d}r$$

$$= \int_0^s T(s-r)B\sum_{k=0}^{n}\left[V_B^kT\right](r)\left[V_B^{n-k}T\right](t)x\,\mathrm{d}r + T(s)\left[V_B^{n+1}T\right](t)$$

$$= \sum_{k=0}^{n}\left[V_B^kT\right](s)\left[V_B^{n-k}T\right](r) + T(s)\left[V_B^{n+1}T\right](t)$$

$$= \sum_{k=0}^{n+1}\left[V_B^kT\right](s)\left[V_B^{n+1-k}T\right](t),$$

which completes the induction. Since $S_{t_0} = S_{|[0,t_0]} \in \mathfrak{X}_{t_0}$ we see immediately that $(S(t))_{t\geq0}$ is a bi-continuous semigroup and from the definition it is straightforward that $T_n(t) = \left(V_B^nT_{|[0,t_0]}\right)(t)$ and hence (7.2.3) is satisfied. The uniform convergence on compact intervals follows from the continuity of $\delta_t$.

In the last step, we will show that the generator of $(S(t))_{t\geq0}$ is given by $(A+B, D(A))$. In what follows, we will denote the generator of the semigroup $(S(t))_{t\geq0}$ by $(C, D(C))$. First of all, we observe that since $(A, D(A))$ is a Hille–Yosida operator also $(A+B, D(A))$ is a Hille–Yosida operator by [17, Thm. 3.5.5], in particular, $\rho(A+B) \neq \varnothing$. For arbitrary $x \in X$ we have

$$\frac{[V_BT](h)x}{h} = \frac{1}{h}\int_0^h T(h-s)BT(s)x\,\mathrm{d}s.$$

Let $p \in \mathcal{P}$ and $\varepsilon > 0$, then by the $\tau$-continuity of $s \mapsto T(s)Bx$ there exists $\delta \in (0, t_0)$ such that

$$p(T(h-s)Bx - Bx) < \frac{\varepsilon}{2},$$

whenever $h \in [0, \delta]$ and $s \in [0, h]$. Further, by taking $\delta > 0$ possibly smaller, we obtain by the bi-equicontinuity of the set $\{T(s) : s \in [0, t_0]\}$ and the $\tau$-continuity of $s \mapsto BT(s)x$ that

$$p\left(\frac{1}{h}\int_0^h T(h-s)BT(s)x\,\mathrm{d}s - Bx\right)$$

$$\leq \frac{1}{h}\int_0^h p(T(h-s)Bx - Bx)\,\mathrm{d}s + \frac{1}{h}\int_0^h p(T(h-s)(BT(s)x - Bx))\,\mathrm{d}s < \frac{\varepsilon}{2} + \frac{\varepsilon}{2},$$

showing that

$$\frac{[V_B T]\,(h)x}{h} \xrightarrow{\tau} Bx. \tag{7.2.6}$$

Furthermore, for $x \in D(A)$ one has

$$\frac{S(h)x - x}{h} = \frac{T(h)x - x}{h} + \frac{[V_B T]\,(h)x}{h} + \sum_{n=2}^{\infty} \frac{\left[V_B^n T\right](h)x}{h}. \tag{7.2.7}$$

By making use of (7.2.2) we obtain

$$\left\| \sum_{n=2}^{\infty} \frac{\left[V_B^n T\right](h)x}{h} \right\| \le h \sum_{n=2}^{\infty} \frac{h^{n-2}\left(\|T(t)\|\,\|B\|\right)^n \|x\|}{n!} \le hK\,\|x\|,$$

for all $x \in X$ and some constant $K > 0$, showing that the last term in (7.2.7) vanishes if one takes the $\tau$-limit for $h \to 0$. All other terms on (7.2.7) stay bounded as $h \to 0$. Putting (7.2.6) and (7.2.7) together yields that

$$\frac{S(h)x - x}{h} \xrightarrow{\tau} Ax + Bx,$$

so that $A + B \subseteq C$. This together with the above remark on the non-empty resolvent set of $A + B$ implies that $A + B = C$. $\qquad\square$

As a consequence of Theorem 7.9, we can formulate the following result which is a consequence of the definition of the weak*-topology, see Sect. 2.1.2. This result will be applicable in Chap. 12.

**Proposition 7.10** *Let $(X, \|\cdot\|)$ be a Banach space and assume that $(T(t))_{t \ge 0}$ is a bi-continuous semigroup on the bi-admissible space $(X', \|\cdot\|_{X'}, \tau_{w*})$ with generator $(A, D(A))$. Then for every $B \in \mathscr{L}(X)$, the operator $(A + B', D(A))$ is the generator of a bi-continuous semigroup on the bi-admissible space $(X', \|\cdot\|_{X'}, \tau_{w*})$.*

## 7.2.1　Counterexamples

Next we will show that the assumption that $B$ is $\tau$-sequentially continuous cannot be dropped in general in Theorem 7.9. In particular, we show that the perturbation of the generator of a bi-continuous semigroup by a norm-bounded operator does not generate a bi-continuous semigroup in general. In fact, it suffices to show that a norm-bounded operator is not necessarily the generator of a bi-continuous semigroup. We will use the fact that whenever $(A, D(A))$ generates a $C_0$-semigroup and a bi-continuous semigroup as well, then the two semigroups must coincide.

**Example 7.11** Consider the Banach space $X := C([0, 1])$ and its dual $X' = \mathcal{M}([0, 1])$, the space of all complex Borel measures on the unit interval. On $X'$ we consider the operator given by

$$(B\mu)(\Omega) := \mu\left((\Omega - \tfrac{3}{4}) \cap [0, 1]\right), \quad \mu \in \mathcal{M}([0, 1]),$$

where $\Omega$ is a Borel subset of $[0, 1]$. Then $B$ is a norm-continuous operator on $\mathcal{M}([0, 1])$, and clearly $B^2 = 0$. Now, consider the sequence of Dirac measures $(\delta_{\frac{3}{4} - \frac{1}{n+1}})_{n \in \mathbb{N}}$. This sequences converges to $\delta_{\frac{3}{4}}$ with respect to the weak*-topology. Furthermore, we observe that $B\delta_{\frac{3}{4} - \frac{1}{n+1}} = 0$, whereas we have $B\delta_{\frac{3}{4}} = \delta_0$, showing that $B$ is not continuous for the weak*-topology. The $C_0$-semigroup $(T(t))_{t \geq 0}$ generated by $B$ is given by

$$T(t) = I + tB, \quad t \geq 0,$$

showing that $(T(t))_{t \geq 0}$ is not bi-continuous with respect to the weak*-topology. In particular, we showed that not every bounded operator on a dual space is a generator of a weak*-bi-continuous semigroup.

**Example 7.12** Let $X := C_b(\mathbb{R})$ and denote by $\beta\mathbb{R}$ the Stone–Čech compactification of $\mathbb{R}$. Take $x \in \beta\mathbb{R} \setminus \mathbb{R}$ and consider the operator $B := \mathbf{1} \oplus \delta_x$ on $X$. Then $B$ is a contractive projection by construction. As a matter of fact, the operator $B$ is not strongly continuous with respect to the compact-open topology $\tau_{co}$ on $C_b(\mathbb{R})$. Indeed, let $(f_n)_{n \in \mathbb{N}}$ be a sequence of continuous functions such that $f_{n|[-n,n]} = 1$ and $\mathrm{supp}(f_n) \subseteq [-n - 1, n + 1]$. Then $f_n \to \mathbf{1}$ with respect to $\tau_{co}$ but $Bf_n = 0$ while $B\mathbf{1} = \mathbf{1}$, hence $B$ is not continuous with respect to the compact-open topology $\tau_{co}$. Moreover, the $C_0$-semigroup $(T(t))_{t \geq 0}$ generated by $B$ is, due to the fact that $B$ is idempotent, given by

$$T(t) = \sum_{n=0}^{\infty} \frac{(tB)^n}{n!} = I + \sum_{n=1}^{\infty} \frac{t^n}{n!} B = I - B + e^t B, \quad t \geq 0.$$

This implies that $(T(t))_{t \geq 0}$ is not sequentially $\tau_{co}$-continuous on bounded sets unless $t = 0$. Therefore, $B$ is not the generator of a bi-continuous semigroup with respect to the compact-open topology $\tau_{co}$.

**Example 7.13** Consider again the Banach space of bounded continuous functions $X = C_b(\mathbb{R})$. We claim that there exists $\varphi \in X'$ such that $\|\varphi\| = 1$, $\varphi(\mathbf{1}) = 1$ and $\varphi(f(\cdot + r)) = \varphi(f)$ and all $f \in C_b(\mathbb{R})$ and $r \in \mathbb{R}$. To see that such a $\varphi \in X'$ indeed exists, define for $t > 0$ a linear functional on $X$ by

$$\varphi_t(f) := \frac{1}{2t} \int_{-t}^{t} f(s)\, ds, \quad f \in C_b(\mathbb{R}).$$

Obviously, $\varphi \in X'$ and $\|\varphi_t\| = 1$. Hence, the sequence $(\varphi_n)_{n \in \mathbb{N}}$ is relatively weakly*-compact, thus it has a weak*-accumulation point $\varphi \in X'$. That $\|\varphi\| = 1$ and $\varphi(\mathbf{1}) = 1$ are satisfied follows by construction. That the last property is satisfied, i.e., $\varphi(f(\cdot + r)) = \varphi(f)$ and all $f \in C_b(\mathbb{R})$ and $r \in \mathbb{R}$ can be seen as follows:

$$\frac{1}{2n}\int_{-n}^{n} f(s+r)\,dr = \frac{1}{2n}\int_{-n+r}^{n+r} f(s)\,ds = \frac{1}{2n}\int_{-n}^{n} f(s)\,ds + \frac{1}{2n}\int_{-n+r}^{-n} f(s)\,ds + \frac{1}{2n}\int_{n}^{n+r} f(s)\,ds.$$

Let $\varepsilon > 0$ be arbitrary, then

$$|\langle f - f(\cdot + r), \varphi\rangle| \leq |\langle f - f(\cdot + r), \varphi - \varphi_n\rangle| + |\langle f - f(\cdot + r), \varphi_n\rangle|$$

$$\leq |\langle f - f(\cdot + r), \varphi - \varphi_n\rangle| + \frac{1}{2n}\int_{-n+r}^{-n} f(s)\,ds + \frac{1}{2n}\int_{n}^{n+r} f(s)\,ds.$$

$$(7.2.8)$$

Since the weak*-neighborhood $U$ of $\varphi$ determined by $\frac{\varepsilon}{2} > 0$ and $f - f(\cdot + r) \in C_b(\mathbb{R})$ contains an infinite sequence $\varphi_{n_k}$ we can make the first term on the right-hand side of (7.2.8) smaller than $\frac{\varepsilon}{2}$. Moreover,

$$\left|\frac{1}{2n}\int_{-n+r}^{-n} f(s)\,ds + \frac{1}{2n}\int_{n}^{n+r} f(s)\,ds\right| \leq \frac{r}{n}\|f\| \to 0,$$

as $n \to \infty$, showing that the last two terms in (7.2.8) together become smaller than $\frac{\varepsilon}{2}$. This proves the third property as desired. We remark that the above properties of $\varphi$ mean that it is indeed an invariant probability measure on $\beta\mathbb{R}$ for the shift semigroup.

Taking the functional $\varphi$ as constructed above, we define a $\|\cdot\|$-bounded linear operator by $B := \mathbf{1} \oplus \varphi$. As in Example 7.12 one is able to show that $B$ is not continuous with respect to the compact-open topology $\tau_{co}$. Hence, the operator $B$ is not the generator of a bi-continuous semigroup with respect to $\tau_{co}$.

However, there holds even more. One can show that there exists a nontrivial bi-continuous semigroup $(S(t))_{t \geq 0}$ with some generator $(A, D(A))$ such that $(A + B, D(A))$ is not the generator of a bi-continuous semigroup. In Sect. 2.2.1, we saw that the translation semigroup on $C_b(\mathbb{R})$ is bi-continuous with respect to $\tau_{co}$. Denote its generator by $(A, D(A))$ (as a matter of fact, we determined the generator explicitly in Proposition 3.4). Now take the operator $B$ from above and observe that by the translation invariance of $\varphi$ the operator $B$ commutes with the operator semigroup $(S(t))_{t \geq 0}$. We define

$$T(t) := S(t)(I - B + e^t B) = (I - B + e^t B)S(t), \quad t \geq 0.$$

Then $(T(t))_{t \geq 0}$ is an operator semigroup which is strongly continuous with respect to $\tau_{co}$. Moreover, a straightforward computation shows that for all $x \in D(A)$ the orbits are differentiable with respect to $\tau_{co}$ and its derivative is given by $(A + B)T(t)x$. If $(A + B, D(A))$ would be the generator of a bi-continuous semigroup, then this semigroup has to coincide with $(T(t))_{t\geq 0}$. However, the operator semigroup $(T(t))_{t\geq 0}$ is not locally bi-equicontinuous which can be seen by taking the sequence of functions as before in Example 7.12. Therefore $(A + B, D(A))$ cannot be the generator of a bi-continuous semigroup.

## 7.3   Miyadera–Voigt Perturbations

We continue with perturbations of bi-continuous semigroups and try to relax the assumptions on the operator $B$. In the case of the bounded perturbations, cf. Theorem 7.9, the operator $B$ was defined on the whole Banach space $X$ subjected to some additional continuity assumption. Now we would like to consider perturbing operators which are defined on the domain of the generator $D(A)$ and satisfy continuity assumptions with respect to the graph topology. In the case of $C_0$-semigroups it suffices that $B \in \mathcal{L}(D(A), X)$. We will provide a similar result for bi-continuous semigroups which can be applied, for example, to perturbations of second-order differential operators with first-order terms. We try to imitate the proof of Theorem 7.9, whereas more subtle arguments are needed as $B$ is not defined for all $x \in X$ yielding discontinuities of the perturbing operator $B$ for the topology $\tau$. In order to proceed with unbounded perturbations, we need the following definition.

**Definition 7.14** Let $(X, \|\cdot\|, \tau)$ be a bi-admissible space and $D \subseteq X$ an arbitrary subset and $\eta > 1$. We say that $D$ is $\eta$-*bi-dense* in $X$ with respect to the topology $\tau$ if for all $x \in X$ there exists a sequence $(x_n)_{n\in\mathbb{N}}$ in $D$ such that $x_n \xrightarrow{\tau} x$ and $\|x_n\| \leq \eta \|x\|$ for all $n \in \mathbb{N}$.

As a matter of fact, the assumption of $\eta$-bi-denseness is not too restrictive.

**Remark 7.15** Suppose that the topology $\tau$ is metrizable, then any bi-dense set $D \subseteq X$ which contains 0 is $\eta$-bi-dense for arbitrary $\eta > 1$. Indeed, let $x \in X \setminus \{0\}$ and assume that $p_1 \leq p_2 \leq \cdots$ is co-final in $\mathcal{P}$. Let $n \in \mathbb{N}$ and choose $x_n \in D$ such that $p_n(x_n - x) < \frac{1}{n}$. Let $p \in \mathcal{P}$ and $0 < \varepsilon < \eta - 1$ and take $N \in \mathbb{N}$ such that $p \leq p_N$ and

$$p(x_n - x) \leq p_N(x_n - x) \leq p_n(x_n - x) \leq \frac{1}{n} < \varepsilon,$$

for all $n \geq N$, showing that $x_n \xrightarrow{\tau} x$. Moreover, since $p(x) \leq \|x\|$ we obtain $p(x_n) \leq \varepsilon + \|x\|$ so that $\|x_n\| \leq \eta \|x\|$ for $n \geq N$ and arbitrary $\eta > 1$.

**Proposition 7.16** *Let $(T(t))_{t\geq 0}$ be a bi-continuous semigroup on a bi-admissible space $(X, \|\cdot\|, \tau)$ with generator $(A, D(A))$. Then, for $k \in \mathbb{N}$, there exists $\eta > 1$ such that $D(A^k)$ is $\eta$-bi-dense in $X$.*

**Proof** Let $x \in X$, then by Theorem 3.15(d) one has that $\tau\text{-}\lim_{n\to\infty} n^k R(n, A)^k x = x$. Moreover, by the definition of bi-continuous semigroups, cf. Definition 2.4, there exists $M > 0$

and $\omega \in \mathbb{R}$ such that $\|T(t)\| \leq Me^{\omega t}$ for all $t \geq 0$. Moreover, by the Hille–Yosida generation theorem for bi-continuous semigroups, cf. Theorem 3.18 or Theorem 5.1, one has that

$$\left\| n^k R(n, A)^k x \right\| \leq \frac{Mn^k}{(n - \omega)^k} \|x\|,$$

for all $n > \omega$. Define a sequence $(x_n)_{n \in \mathbb{N}}$ by $x_n := n^k R(n, A)^k x, n \in \mathbb{N}$, and take any $\eta > M$, then $x_n \in D(A^k)$ for all $n \in \mathbb{N}$ and by all previous observations $x_n \xrightarrow{\tau} x$ and $\|x_n\| \leq \eta \|x\|$ for $n \in \mathbb{N}$ large enough.                                                                   $\square$

**Definition 7.17**  Let $(T(t))_{t \geq 0}$ be a bi-continuous semigroup on a bi-admissible space $(X, \| \cdot \|, \tau)$ and let $(A, D(A))$ be the generator such that $D(A)$ is $\eta$-bi-dense. Assume that $\tau$ is generated by a family of seminorms $\mathcal{P}$. The *graph topology* $\tau_A$ is defined to be the locally convex topology determined by the family of seminorms

$$\mathcal{P}_A := \{p(\cdot) + q(A\cdot) : \ p, q \in \mathcal{P}\}.$$

The following definition will be important for our perturbation result. In what follows $\|\cdot\|_A$ denotes the graph norm as defined by (4.1.1).

**Definition 7.18**  Let $Y$ be a closed subspace of $\mathfrak{X}_{t_0}$. We say that a linear operator $B : (D(A), \tau_A) \to (X, \tau)$ which is continuous on $\|\cdot\|_A$-bounded sets is *Miyadera–Voigt admissible* on $Y$, if there exists $t_0 > 0$ such that $T_{|[0,t_0]} \in Y$ and the following conditions are satisfied:

(i)  For all $x \in D(A)$ the map $s \mapsto \|BT(s)x\|, s \in [0, t_0]$, is bounded.
(ii)  The operator

$$B(F, t)x := \int_0^t F(t - s)BT(s)x \, \mathrm{d}s \tag{7.3.1}$$

defined on $D(A)$ extends to a linear operator $\overline{B}(F, t)$ on $X$ which is $\tau$-continuous on $\|\cdot\|$-bounded sets for all $t \in [0, t_0]$ and $F \in Y$. Moreover, we require that the operator $\overline{B}(F, t)$ is $\|\cdot\|$-bounded.
(iii)  The operator $V_{t_0}$ on $Y$ defined by $\left[V_{t_0} F\right](t)x := \overline{B}(F, t)x, \ x \in X, \ t \in [0, t_0]$ is a bounded operator on $Y$ and $\left\| V_{t_0} \right\| < \frac{1}{\eta}$.

We are now able to state and to prove a Miyadera–Voigt-type perturbation theorem.

**Theorem 7.19**  *Let $(T(t))_{t \geq 0}$ be a bi-continuous semigroup on a bi-admissible space $(X, \| \cdot \|, \tau)$ with generator $(A, D(A))$ whose domain $D(A)$ is $\eta$-bi-dense. Suppose that*

*B is Miyadera–Voigt admissible on* $\mathfrak{X}_{t_0}$. *Then* $(A + B, D(A))$ *generates a bi-continuous semigroup* $(S(t))_{t \geq 0}$ *on* $X$ *satisfying*

$$S(t)x = T(t)x + \int_0^t S(t - s)BT(s)x \, ds, \tag{7.3.2}$$

*for all* $x \in D(A)$ *and* $t \geq 0$.

**Proof** For the sake of simplicity we will denote the restriction of the semigroup $(T(t))_{t \geq 0}$ to $[0, t_0]$ just by $T$. Define the abstract Volterra operator $V_{t_0} \in \mathscr{L}(\mathfrak{X}_{t_0})$ as by (7.2.1). By the assumption of Definition 7.18 we have $1 \in \rho(V_{t_0})$. Let $t > 0$ and write $t = nt_0 + t_1$ for some $n \in \mathbb{N}$ and $t_1 \in [0, t_0)$. We define

$$S(t) = (\delta_{t_0}(R(1, V_{t_0})T))^n \cdot \delta_{t_1}(R(1, V_{t_0})T).$$

In the first step we show that $(S(t))_{t \geq 0}$ is a semigroup. To do so, let first of all $0 \leq s, t \leq t + s \leq t_0$. By following the lines of the proof of the bounded perturbation result, cf. Theorem 7.9, we conclude that $S(t + s) = S(t)S(s)$. Now, let $t, s > 0$ be arbitrary and find $n, m \in \mathbb{N}$ and $t_1, t_2 \in [0, t_0)$ such that $t = nt_0 + t_1$ and $s = mt_0 + t_2$. Then, by using the previous observation, one obtains

$$\begin{aligned}
S(t)S(s) &= S(t_0)^n S(t_1) S(t_0)^m S(t_2) \\
&= S(t_0)^n S(t_1) S(t_0) S(t_0)^{m-1} S(t_2) \\
&= S(t_0)^n S(t_1) S(t_0 - t_1) S(t_1) S(t_0)^{m-1} S(t_2) \\
&= \cdots \\
&= S(t_0)^n S(t_0)^m S(t_1) S(t_2) \\
&= \begin{cases} S(t_0)^{n+m} S(t_1 + t_2), & t_1 + t_2 < t_0, \\ S(t_0)^{n+m+1} S(t_2 - (t_0 - t_1)), & t_1 + t_2 \geq t_0. \end{cases}
\end{aligned}$$

We observe that both cases by definition equal $S(t + s)$. As $S(0) = I$ is clear we can conclude that $(S(t))_{t \geq 0}$ is a semigroup. We observe that

$$S_{|[0,t_0]} = R(1, V_{t_0})T_{|[0,t_0]},$$

so that $(S(t))_{t \geq 0}$ is locally bounded and $\{S(t) : t \in [0, t_0]\}$ is bi-equicontinuous. The next step is to show that for $t > 0$ one has the bi-equicontinuity of $\{S(s) : s \in [0, t]\}$. For this, let $m := \left\lfloor \frac{t}{t_0} \right\rfloor$ and we observe that $\{S(t_0)^k : 0 \leq k \leq m\}$ is bi-equicontinuous on $\|\cdot\|$-bounded sets. This yields the bi-equicontinuity of

$$\left\{ S(t_0)^k : 0 \leq k \leq m \right\} \cdot \{S(t) : t \in [0, t_0]\}.$$

Furthermore, since $S_{|[0,t_0]}$ is strongly continuous with respect to $\tau$ also $(S(t))_{t\geq 0}$ is strongly continuous with respect to $\tau$ as we already showed that $S(t+s) = S(t)S(s)$ whenever $0 \leq s, t \leq t+s \leq t_0$. Hence, $(S(t))_{t\geq 0}$ is indeed a bi-continuous semigroup and from the construction of $(S(t))_{t\geq 0}$ we immediately obtain (7.3.2) but only for $x \in D(A)$ and $t \in [0, t_0]$. Next, we show that (7.3.2) indeed holds for all $t > 0$. Again, find $n \in \mathbb{N}$ and $t_1 \in [0, t_0)$ such that $t = nt_0 + t_1$ then we can write for all $x \in D(A)$ the following:

$$\int_0^t S(t-s)BT(s)x\,\mathrm{d}s$$

$$= \sum_{k=0}^{n-1} \int_{kt_0}^{(k+1)t_0} S(t-s)BT(s)x\,\mathrm{d}s + \int_{nt_0}^t S(t-s)BT(s)x\,\mathrm{d}s$$

$$= \sum_{k=0}^{n-1} \int_0^{t_0} S(t-(s+kt_0))BT(s+kt_0)x\,\mathrm{d}s + \int_0^{t-nt_0} S(t-(s+nt_0))BT(s+nt_0)x\,\mathrm{d}s.$$

By taking the definition of $(S(t))_{t\geq 0}$ into account this yields

$$\sum_{k=0}^{n-1} S(t-(k+1)t_0) \int_0^{t_0} S(t_0-s)BT(s)T(kt_0)x\,\mathrm{d}s + \int_0^{t-nt_0} S((t-nt_0)-s)BT(s)T(nt_0)x\,\mathrm{d}s$$

$$= \sum_{k=0}^{n-1} S(t-(k+1)t_0)\,[S(t_0)-T(t_0)]\,T(kt_0)x + [S(t-nt_0)-T(t-nt_0)]\,T(nt_0)x$$

$$= S(t)x - S(t-t_0)T(t_0)x + \sum_{k=1}^{n-1} S(t-kt_0)T(kt_0)x$$

$$\quad - \sum_{k=1}^{n-1} S(t-(k+1)t_0)T((k+1)t_0)x + S(t-nt_0)T(nt_0)x - T(t)x$$

$$= [S(t)-T(t)]\,x.$$

In the last step, we show that $(S(t))_{t\geq 0}$ is indeed generated by the operator $(A+B, D(A))$. As a first step we show that $\rho(A+B) \neq \emptyset$. To do so, let $\lambda \in \rho(A)$ and observe that by taking the Laplace transform of the semigroup $(T(t))_{t\geq 0}$ one obtains that for all $x \in D(A)$ holds that

$$R(\lambda, A)x = \sum_{k=0}^{\infty} e^{-\lambda kt_0} \int_0^{t_0} e^{-\lambda t} T(t)T(kt_0)x\,\mathrm{d}t,$$

where the series and the integral converge in the locally convex topology $\tau$. Moreover, in this case, the convergence also in the graph topology $\tau_A$, cf. Definition 7.17, Lemma 3.10, and Theorem 3.14. Hence,

$$BR(\lambda, A)x = \sum_{k=0}^{\infty} e^{-\lambda k t_0} \int_0^{t_0} e^{-\lambda t} BT(t) T(kt_0)x \, dt$$

$$= \sum_{k=0}^{\infty} e^{-\lambda k t_0} \left[ V_{t_0} F_\lambda \right] (t_0) T(kt_0)x,$$

for all $x \in D(A)$ and where $F_\lambda \in \mathfrak{X}_{t_0}$ with $F_\lambda(t) := e^{-\lambda(t_0 - t)} I$, $t \in [0, t_0]$. The previous expression yields for $x \in D(A)$ the following norm estimate:

$$\|BR(\lambda, A)x\| \le \sum_{k=0}^{\infty} e^{-\lambda k t_0} \left\| \left[ V_{t_0} F_\lambda \right] (t_0) T(kt_0)x \right\|$$

$$\le \sum_{k=0}^{\infty} e^{-\lambda k t_0} \left\| V_{t_0} \right\| \, \|F_\lambda\| \, \|T(kt_0)\| \, \|x\|$$

$$\le \left\| V_{t_0} \right\| \|x\| + M \left\| V_{t_0} \right\| \sum_{k=1}^{\infty} e^{-\lambda k t_0 + \omega k t_0} \|x\|$$

$$\le \left\| V_{t_0} \right\| \|x\| + \frac{M}{\eta} \cdot \frac{e^{(\omega - \lambda)t_0}}{1 - e^{(\omega - \lambda)t_0}} \|x\|,$$

where $M > 0$ and $\omega \in \mathbb{R}$ are chosen in such a way that $\|T(t)\| \le M e^{\omega t}$ for all $t \ge 0$. Let $x \in X$ now be arbitrary. As $D(A)$ is assumed to be $\eta$-bi-dense, by Definition 7.14 there exists a sequence $(x_n)_{n \in \mathbb{N}}$ in $D(A)$ such that $x_n \overset{\tau}{\to} x$ and $\|x_n\| \le \eta \|x\|$ for all $n \in \mathbb{N}$. In this case, by the continuity assumption on the operator $B$ one has $BR(\lambda, A)x_n \overset{\tau}{\to} BR(\lambda, A)x$ for $n \to \infty$. As $\|x\| = \sup_{p \in \mathcal{P}} p(x)$ one also has that the norm $\|\cdot\|$ is $\tau$-lower semicontinuous, and that the unit ball in $X$ is $\tau$-closed. Therefore,

$$\|BR(\lambda, A)x\| \le \limsup_{n \to \infty} \|BR(\lambda, A)x_n\|$$

$$\le \limsup_{n \to \infty} \left( \left\| V_{t_0} \right\| \|x_n\| + \frac{M}{\eta} \cdot \frac{e^{(\omega - \lambda)t_0}}{1 - e^{(\omega - \lambda)t_0}} \|x_n\| \right)$$

$$\le \eta \left\| V_{t_0} \right\| \|x\| + M \cdot \frac{e^{(\omega - \lambda)t_0}}{1 - e^{(\omega - \lambda)t_0}} \|x\|.$$

By taking $\lambda$ large enough, we obtain $\|BR(\lambda, A)\| < 1$ and therefore

$$R(\lambda, A + B) = R(\lambda, A) \left[ I - BR(\lambda, A) \right]^{-1},$$

showing that indeed $\rho(A + B) \ne \varnothing$. As a final step, let $x \in D(A)$ and $t < t_0$. We already proved that (7.3.2) is valid. Hence, by using (7.3.2) one obtains

$$\frac{S(t)x - x}{t} = \underbrace{\frac{T(t)x - x}{t}}_{\overset{\tau}{\to} Ax} + \frac{1}{t} \int_0^t S(t - s) BT(s)x \, \mathrm{d}s.$$

Let $\varepsilon > 0$, then by the $\tau$-continuity of the orbits $s \mapsto S(t)Bx$ we can find $\delta \in (0, t_0)$ such that

$$p(S(t - s)Bx - Bx) < \frac{\varepsilon}{2},$$

for $t \in [0, \delta]$ and $s \in [0, t]$. Furthermore, by taking $\delta$ possibly smaller we can make use of the bi-equicontinuity of $\{S(s) : s \in [0, t_0]\}$ and the $\tau$-continuity of $s \mapsto BT(s)x$ to see that

$$p\left(\frac{1}{t}\int_0^t S(t - s)BT(s)x \, \mathrm{d}s - Bx\right)$$
$$\leq \frac{1}{t}\int_0^t p(S(t - s)Bx - Bx) \, \mathrm{d}s + \frac{1}{t}\int_0^t p(S(t - s)(BT(s)x - Bx)) \, \mathrm{d}s < \varepsilon.$$

This shows that the operator $(A + B, D(A))$ is a restriction of the generator of $(S(t))_{t \geq 0}$, hence $(A + B, D(A))$ is indeed its generator. $\qquad\square$

The next result provides sufficient conditions on the operator $B$ such that it becomes Miyadera–Voigt admissible in the sense of Definition 7.18. This is useful as the conditions in Definition 7.18 are unnatural and difficult to check is explicit situations. The next result is similar to that of [231] and [101, Chap. III, Cor. 3.16] but requires more assumptions on the topology and the operators as one has to deal with the norm and the additional locally convex topology simultaneously.

**Theorem 7.20** *Let $(T(t))_{t \geq 0}$ be a bi-continuous semigroup on a bi-admissible space $(X, \|\cdot\|, \tau)$ with generator $(A, D(A))$. Suppose that $D(A)$ is $\eta$-bi-dense in $X$ and that $B : (D(A), \tau_A) \to (X, \tau)$ is continuous on $\|\cdot\|_A$-bounded sets. Suppose that there exists $t_0 > 0$ and $0 < K < \frac{1}{\eta}$ such that*

*(i)   the map $s \mapsto \|BT(s)x\|$ is bounded on $[0, t_0]$ and for each $x \in D(A)$;*

*(ii)  $\displaystyle\int_0^t \|BT(s)x\| \, \mathrm{d}s < K \|x\|$ for each $t \in [0, t_0]$ and $x \in D(A)$;*

*(iii) for all $\varepsilon > 0$ and $p \in \mathcal{P}$ there exists $q \in \mathcal{P}$ and $M > 0$ such that*

$$\int_0^{t_0} p(BT(s)x) \, \mathrm{d}s < Mq(x) + \varepsilon \|x\|,$$

*for each $x \in D(A)$.*

*Then $B$ is Miyadera–Voigt admissible on $\mathfrak{X}_{t_0}^{\mathrm{loc}}$. In particular, $(A + B, D(A + B))$ generates a bi-continuous semigroup $(S(t))_{t \geq 0}$ which satisfies the variation of parameter formula (7.3.2).*

***Proof*** The goal is to show that $\mathfrak{X}^{loc}_{t_0}$ is invariant under $V_{t_0}$ and simultaneously to verify the three conditions of Definition 7.18. We refer to Definition 7.4 for the definition of $\mathfrak{X}^{loc}_{t_0}$. For $x \in X$ there exists a sequence $(x_n)_{n \in \mathbb{N}}$ in $D(A)$ such that $x_n \overset{\tau}{\to} x$ and there exists $C > 0$ satisfying $\|x_n\| \leq C$ for all $n \in \mathbb{N}$. Let us show that $(B(F,t)x_n)_{n \in \mathbb{N}}$ is a Cauchy sequence with respect to $\tau$. For $p \in \mathcal{P}$ and $F \in \mathfrak{X}^{loc}_{t_0}$ one has

$$
p(B(F,t)(x_n - x_m)) = p\left( \int_0^t F(t-s)BT(s)(x_n - x_m)\, ds \right)
$$

$$
\leq \int_0^t p(F(t-s)BT(s)(x_n - x_m))\, ds
$$

$$
\leq L \int_0^t q_1(BT(s)(x_n - x_m))\, ds + \varepsilon \int_0^t \|BT(s)(x_n - x_m)\|\, ds
$$

$$
\leq LM q_2(x_n - x_m) + (L\varepsilon' + K\varepsilon)\|x_n - x_m\|
$$

$$
\leq LM q_2(x_n - x_m) + 2C(L\varepsilon' + K\varepsilon),
$$

for certain $L, M > 0$ and $q_1, q_2 \in \mathcal{P}$, see Definition 7.4. By taking $\varepsilon > 0$ very small and accordingly $\varepsilon' > 0$ appropriately small, one observes that $(B(F,t)x_n)_{n \in \mathbb{N}}$ is indeed a Cauchy sequence with respect to $\tau$. As the sequence is also $\|\cdot\|$-bounded and since one has sequential completeness on $\|\cdot\|$-bounded sets, we obtain that $(B(F,t)x_n)_{n \in \mathbb{N}}$ is actually convergent with respect to $\tau$. Its limit will be denoted by $\overline{B}(F,t)x$. Observe that the definition of $\overline{B}(F,t)x$ is indeed independent of the particular choice of the approximating sequence $(x_n)_{n \in \mathbb{N}}$ and that $\overline{B}(F,t)$ yields a linear map. Moreover, one has

$$
p(\overline{B}(F,t)x) \leq LM q_2(x) + 2C(L\varepsilon' + K\varepsilon).
$$

As $D(A)$ is actually supposed to be $\eta$-bi-dense there exists a sequence $(x_n)_{n \in \mathbb{N}}$ in $D(A)$ such that $x_n \overset{\tau}{\to} x$ and $\|x_n\| \leq \eta \|x\|$ for all $n \in \mathbb{N}$, cf. Definition 7.14. According to this, the previous estimate can be amended to

$$
p(\overline{B}(F,t)x) \leq LM q_2(x) + 2\eta(L\varepsilon' + K\varepsilon). \tag{7.3.3}
$$

Let us now verify the conditions of Definition 7.18. Observe that the first assumption of Definition 7.18 appears as condition in the theorem itself. Let us continue with the second assumption that appears in Definition 7.18. Take $\eta > 0$ as it appears in Definition 7.18 and let $x \in X$ be arbitrary. As $D(A)$ is assumed to be $\eta$-bi-dense in $X$ there exists a sequence $(x_n)_{n \in \mathbb{N}}$ in $D(A)$ such that $x_n \overset{\tau}{\to} x$ and $\|x_n\| \leq \eta \|x\|$ for all $n \in \mathbb{N}$, cf. Definition 7.14. Therefore

$$
\begin{aligned}
\|B(F,t)x_n\| &= \sup_{\substack{\varphi\in(X,\tau)' \\ \|\varphi\|\le 1}} \left\| \left\langle \int_0^t F(t-s)BT(s)x_n \, ds, \varphi \right\rangle \right\| \\
&\le \sup_{\substack{\varphi\in(X,\tau)' \\ \|\varphi\|\le 1}} \int_0^t |\langle F(t-s)BT(s)x_n, \varphi\rangle| \, ds \\
&\le \sup_{\substack{\varphi\in(X,\tau)' \\ \|\varphi\|\le 1}} \|F\|\,\|\varphi\| \int_0^t \|BT(s)x_n\| \, ds \\
&\le K\,\|F\|\,\|x_n\| \\
&\le \eta K\,\|F\|\,\|x\|.
\end{aligned}
$$

Hence, the sequence $(B(F,t)x_n)_{n\in\mathbb{N}}$ is $\|\cdot\|$-bounded and $\tau$-convergent to $\overline{B}(F,t,x)$. As already mentioned previously, the norm $\|\cdot\|$ is $\tau$-lower semicontinuous so that $\|\overline{B}(F,t)x\| \le \eta K\,\|F\|\,\|x\|$ showing that $\overline{B}(F,t) \in \mathscr{L}(X)$. That the operator $\overline{B}(F,t)$ is even local and that $\{\overline{B}(F,t) : t \in [0,t_0]\}$ is bi-equicontinuous follows from (7.3.3). Now, let $x \in D(A)$ be arbitrary, then for $p \in \mathcal{P}$ and $h > 0$ one has

$$
\begin{aligned}
&p\left(B(F,t+h)x - B(F,t)x\right) \\
&= p\left( \int_0^{t+h} F(t+h-s)BT(s)x \, ds - \int_0^t F(t-s)BT(s)x \, ds \right) \\
&\le p\left( \int_0^t (F(t+h-s) - F(t-s))BT(s)x \, ds \right) + p\left( \int_t^{t+h} F(t+h-s)BT(s)x \, ds \right) \\
&\le \int_0^t p\left((F(t+h-s) - F(t-s))BT(s)x\right) ds + \int_t^{t+h} p\left(F(t+h-s)BT(s)x\right) ds.
\end{aligned}
$$

The second term automatically converges to 0 whenever $h \to 0$. The same holds for the first time by an application of Lebesgue's dominated convergence theorem. The case $h < 0$ can be treated similarly. In order to conclude that the map $t \mapsto \overline{B}(F,t)$ is strongly continuous with respect to the locally convex topology $\tau$ let $x \in X$, $\varepsilon > 0$ and $p \in \mathcal{P}$ be arbitrary. We can choose a sequence $(x_n)_{n\in\mathbb{N}}$ in $D(A)$ such that $x_n \xrightarrow{\tau} x$ and $\|x_n\| \le \eta\,\|x\|$ for all $n \in \mathbb{N}$. Then for sufficiently small $h > 0$

$$
\begin{aligned}
&p\left(\overline{B}(F,t+h)x - \overline{B}(F,t)x\right) \\
&= p\left(\overline{B}(F,t+h)(x-x_n) + \overline{B}(F,t+h)x_n - \overline{B}(F,t)x_n - \overline{B}(F,t)(x-x_n)\right) \\
&\le p\left(\overline{B}(F,t+h)(x-x_n)\right) + p\left(\overline{B}(F,t+h)x_n - \overline{B}(F,t)x_n\right) - p\left(\overline{B}(F,t)(x-x_n)\right) < \varepsilon,
\end{aligned}
$$

by using the bi-equicontinuity of the family $\{\overline{B}(F,t) : t \in [0,t_0]\}$ and choosing $n \in \mathbb{N}$ sufficiently large, which then yields the desired strong continuity with respect to $\tau$. We can therefore conclude that $\overline{B}(F,t) \in \mathfrak{X}_{t_0}^{\mathrm{loc}}$ and that the Volterra operator given by $(V_{t_0}F)(t)x :=$

$\overline{B}(F, t)x$ is a bounded operator on $\mathcal{X}_{t_0}^{\mathrm{loc}}$ satisfying $\left\| (V_{t_0} F)(t)x \right\| \leq \eta K \left\| F \right\| \left\| x \right\|$ so that $\left\| V_{t_0} \right\| < \frac{1}{\eta}$. $\qquad\qquad\square$

### 7.3.1   Transition Semigroups on $C_b(\mathcal{H})$

As application of the Miyadera–Voigt-type perturbation theorem for bi-continuous semigroups, cf. Theorem 7.20, we want to investigate stochastic differential equations. Our setting will be the one as already described in Sect. 2.2.5: let $\mathcal{H}$ be a separable Hilbert space and $(A, D(A))$ the generator of a $C_0$-semigroup $(S(t))_{t \geq 0}$ on $\mathcal{H}$. Furthermore, we assume that $W$ is a $\mathcal{H}$-valued cylindrical Wiener process and that $Q$ is a bounded, positive, self-adjoint operator on $\mathcal{H}$ with trivial kernel. From this we define

$$Q(t) := \int_0^t S(s) Q S^*(s)\, \mathrm{d}s, \tag{7.3.4}$$

where we understand the integral in the strong sense. Another additional assumption we make is that $Q(t)$ is a trace-class operator for each $t \geq 0$. Now, we consider for $F \in C_b(\mathcal{H})$ the equation given by

$$\begin{cases} \mathrm{d}X(t) = (AX(t) + F(X(t)))\mathrm{d}t + Q^{\frac{1}{2}}\mathrm{d}W(t), & t \geq 0, \\ X(0) = x \in \mathcal{H}. \end{cases} \tag{SDE}$$

For more information regarding this equation, we refer to [78]. The goal is to construct a transition semigroup on $C_b(\mathcal{H})$ which represents the solution of (SDE). For this we will use the previously developed perturbation theory. Such an approach is appropriate as the Ornstein–Uhlenbeck semigroup $(P(t))_{t \geq 0}$ (for the definition we refer to Sect. 2.2.5 and especially to (2.2.4)) actually corresponds to the equation

$$\begin{cases} \mathrm{d}X(t) = AX(t)\mathrm{d}t + Q^{\frac{1}{2}}\mathrm{d}W(t), & t \geq 0, \\ X(0) = x \in \mathcal{H}, \end{cases}$$

as discussed in Sect. 2.2.5, see also (2.2.1). In what follows, we will denote by $(L, D(L))$ the generator of the bi-continuous Ornstein–Uhlenbeck semigroup. Let us collect some results about the Ornstein–Uhlenbeck semigroup $(P(t))_{t \geq 0}$ as well as its generator $(L, D(L))$.

**Remark 7.21** By Proposition 2.10, the semigroup $(P(t))_{t \geq 0}$ is a bi-continuous semigroup on $(C_b(\mathcal{H}), \| \cdot \|_\infty, \tau_{\mathrm{co}})$. Moreover, by the definition of $(P(t))_{t \geq 0}$ it immediately follows that each operator $P(t), t \geq 0$, is $\tau_{\mathrm{co}}$-continuous on $\|\cdot\|$-bounded sets. Hence, Theorem 2.16 is applicable and (2.2.5) holds. From this, we can deduce a local estimate for the resolvent of $(L, D(L))$. Let $\lambda > 0, \varepsilon > 0$, and $K \subseteq \mathcal{H}$ compact. By taking the Laplace transform for the semigroup $(P(t))_{t \geq 0}$ one obtains for a suitable compact set $K' \subseteq \mathcal{H}$ that

$$\sup_{x\in K}|R(\lambda,L)f(x)| \le \sup_{x\in K}\int_0^\infty e^{-\lambda s}\,|P(s)f(x)|\,\mathrm{d}s$$

$$\le \sup_{x\in K}\int_0^t e^{-\lambda s}\,|P(s)f(x)|\,\mathrm{d}s + \sup_{x\in K}\int_t^\infty e^{-\lambda s}\,|P(s)f(x)|\,\mathrm{d}s$$

$$\le \sup_{x\in K}\int_0^t e^{-\lambda s}\,|P(s)f(x)|\,\mathrm{d}s + \sup_{x\in K}\int_t^\infty e^{-\lambda s}\,\|f\|_\infty\,\mathrm{d}s$$

$$= \sup_{x\in K}\int_0^t e^{-\lambda s}\,|P(s)f(x)|\,\mathrm{d}s + \frac{e^{-\lambda t}}{\lambda}\,\|f\|_\infty$$

$$\le \int_0^t e^{-\lambda s}\,\sup_{x\in K'}|f(x)| + \varepsilon'\,\|f\|_\infty\,\mathrm{d}s + \frac{e^{-\lambda t}}{\lambda}\,\|f\|_\infty$$

$$\le \frac{1}{\lambda}\left(\sup_{x\in K'}|f(x)| + \varepsilon'\,\|f\|_\infty\right) + \frac{e^{-\lambda t}}{\lambda}\,\|f\|_\infty$$

$$\le \frac{1}{\lambda}\sup_{x\in K'}|f(x)| + \varepsilon\,\|f\|_\infty\,.$$

Remember that we assumed that $F \in C_b(\mathcal{H})$. By $Df$ we denote the derivative of a Fréchet differentiable function. Moreover, we will make the following additional assumption.

**Assumption 7.22** $\operatorname{Ran}(S(t)) \subseteq \operatorname{Ran}(Q^{1/2}(t))$ for all $t \ge 0$.

Under Assumption 7.22 we are able to define the following function:

$$\Gamma(t) := Q(t)^{-1/2}S(t), \quad t \ge 0.$$

The following was proven in [75, Prop. 2.3], however, it is also valid for functions in $C_b(\mathcal{H})$

**Lemma 7.23** *For all $f \in C_b(\mathcal{H})$ and $t > 0$ we have $P(t)f \in C_b^1(\mathcal{H})$ and*

$$\langle (DP(t)f)(x), h \rangle = \int_{\mathcal{H}} \langle \Gamma(t)h, Q^{-1/2}\cdot \rangle\, f(S(t)x + \cdot)\,\mathrm{d}\mu_{t,0}, \tag{7.3.5}$$

*so that*

$$\|(DP(t)f)(x)\| \le \|\Gamma(t)\|^2 \cdot \left|(P(t)f^2)(x)\right|, \quad x \in \mathcal{H}. \tag{7.3.6}$$

*In particular,*

$$\|(DP(t)f)(x)\| \le \|\Gamma(t)\|^2 \cdot \|f\|_\infty, \quad x \in \mathcal{H}. \tag{7.3.7}$$

**Assumption 7.24** There exists $t_0 > 0$ such that

$$\int_0^{t_0} \|\Gamma(s)\| \, \mathrm{d}s < \infty.$$

Together with Assumption 7.24 we are able to prove a result similar to [75, Prop. 2.4]

**Lemma 7.25** *For $\lambda > 0$ and $f \in C_b(\mathcal{H})$ we have $R(\lambda, L)f \in C_b^1(\mathcal{H})$ and*

$$\|(DR(\lambda, L)f)(x)\| \le M \, \|f\|_\infty, \tag{7.3.8}$$

*where $M \ge 0$ is independent of $f$. Moreover, for all $K \subseteq \mathcal{H}$ compact and $\varepsilon > 0$ there exists a constant $M > 0$ and $K' \subseteq \mathcal{H}$ compact such that*

$$\sup_{x \in K} \|(DR(\lambda, L)f)(x)\| \le M \cdot \sup_{x \in K'} |f(x)| + \varepsilon \, \|f\|_\infty. \tag{7.3.9}$$

*Proof* Let $t_0 > 0$ as in Assumption 7.24 and $0 < \lambda \in \rho(L)$. Then

$$R(\lambda, L)f = \mathrm{e}^{-\lambda t_0} P(t_0) R(\lambda, L)f + \int_0^{t_0} \mathrm{e}^{-\lambda s} P(s) f \, \mathrm{d}s$$

for all $f \in C_b(\mathcal{H})$. By (7.3.6) we obtain

$$\|(DR(\lambda, L)f)(x)\| \le \mathrm{e}^{-\lambda t_0} \|(DP(t_0)R(\lambda, L)f)(x)\| + \int_0^{t_0} \mathrm{e}^{-\lambda s} \|(DP(s)f)(s)\| \, \mathrm{d}s$$

$$\le \mathrm{e}^{-\lambda t_0} \|\Gamma(t_0)\| \cdot \left| P(t_0)(R(\lambda, L)f)^2(x) \right|^{1/2} + \int_0^{t_0} \|\Gamma(s)\| \cdot \left| (P(s)f^2)(x) \right|^{1/2} \, \mathrm{d}s.$$

This yields

$$\|(DR(\lambda, L)f)(x)\| \le \left( \mathrm{e}^{-\lambda t_0} \|\Gamma(t_0)\| \cdot \|R(\lambda, L)\| + \int_0^{t_0} \|\Gamma(s)\| \, \mathrm{d}s \right) \|f\|_\infty$$

which therefore proves (7.3.8). Furthermore, from Remark 7.21 one can derive (7.3.9) as follows:

$$\sup_{x \in K} \|(DR(\lambda, L)f)(x)\|$$

$$\le \mathrm{e}^{-\lambda t_0} \|\Gamma(t_0)\| \cdot \left( \sup_{x \in K''} |(R(\lambda, L)f)(x)| + \varepsilon \, \|R(\lambda, L)\| \cdot \|f\| \right)$$

$$+ \int_0^{t_0} \|\Gamma(s)\| \, \mathrm{d}s \cdot \left( \sup_{x \in K''} |f(x)| + \varepsilon \, \|f\|_\infty \right)$$

$$\le \mathrm{e}^{-\lambda t_0} \|\Gamma(t_0)\| \cdot \left( \frac{1}{\lambda} \sup_{x \in K'} |f(x)| + \varepsilon \, \|f\|_\infty + \varepsilon \, \|R(\lambda, L)\| \cdot \|f\|_\infty \right)$$

$$+ \int_0^{t_0} \|\Gamma(s)\| \, ds \cdot \left( \sup_{x \in K''} |f(x)| + \varepsilon \|f\|_\infty \right)$$

$$\leq M \sup_{x \in K'} |f(x)| + M'\varepsilon \|f\|_\infty.$$

$\square$

With the previous results in hand, we are now ready to investigate the promised perturbation result for the Ornstein–Uhlenbeck semigroup which eventually leads to a solution associated to (SDE). To apply the Miyadera–Voigt perturbations for bi-continuous semigroups, cf. Theorem 7.20, we introduce the operator $(B, D(B))$ defined by

$$(Bf)(x) := \langle F(x), Df(x) \rangle, \quad D(B) := C_b^1(\mathcal{H}). \tag{7.3.10}$$

The goal is to show that this operator $(B, D(B))$ satisfies the assumptions of Theorem 7.20.

**Proposition 7.26** *Let $(B, D(B))$ be the operator defined by (7.3.10). Then for all $f \in D(L)$ the orbit $t \mapsto BP(t)f$ is locally $\|\cdot\|$-bounded.*

**Proof** We notice that by Lemma 7.25 one has that $B \in \mathscr{L}(D(L), C_b(\mathcal{H}))$. Moreover, we observe that the set $\{P(t)f : t \in [0, t_0]\}$ is $\|\cdot\|_L$-bounded for all $f \in D(L)$ which yields immediately the assertion (here $\|\cdot\|_L$ denotes the graph norm of the operator $(L, D(L))$, see also (4.1.1)). $\square$

**Proposition 7.27** *Let $(B, D(B))$ be the operator defined by (7.3.10). Then $B : (D(L), \tau_L) \to (C_b(\mathcal{H}), \tau_{co})$ is continuous on $\|\cdot\|_L$-bounded sets. In particular, the orbits $t \mapsto BP(t)f$, $t \geq 0$, are $\tau_{co}$-continuous for all $f \in D(L)$ (here $\tau_L$ denotes the graph topology corresponding to the operator $(L, D(L))$, cf. Definition 7.17).*

**Proof** The continuity of $B$ is a direct consequence of (7.3.9). The continuity of the orbits follows by Proposition 7.26. $\square$

We are now able to state the main result.

**Theorem 7.28** *Let $(P(t))_{t \geq 0}$ be the Ornstein–Uhlenbeck semigroup on $C_b(\mathcal{H})$ with generator $(L, D(L))$. Let $(B, D(B))$ be the operator defined by (7.3.10). Then $(L + B, D(L))$ is the generator of a bi-continuous semigroup $(R(t))_{t \geq 0}$ on $C_b(\mathcal{H})$.*

**Proof** As promised, the want to make use of the Miyadera–Voigt perturbation result, cf. Theorem 7.20. For this, we have to check the conditions of Theorem 7.20 carefully. We observe that condition (i) of Theorem 7.20 is fulfilled by means of Proposition 7.26. From (7.3.7) and (7.3.8), we see that also the second condition of Theorem 7.20 is satisfied.

Condition (iii) in Theorem 7.20 is justified by an integrated version of (7.3.6). Hence, we can apply Theorem 7.20 which yields the desired result.                    □

## 7.4  Positive Miyadera–Voigt Perturbations

Various models of physical processes ask for positive solutions in order to have a reasonable interpretation, e.g., consider solutions containing the absolute temperature or a density. The maximum principle for elliptic and parabolic partial differential equations guarantees positive solutions under positive initial data. This demonstrates the importance of positivity in the theory of operator semigroups on Banach lattices, hence in the theory of linear evolution equations.

**Definition 7.29** Let $(X, \|\cdot\|, \tau)$ be a Banach lattice with a locally convex topology $\tau$ generated by a family $\mathcal{P}$ of seminorms which satisfies Assumption 2.1. We say that $X$ is a *bi*-AL *space* if the following conditions are satisfied:

1. $X$ is a AL-space, i.e., for all $x, y \in X_+$ the equality $\|x + y\| = \|x\| + \|y\|$ holds, cf. Definition C.4(a).
2. There exists $\mathcal{P}_+ \subseteq \mathcal{P}$ such that $\mathcal{P}_+$ still generates the locally convex topology $\tau$ and for all $x, y \in X_+$ and for all $p \in \mathcal{P}_+$ one has $p(x + y) = p(x) + p(y)$.
3. All $p \in \mathcal{P}_+$ are Riesz seminorms, i.e., for all $x, y \in X$ one has that $|x| \leq |y|$ implies $p(x) \leq p(y)$.

**Remark 7.30** We notice that the notion of Riesz seminorms are not new, see, for example, [207, Chapter II, Def. 5.1] or [177, Chap. 9, Sect. 92]. Another characterization of a Riesz seminorm is the following: $p(x) \leq p(y)$ for $0 \leq x \leq y$ and $p(x) = p(|x|)$ for all $x \in X$. This will be important later on.

In Sect. 7.4.1.2, we will consider an explicit example of a space which satisfies the properties of Definition 7.29. Locally convex spaces satisfying such additivity conditions as in Definition 7.29 are also mentioned in [74] to discuss regular operators on vector lattices.

The mixed topology $\gamma$ will now play an important role. Recall from either the previous chapters or directly from Chap. A the definition of the mixed topology. Let $(X, \|\cdot\|, \tau)$ be a bi-admissible space and consider the mixed topology $\gamma = \gamma(\|\cdot\|, \tau)$. We also kindly refer the reader to recall the definition of the sub-mixed topology $\tilde{\gamma}$, see also Definition B.5.

The main result is the following theorem.

**Theorem 7.31** *Let* $(A, D(A))$ *be the generator of a positive, local bi-continuous semigroup* $(T(t))_{t \geq 0}$ *on a bi-AL space* $X$ *with* $\eta$-*bi-dense domain* $D(A)$ *for some* $1 < \eta < 2$. *Assume that* $(X, \gamma)$ *is C-sequential and* $\gamma = \tilde{\gamma}$. *Let* $B : D(A) \to X$ *be a positive operator, i.e.,*

*$Bx \geq 0$ for each $x \in D(A) \cap X_+$, and assume that $BR(\lambda, A)$ is local and $(A + B, D(A))$ is resolvent positive. Then $(A + B, D(A))$ is the generator of a positive bi-continuous semigroup.*

**Remark 7.32** (a)  As compared with [232, Thm. 0.1] we need the additional technical assumption of $\eta$-bi-denseness of the domain and the localness of the operator $BR(\lambda, A)$ in Theorem 7.31 due to the Miyadera–Voigt perturbation theorem for bi-continuous semigroups, which we will recall as Theorem 7.20.

(b)  We mention that the assumption $1 < \eta < 2$ does not appear in the Miyadera–Voigt perturbation theorem for bi-continuous semigroups (Theorem 7.20), however we will need this assumption to prove the main result Theorem 7.31. Nevertheless, this additional assumption is not too restrictive. In fact, by renorming the semigroup, see Sect. 3.2, one can achieve that $D(A)$ is $\eta$-bi-dense for all $\eta > 1$ by using Theorem 3.15(d) and Theorem 3.18.

The following results will be useful for proving the main theorem. The first one is actually related to [23, Lemma 4.15].

**Lemma 7.33**  *Let $(T(t))_{t\geq 0}$ be a bi-continuous semigroup on a bi-admissible space $(X, \|\cdot\|, \tau)$ with generator $(A, D(A))$. Suppose that $D(A)$ is $\eta$-bi-dense in $X$ and that $B : (D(A), \tau_A) \to (X, \tau)$ is continuous on $\|\cdot\|_A$-bounded sets. The following assertions are equivalent:*

*(a)  There exist $t_0 > 0$ and $0 < K < \frac{1}{\eta}$ such that for each $t \in [0, t_0]$ and $x \in D(A)$ one has*

$$\int_0^t \|BT(s)x\| \; ds < K \|x\|.$$

*(b)  Let $\lambda \in \mathbb{R}$, then there exist $t_0' > 0$ and $0 < K' < \frac{1}{\eta}$ such that for each $t \in [0, t_0']$ and $x \in D(A)$ one has*

$$\int_0^t \left\| e^{-\lambda s} BT(s)x \right\| \; ds < K' \|x\|.$$

***Proof***  Firstly, we mention that if $(A, D(A))$ generates a bi-continuous semigroup $(T(t))_{t\geq 0}$, then $(A - \lambda I, D(A))$ is the generator of the rescaled bi-continuous semigroup $(e^{-\lambda t} T(t))_{t\geq 0}$, see also Sect. 3.2.2. Moreover, since $A = (A - \lambda I) + \lambda I = (A + \lambda I) - \lambda I$ it suffices to prove that the assertion is true for $\lambda > 0$. To prove (b) $\Rightarrow$ (a) assume that there exist $t_0' > 0$ and $0 < K' < \frac{1}{\eta}$ such that for all $t \in [0, t_0']$ and $x \in D(A)$ one has that

$$\int_0^t \left\| e^{-\lambda s} BT(s)x \right\| \; ds < K' \|x\|.$$

One observes that

$$\int_0^t \|BT(s)x\| \ \mathrm{d}s \le e^{\lambda t_0} \int_0^t \left\|e^{-\lambda s} BT(s)x\right\| \ \mathrm{d}s < e^{\lambda t_0'} K' \|x\| .$$

If $e^{\lambda t_0'} K' < \frac{1}{\eta}$, then we choose $K := e^{\lambda t_0'} K' < \frac{1}{\eta}$ and we are done. If not, then choose $0 < t_0 < t_0'$ such that $e^{\lambda t_0} K' < 1$ which is possible since $0 < K' < \frac{1}{\eta} < 1$ and we are done. Conversely, for (a) $\Rightarrow$ (b) assume that there exist $t_0 > 0$ and $0 < K < \frac{1}{\eta}$ such that for all $t \in [0, t_0]$ and $x \in D(A)$ one has that

$$\int_0^t \|BT(s)x\| \ \mathrm{d}s < K \|x\| .$$

Since $\lambda > 0$, we directly conclude that

$$\int_0^t e^{-\lambda s} \|BT(s)x\| \ \mathrm{d}s \le \int_0^t \|BT(s)x\| \ \mathrm{d}s < K \|x\| .$$

$\square$

A similar result can be proven for the third condition occurring in Theorem 7.20.

**Lemma 7.34** *Let $(T(t))_{t\ge 0}$ be a bi-continuous semigroup on a bi-admissible space $(X, \|\cdot\|, \tau)$ with generator $(A, D(A))$. Suppose that $D(A)$ is $\eta$-bi-dense in $X$ and that $B : (D(A), \tau_A) \to (X, \tau)$ is continuous on $\|\cdot\|_A$-bounded sets. Fix $t_0 > 0$. Then, the following assertions are equivalent:*

(a) *For all $\varepsilon > 0$ and $p \in \mathcal{P}$ there exist $M \ge 0$ and $q \in \mathcal{P}$ such that for all $x \in D(A)$ one has*

$$\int_0^{t_0} p(BT(s)x) \ \mathrm{d}s \le Mq(x) + \varepsilon \|x\|.$$

(b) *For all $\varepsilon > 0$ and $p \in \mathcal{P}$ there exist $M' \ge 0$ and $q' \in \mathcal{P}$ such that for all $x \in D(A)$ one has*

$$\int_0^{t_0} p(e^{-\lambda s} BT(s)x) \ \mathrm{d}s \le M'q'(x) + \varepsilon \|x\|.$$

***Proof*** The same argument as at the beginning of the proof of Lemma 7.33 applies, hence we only have to consider $\lambda > 0$. So fix $t_0 > 0$. We first prove (b) $\Rightarrow$ (a), i.e., we assume that there exist $K' \ge 0$ and $q' \in \mathcal{P}$ such that for all $x \in D(A)$ one has

$$\int_0^{t_0} p(e^{-\lambda s} BT(s)x) \ \mathrm{d}s \le K'q'(x) + \varepsilon \|x\|.$$

Observe that

$$\int_0^{t_0} p(BT(s)x)\,\mathrm{d}s \le \mathrm{e}^{\lambda t_0}\int_0^{t_0} p(\mathrm{e}^{-\lambda s}BT(s)x)\,\mathrm{d}s \le \mathrm{e}^{\lambda t_0}M'q'(x) + \mathrm{e}^{\lambda t_0}\varepsilon\,\|x\|.$$

Since $\lambda > 0$ we can find $\alpha > 0$ such that $\mathrm{e}^{\lambda t_0}M'q'(x) + \mathrm{e}^{\lambda t_0}\varepsilon\,\|x\| < \mathrm{e}^{\lambda\alpha}M'q'(x) + \varepsilon\,\|x\|$, which concludes the proof of this implication. For the converse, (a) $\Rightarrow$ (b), observe that since $\lambda > 0$, one directly obtains

$$\int_0^{t_0} \mathrm{e}^{-\lambda s}p(BT(s)x)\,\mathrm{d}s \le \int_0^{t_0} p(BT(s)x)\,\mathrm{d}s \le Mq(x) + \varepsilon\,\|x\|.$$

$\square$

Summarizing Lemmas 7.33 and 7.34 one can say that an operator $B : (D(A), \tau_A) \to (X, \tau)$ which is continuous on $\|\cdot\|_A$-bounded sets is a Miyadera–Voigt perturbation for $(A, D(A))$ if and only if it is a Miyadera–Voigt perturbation of $(A - \lambda I, D(A))$.

Recall from [101, Chap. II, Def. 1.12] that the spectral bound $s(A)$ of a linear operator is defined by

$$s(A) := \sup\{\mathrm{Re}(\lambda) : \lambda \in \sigma(A)\},$$

where $\sigma(A)$ denotes the spectrum of the operator $(A, D(A))$. For the proof of Theorem 7.31 we need the following lemma.

**Lemma 7.35** *Let $1 < \eta < 2$ and let $(A, D(A))$ be the $\eta$-bi-densely defined generator of a positive local bi-continuous semigroup $(T(t))_{t\ge 0}$ on a bi-AL space $(X, \|\cdot\|, \tau)$. Let $B : D(A) \to X$ be a positive operator such that there exists $\lambda > s(A)$ such that the operator $BR(\lambda, A)$ is local and $\|BR(\lambda, A)\| < \frac{1}{2\eta} < 1$. Then $(A + B, D(A))$ is the generator of a positive bi-continuous semigroup.*

**Proof** We divide the proof into three steps.

**Step 1:** We show that conditions (i)–(iii) of Theorem 7.20 hold for each $x \in D(A)_+$.

We start by establishing condition (ii) in Theorem 7.20. By Lemma 7.33 it suffices to show the following estimate for each $x \in D(A)_+$

$$\int_0^t \left\| Be^{-\lambda s}T(s)x \right\|\,\mathrm{d}s = \left\| \int_0^t Be^{-\lambda s}T(s)x\,\mathrm{d}s \right\| = \left\| B\int_0^t e^{-\lambda s}T(s)x\,\mathrm{d}s \right\| \le \|BR(\lambda, A)x\| < \frac{1}{2\eta}\,\|x\|,$$

where the first equality is justified by the AL-property of the space and the fact that if $x \in D(A)_+$ then it is an element of the space of strong continuity (which by [59, Lemma 4.5] coincides with $\underline{X}_0 := \overline{D(A)}^{\|\cdot\|}$). The second equality follows by the fact that the operator $B$ is a continuous map from $(D(A), \tau_A)$ to $(X, \tau)$ which is $A$-bounded by [23, Lemma 4.1], i.e., there exists $a, b \ge 0$ such that $\|Bx\| \le a\,\|Ax\| + b\,\|x\|$ for each $x \in D(A)$. The first inequality follows by the Laplace transform representation of the resolvent $R(\lambda, A)$ and the positivity of the semigroup. As a consequence we conclude that condition (ii) of Theorem 7.20 holds.

Now let $\varepsilon > 0$ and $p \in \mathcal{P}_+$ be arbitrary. Then

$$\int_0^{t_0} p\left(e^{-\lambda s} B T(s) x\right) \, ds = p\left(\int_0^{t_0} e^{-\lambda s} B T(s) x \, ds\right) = p\left(B \int_0^{t_0} e^{-\lambda s} T(s) x \, ds\right)$$

$$\leq p\left(B R(\lambda, A) x\right) \leq K q(x) + \varepsilon \|x\|.$$

Here the first identity follows by the properties of the bi-AL space. The second identity follows by the same argument as before and the last inequality follows by the assumption of localness of the operator $B R(\lambda, A)$. Hence condition (iii) of Theorem 7.20 is fulfilled. Moreover also (i) holds since $(B, D(B))$ is $A$-bounded by [23, Lemma 4.1].

**Step 2:** We show that conditions (i)–(ii) of Theorem 7.20 hold for each $x \in D(A)$.

Take $x_+ := x \vee 0$ and $x_- := -(x \wedge 0)$ and observe that $x_+ + x_- = |x|$ and $x_+ - x_- = x$. Using this decomposition, one directly concludes that condition (i) of Theorem 7.20 holds. To show that also condition (ii) holds, recall that the semigroup $(T(t))_{t \geq 0}$ is exponentially bounded, i.e., there exists $\omega \in \mathbb{R}$ and $M \geq 1$ such that $\|T(t)\| \leq M e^{\omega t}$ for all $t \geq 0$. For $n \in \mathbb{N}$ define

$$x_{n,\pm} := \begin{cases} n R(n, A) x_\pm, & n > \omega, \\ 0, & \text{otherwise,} \end{cases}$$

and notice that $x_{n,\pm} \xrightarrow{\tau} x_\pm$ in $X$ and $x_{n,+} - x_{n,-} \xrightarrow{\tau_A} x$. The sequences $(x_{n,+})_{n \in \mathbb{N}}$ and $(x_{n,-})_{n \in \mathbb{N}}$, and hence also $(x_{n,+} - x_{n,-})_{n \in \mathbb{N}}$, are norm-bounded, see also Theorem 3.15(a). Furthermore, we notice that the sequence $(x_n)_{n \in \mathbb{N}}$ defined by $x_n := x_{n,+} - x_{n,-}$, $n \in \mathbb{N}$, actually satisfies $\|x_n\| \leq \eta \|x\|$ for all $n \in \mathbb{N}$. Hence we obtain

$$\int_0^t \left\| e^{-\lambda s} B T(s)(x_{n,+} - x_{n,-}) \right\| \, ds < \frac{1}{2\eta} \left(\|x_{n,+}\| + \|x_{n,-}\|\right) = \frac{1}{2\eta} \|x_n\| \leq \frac{1}{2} \|x\| < \frac{1}{\eta} \|x\|. \tag{7.4.1}$$

The last inequality follows from the assumption that $(A, D(A))$ is an $\eta$-bi-densely defined generator with $1 < \eta < 2$ and an application of Remark 7.32(b). Furthermore, we have to show that the integrand converges to $\left\| e^{-\lambda s} B T(s) x \right\|$ in order to conclude the desired estimate. Observe that for each $\varphi \in (X, \tau)'$

$$\varphi\left(\int_0^t e^{-\lambda s}BT(s)x\,\mathrm{d}s\right) = \int_0^t \varphi\left(e^{-\lambda s}BT(s)x\right)\,\mathrm{d}s$$

$$= \int_0^t \lim_{n\to\infty}\varphi\left(e^{-\lambda s}BT(s)(x_{n,+}-x_{n,-})\right)\,\mathrm{d}s$$

$$\leq \liminf_{n\to\infty}\int_0^t \left\|e^{-\lambda s}BT(s)(x_{n,+}-x_{n,-})\right\|\,\mathrm{d}s$$

$$\leq \frac{1}{\eta}\,\|x\|.$$

Due to the norming property of the locally convex topology (cf. Assumption 2.1) and the AL-property we obtain by taking the supremum over all $\varphi \in (X,\tau)'$ with $\|\varphi\| \leq 1$ the following inequality:

$$\int_0^t \left\|e^{-\lambda s}BT(s)x\right\|\,\mathrm{d}s \leq \frac{1}{\eta}\,\|x\|.$$

**Step 3:** We show that also condition (iii) of Theorem 7.20 holds for each $x \in D(A)$. By doing so, we conclude by Theorem 7.20 that indeed $(A+B, D(A))$ generates a positive bi-continuous semigroup.

To show that condition (iii) of Theorem 7.20 holds, let $\varepsilon > 0$ and $p \in \mathcal{P}_+$ be arbitrary and observe that there exist $K > 0$ and $q \in \mathcal{P}_+$ such that for each $n \in \mathbb{N}$

$$\int_0^{t_0} p(e^{-\lambda s}BT(s)(x_{n,+}-x_{n,-}))\,\mathrm{d}s \leq \int_0^{t_0} p(e^{-\lambda s}BT(s)x_{n,+})\,\mathrm{d}s + \int_0^{t_0} p(e^{-\lambda s}BT(s)x_{n,-})\,\mathrm{d}s$$

$$\leq K\left(q(x_{n,+})+q(x_{n,-})\right) + \varepsilon\left(\|x_{n,+}\| + \|x_{n,-}\|\right)$$

$$\leq K\left(q(x_{n,+})+q(x_{n,-})\right) + 2L\varepsilon\,\|x\|.$$

Since by construction $x_{n,+} \to x_+$ and $x_{n,-} \to x_-$ we conclude that $q(x_{n,+})+q(x_{n,-}) \to q(|x|) = q(x)$, cf. Remark 7.30. To see that the integrand on the left-hand side of the previous inequality converges, one notices that $x_{n,+}-x_{n,-} \to x$ with respect to the topology $\tau_A$. Since $T(s)$, $s \geq 0$, is $\tau_A$-continuous and $B : (D(A), \tau_A) \to (X,\tau)$ is continuous as well, we conclude that the integrand is converging pointwise to $p(e^{-\lambda s}BT(s)x)$. By an application of the Lebesgue dominated convergence theorem we conclude the proof. $\qquad\square$

The main work for proving Theorem 7.31 is included in Lemma 7.35. Now we are able to accomplish the proof of Theorem 7.31 which runs in fact parallel to the proof of [232, Thm. 0.1].

***Proof of Theorem*** 7.31 By [232, Thm. 1.1] there exists a $\lambda \in \rho(A+B)$ such that $r(BR(\lambda,A)) < 1$ and

$$R(\lambda,A) \leq R(\lambda,A)\sum_{n=0}^{\infty}(BR(\lambda,A))^n = R(\lambda,A+B).$$

Now, by taking $sB$ instead of $B$ for $s \in [0, 1]$ we obtain by the above

$$R(\lambda, A) \leq R(\lambda, A + sB) \leq R(\lambda, A + B).$$

Since $B$ is positive and $\mathrm{Ran}(R(\lambda, A + B)) = D(A)$, we conclude that $BR(\lambda, A + B) \in \mathscr{L}(X)$. Therefore, also $2\eta BR(\lambda, A + B) \in \mathscr{L}(X)$ and there exists $n \in \mathbb{N}$ such that

$$\|2\eta BR(\lambda, A + B)\| < n.$$

Hence

$$\left\| \frac{1}{n} BR(\lambda, A + sB) \right\| < \frac{1}{2\eta},$$

for each $s \in [0, 1]$. In particular, one has

$$\left\| \frac{1}{n} BR\left(\lambda, A + \frac{j}{n}B\right) \right\| < \frac{1}{2\eta},$$

for $0 \leq j \leq n - 1$. We observe that the operator $BR(\lambda, A + sB)$ is local for $s \in [0, 1]$. In fact, let $\varepsilon > 0$ and $p \in \mathcal{P}$ be arbitrary. We have to show that there exists $q \in \mathcal{P}$ and $L > 0$ such that

$$p(BR(\lambda, A + sB)x) \leq Lq(x) + \varepsilon \|x\|$$

for each $x \in X$. We notice that $BR(\lambda, A + sB) = BR(\lambda, A)R(1, sBR(\lambda, A))$. By assumption the operator $BR(\lambda, A)$ is local, hence, for $\varepsilon' := \varepsilon/(2\|R(1, sBR(\lambda, A))\|)$ one obtains that there exists $K > 0$ and $q' \in \mathcal{P}$ such that

$$p(BR(\lambda, A)y) \leq Kq'(y) + \varepsilon' \|y\|, \tag{7.4.2}$$

for each $y \in X$. Therefore, by using (7.4.2) with $y := R(1, sBR(\lambda, A))x$ one gets

$$p\left(BR(\lambda, A)R(1, sBR(\lambda, A))x\right) \leq Kq'(R(1, sBR(\lambda, A))x) + \frac{\varepsilon}{2} \|x\|. \tag{7.4.3}$$

Furthermore, since the operator $sBR(\lambda, A)$ is a bounded operator and local, it obviously generates a bi-continuous semigroup $(E(t))_{t\geq 0}$, by taking the exponential of the operator, i.e.:

$$E(t) = e^{tsBR(\lambda, A)} = \sum_{n=0}^{\infty} \frac{t^n(sBR(\lambda, A))^n}{n!}, \quad t \geq 0.$$

Therefore, by Lemma 7.6 as well as the assumption that $(X, \gamma)$ is $C$-sequential, the operator $R(1, sBR(\lambda, A)$ is local. This in fact is an application of [154, Thm. 3.17(b)]. For $\varepsilon'' := \frac{\varepsilon}{2K}$ there exists $K' > 0$ and $q'' \in \mathcal{P}$ such that

$$q'(R(1, sBR(\lambda, A))x) \leq K'q''(x) + \varepsilon'' \|x\|, \tag{7.4.4}$$

for each $x \in X$. Combining (7.4.3) and (7.4.4) one obtains that for each $x \in X$ holds that

$$p(BR(\lambda, A + sB)x) = p\left(BR(\lambda, A)R(1, sBR(\lambda, A))x\right) \leq KK'q''(x) + \varepsilon \|x\|.$$

By taking $L := KK'$ and $q := q''$ we conclude that $BR(\lambda, A + sB)$ is local for $s \in [0, 1]$. Now we apply Lemma 7.35 for the perturbation $\frac{1}{n}B$ repeatedly for $A, A + \frac{1}{n}B, \ldots, A + \frac{n-1}{n}B$ and obtain the generation by $A + B$ in the last step.  $\square$

### 7.4.1  Examples

#### 7.4.1.1 Rank-One Perturbations

As a first example we consider rank-one perturbations as they are treated for $C_0$-semigroups by Arendt in Rhandi [21, Thm. 2.2]. Let $(T(t))_{t \geq 0}$ be a positive bi-continuous semigroup on $X$ with respect to $\tau$ with generator $(A, D(A))$. Furthermore, assume that $(X, \gamma)$ is $C$-sequential and $\gamma = \tilde{\gamma}$ (see also the conditions of Theorem 7.31). For a positive $\tau_A$-continuous linear functional $\varphi : D(A) \to \mathbb{R}$ and for $y \in X_+$ we define the rank-one perturbation $B : D(A) \to X$ by

$$Bx := \varphi(x)y, \quad x \in D(A).$$

The operator $(B, D(A))$ satisfies the assumptions of Theorem 7.31, and hence it is a Miyadera–Voigt perturbation. To see this let $\varepsilon' > 0$ and $p \in \mathscr{P}$ be arbitrary and observe that

$$p(BR(\lambda, A)x) = p\left(\varphi(R(\lambda, A)x)\,y\right) = |\varphi(R(\lambda, A)x)|\,p(y), \quad x \in X.$$

Since $\varphi$ is $\tau_A$-continuous we conclude that there exist $M > 0$ and $p', q' \in \mathcal{P}$ such that

$$|\varphi(R(\lambda, A)x)| \leq M\left(p'(R(\lambda, A)x) + q'(x)\right), \quad x \in X.$$

By Lemma 7.6 the operator $R(\lambda, A)$, $\lambda \in \rho(A)$, is local, i.e., for each $\varepsilon' > 0$ there exist $K' > 0$ and $q'' \in \mathcal{P}$ such that

$$p'(R(\lambda, A)x) \leq K'q''(x) + \varepsilon'\|x\|, \quad x \in X.$$

With $K'' := p(y)MK'$ this leads to the following inequality:

$$p(BR(\lambda, A)x) \leq K''\left(q''(x) + \frac{1}{K'}q'(x)\right) + \varepsilon\|x\|, \quad x \in X,$$

where $\varepsilon = Mp(y)\varepsilon'$. We conclude from [72, Prop. 7.1.4] that $q'' + \frac{1}{K'}q' \in \mathcal{P}$, showing that $BR(\lambda, A)$ is local. Since, by Theorem 3.15, $(A, D(A))$ is a Hille–Yosida operator one obtains

$$\|BR(\lambda, A)x\| \leq |\varphi(R(\lambda, A)x| \cdot \|y\| \leq \|\varphi\|\,\|R(\lambda, A)x\|\,\|y\| \leq \|\varphi\|\,\frac{M}{|\lambda - \omega|}\,\|x\|\,\|y\|.$$

Since all constants are fixed from the beginning the expression becomes smaller than 1 by choosing $\lambda > \omega$ large enough, this shows that rank-one perturbations satisfy the required conditions on Theorem 7.31, i.e., $A + B$ generates a bi-continuous semigroup by an application of Theorem 7.31. Observe that the argumentation can be extended to finite rank operators.

### 7.4.1.2 Gauss–Weierstrass Semigroup on $M(\mathbb{R})$

Now we discuss a second example, i.e., we consider the space $M(\Omega)$ of bounded Borel measures on $\Omega$, where $\Omega$ is a Polish space. First of all we show that $M(\Omega)$ satisfies the conditions of Definition 7.29, i.e., we show that the space $M(\Omega)$ of bounded Borel measures is a bi-AL-space with respect to an appropriate locally convex topology. By Proposition 2.26 every bi-continuous semigroup on $C_b(\Omega)$ with respect to the compact-open topology $\tau_{co}$ gives rise to a bi-continuous semigroup on $M(\Omega)$ with respect to $\tau = \sigma(M(\Omega), C_b(\Omega))$ and vice versa. So the locally convex topology on $M(\Omega)$ is given by the weak$^*$-topology generated by the family of seminorms given by

$$\mathcal{P} := \left\{ p_f : f \in C_b(\mathbb{R}) \right\},$$

where

$$p_f(\mu) := \left| \int_\Omega f \, d\mu \right|, \quad \mu \in M(\Omega).$$

The norm on $M(\Omega)$ is the total variation norm, which turns the space into a Banach space. Moreover, the norm satisfies the first condition of Definition 7.29, i.e., $\|\mu + \nu\|_{M(\Omega)} = \|\mu\|_{M(\Omega)} + \|\nu\|_{M(\Omega)}$. Also, the second condition of the additivity of the seminorms is fulfilled. In fact, let

$$\mathcal{P}_+ := \left\{ p_f : f \in C_b(\Omega), \ f \geq 0 \right\},$$

and let $\tau_+$ denote the locally convex topology generated by $\mathcal{P}_+$. We first observe that for each $p = p_f \in \mathcal{P}_+$ and for each $\mu, \nu \geq 0$ one has

$$p_f(\mu) + p_f(\nu) = \left| \int_\Omega f \, d\mu \right| + \left| \int_\Omega f \, d\nu \right| = \int_\Omega f \, d\mu + \int_\Omega f \, d\nu = \int_\Omega f \, d(\mu + \nu)$$

$$= \left| \int_\Omega f \, d(\mu + \nu) \right| = p_f(\mu + \nu).$$

The following result shows that actually the locally convex topology $\tau$ coincides with the originally topology $\tau$. It directly follows from the fact that one can decompose every $f \in C_b(\Omega)$ into a positive and negative part, i.e., $f = f_+ - f_-$ for some positive functions $f_+, f_- \in C_b(\Omega)$, and therefore we omit the proof here.

**Lemma 7.36** *The family $\mathcal{P}_+$ generates the topology $\tau$, i.e., for a net $(\mu_\alpha)_{\alpha \in \Lambda}$ in $M(\Omega)$ and $\mu \in M(\Omega)$:*

$$\mu_\alpha \xrightarrow{\tau} \mu \iff \mu_\alpha \xrightarrow{\tau_+} \mu$$

As special case we now consider the space $M(\mathbb{R})$ of bounded Borel measures on $\mathbb{R}$ with respect to the Borel $\sigma$-algebra $\mathscr{B}(\mathbb{R})$. Notice that since we consider bounded Borel measures on the separable metric space $\mathbb{R}$ every $\mu \in M(\mathbb{R})$ is also a Radon measure. In fact, $M(\mathbb{R})$ in this case coincides with the space $M_t(\mathbb{R})$ of bounded Radon measure. Moreover, it is known that the corresponding mixed topology $\gamma = \gamma\left(\|\cdot\|_{M_t(\mathbb{R})}, \sigma(M_t(\mathbb{R}), C_b(\mathbb{R}))\right)$ is $C$-sequential and $\gamma = \tilde{\gamma}$, cf. [154, Cor. 3.23.]. Here $\|\cdot\|_{M_t(\mathbb{R})}$ denotes the total variation norm. Hence, the technical assumption on the mixed topology in Theorem 7.31 is satisfied.

We first recall the definition of the (centered) Gaussian measure $\gamma_t$, $t > 0$, on $\mathbb{R}$ defined by

$$\gamma_t(C) = \frac{1}{\sqrt{2\pi t}} \int_C e^{-\frac{|x|^2}{2t}} \, d\lambda^1, \quad C \in \mathscr{B}(\mathbb{R}),$$

where $\lambda^1$ denotes the Lebesgue measure on $\mathbb{R}$. This measure $\gamma_t$, $t > 0$, is a strictly positive bounded Borel measure. As a matter of fact, $\gamma_t$, $t > 0$, is absolutely continuous with respect to the Lebesgue measure with density $\varphi_t$ given by

$$\varphi_t(x) = \frac{1}{\sqrt{2\pi t}} e^{-\frac{|x|^2}{2t}}.$$

Next we recall the convolution of measures. For $\mu, \nu \in M(\mathbb{R})$ one defines the convolution of $\mu$ and $\nu$ by

$$(\mu * \nu)(\Omega) = \int_{\mathbb{R}} \int_{\mathbb{R}} \mathbf{1}_\Omega(x + y) \, d\mu(x) \, d\nu(y) = \int_{\mathbb{R}} \nu(\Omega - x) \, d\mu(x).$$

**Remark 7.37** If one of the measures $\mu, \nu \in M(\mathbb{R})$ has some density with respect to the Lebesgue measure, say $\nu = g \cdot \lambda^1$ for some non-negative, integrable function relative to the Lebesgue measure, as it is the case for $\gamma_t$, then by an application of Tonelli's theorem and the translation invariance of the Lebesgue measure, the convolution has density $g * \mu$ defined by

$$(g * \mu)(x) = \int_{\mathbb{R}} g(x - y) \, d\mu(y).$$

Now we define a family of operators $(T(t))_{t\geq 0}$ on $M(\mathbb{R})$ by $T(0)\mu = \mu$ and

$$T(t)\mu = \gamma_t * \mu, \quad t > 0. \tag{7.4.5}$$

Since $\gamma_t \geq 0$ for each $t > 0$ we conclude that the family $(T(t))_{t\geq 0}$ consists of positive operators on $M(\mathbb{R})$. The next result shows that $(T(t))_{t\geq 0}$ is in fact a bi-continuous semigroup on $M(\mathbb{R})$ equipped with the total variation norm and the topology $\sigma(M(\mathbb{R}), C_b(\Omega))$. For that recall that the so-called Gauss–Weierstrass semigroup on $C_b(\mathbb{R})$, here denoted by $(T_*(t))_{t\geq 0}$, is defined by $T(0)f := f$ and

$$T_*(t)f := \varphi_t * f, \quad t > 0.$$

As a matter of fact, $(T_*(t))_{t \geq 0}$ is a bi-continuous semigroup on $C_b(\mathbb{R})$ with respect to the compact-open topology, cf. Sect. 2.2.2. We now relate the semigroup $(T(t))_{t \geq 0}$ to $(T_*(t))_{t \geq 0}$.

**Theorem 7.38** *The semigroup $(T(t))_{t \geq 0}$ is the adjoint of the Gauss–Weierstrass semigroup $(T_*(t))_{t \geq 0}$ on $C_b(\mathbb{R})$.*

***Proof*** We observe that by Remark 7.37 for $f \in C_b(\mathbb{R})$ with $f \geq 0$ and $\mu \in M(\mathbb{R})$ the following holds:

$$
\begin{aligned}
\langle f, T(t)\mu \rangle &= \left| \int_{\mathbb{R}} f(x)\, d(T(t)\mu)(x) \right| \\
&= \left| \int_{\mathbb{R}} f(x)\, d(\gamma_t * \mu)(x) \right| \\
&= \left| \int_{\mathbb{R}} f(x)\, d((\varphi_t \cdot \lambda^1) * \mu)(x) \right| \\
&= \left| \int_{\mathbb{R}} f(x)\, d((\varphi_t * \mu) \cdot \lambda^1)(x) \right| \\
&= \left| \int_{\mathbb{R}} f(x)(\varphi_t * \mu)(x)\, d\lambda^1(x) \right| \\
&= \left| \int_{\mathbb{R}} f(x) \int_{\mathbb{R}} \varphi_t(x - y)\, d\mu(y)\, d\lambda^1(x) \right| \\
&= \left| \int_{\mathbb{R}} (f * \varphi_t)(y)\, d\mu(y) \right| \\
&= \left| \int_{\mathbb{R}} T_*(t)f(y)\, d\mu(y) \right| \\
&= \langle T_*(t)f, \mu \rangle .
\end{aligned}
$$

Here $\langle \cdot, \cdot \rangle$ denotes the duality pairing between $C_b(\mathbb{R})$ and $M(\mathbb{R})$. This proves the assertion. $\qquad\square$

**Remark 7.39** (a)  The notation $(T_*(t))_{t \geq 0}$ is intuitive since $(T_*(t))_{t \geq 0}$ is the preadjoint of $(T(t))_{t \geq 0}$.

(b)  Since $(T_*(t))_{t \geq 0}$ is known to be bi-continuous on $C_b(\mathbb{R})$ with respect to the compact-open topology we conclude by Theorem 2.26 that $(T(t))_{t \geq 0}$ on $M(\mathbb{R})$ is bi-continuous with respect to $\sigma(M(\mathbb{R}), C_b(\Omega))$.

(c)  Recall that the generator $(A_*, D(A_*))$ of the Gauss–Weierstrass semigroup on $C_b(\mathbb{R})$ is given by $(\Delta, C_b^2(\mathbb{R}))$ where $\Delta$ denotes the Laplacian and $C_b^2(\mathbb{R})$ the space of twice continuous differentiable functions with bounded derivatives, cf. [169] or [175].

(d)  We also notice that $\mathbb{C}_+ \subseteq \rho(A_*)$, where $\mathbb{C}_+ = \{z \in \mathbb{C} : \mathrm{Re}(z) > 0\}$.

The next step is to determine the generator $(A, D(A))$ of $(T(t))_{t \geq 0}$. To do so we have to consider differentiable measures. The original study of such measures is due to Fomin [113] and Skorohod [218], [219, Chap. 4, Sect. 21]. For more details, we refer to the work of Bogachev [43].

**Definition 7.40** A Borel measure $\mu \in M(\mathbb{R})$ is called *Skorohod differentiable* or *S-differentiable* if, for every function $f \in C_b(\mathbb{R})$, the function

$$t \mapsto \int_{\mathbb{R}} f(x - t) \, \mathrm{d}\mu(x)$$

is differentiable.

Recall from [43, Thm. 3.6.1] that the previous definition is equivalent to the existence of a Borel measure $\nu$, called the Skorohod derivative of $\mu$, such that for each bounded continuous function on $\mathbb{R}$ one has

$$\lim_{t \to 0} \int_{\mathbb{R}} \frac{f(x - t) - f(x)}{t} \, \mathrm{d}\mu(x) = \int_{\mathbb{R}} f(x) \, \mathrm{d}\nu(x). \tag{7.4.6}$$

**Theorem 7.41** *Let $\mu$ be a Skorohod differentiable Borel measure on $\mathbb{R}$. Then there exists a Borel measure $\nu$ which is its Skorohod derivative, i.e., $\nu$ satisfies (7.4.6) for all bounded continuous functions $f$ on $\mathbb{R}$.*

**Remark 7.42** (a)  The Skorohod derivative $\nu$ of $\mu$ as it appears in (7.4.6) and Theorem 7.41 is denoted by $\mathscr{D}\mu$, i.e., $\nu = \mathscr{D}\mu$.

(b)  By [43, Prop. 3.4.1] one has that for a bounded Borel measure $\mu$ on $\mathbb{R}$ the measure $\mu$ is Skorohod differentiable if and only if it has density of bounded variation, in particular, every Skorohod differentiable measure on $\mathbb{R}$ admits a bounded density. In that case the density of the Skorohod derivative is just the (distributional) derivative of the original density.

(c)  For higher order derivatives we recall from [43, Prop. 3.7.1] that if the map $t \mapsto \int_{\mathbb{R}} f(x - t) \, \mathrm{d}\mu(x)$ is $n$-times differentiable, then $\mu$ is $n$-times Skorohod differentiable, i.e., for all $f \in C_b(\mathbb{R})$ the function

$$(t_1, \ldots, t_n) \mapsto \int_{\mathbb{R}} f(x + t_1 + \cdots + t_n) \, \mathrm{d}\mu(x)$$

has partial derivatives $\partial_{t_1} \cdots \partial_{t_n}$.

(d)  $\mu$ is Skorohod differentiable with Skorohod derivative $\mathscr{D}\mu$ if and only if for each $f \in C_b^\infty(\mathbb{R})$

$$\int_{\mathbb{R}} \frac{\mathrm{d}}{\mathrm{d}x} f(x) \, \mathrm{d}\mu(x) = - \int_{\mathbb{R}} f(x) \, \mathrm{d}(\mathscr{D}\mu)(x),$$

in particular, one concludes that Skorohod differentiability of a measure $\mu$ is equivalent to that its derivative in the sense of distributions is a bounded measure, cf. [43, Prop. 3.4.3].

Now we prove the following result.

**Theorem 7.43** *The generator $(A, D(A))$ of $(T(t))_{t \geq 0}$ is given by*

$$A\mu := \Delta\mu, \quad D(A) := \{\mu \in M(\mathbb{R}) : \mu \text{ is twice Skorohod differentiable}\},$$

*where $\Delta\mu$ denotes the second-order Skorohod derivative of $\mu$ introduced in Remark 7.42(c).*

**Proof** Let us denote the generator of the semigroup $(T(t))_{t \geq 0}$ by $(C, D(C))$. By Lemma 3.7 and Remark 7.39 we conclude that the domain of the generator is of the following form:

$$D(C) = \left\{\mu \in M(\mathbb{R}) : \exists \nu \in M(\mathbb{R}) \, \forall f \in C_b^2(\mathbb{R}) : \langle f, \nu \rangle = \langle \Delta f, \mu \rangle\right\}.$$

Now let $\mu \in M(\mathbb{R})$ be twice Skorohod differentiable and denote its second derivative by $\nu := \Delta\mu$. For $f \in C_b^2(\mathbb{R})$ one has

$$\langle \Delta f, \mu \rangle = \int_{\mathbb{R}} \Delta f \, d\mu = \int_{\mathbb{R}} f \, d\nu = \langle f, \nu \rangle,$$

showing that $D(A) \subseteq D(C)$, cf. Lemma 3.7. For the converse let $\mu \in D(C)$, i.e., there exists $\nu \in M(\mathbb{R})$ such that for each $f \in C_b^2(\mathbb{R})$ holds that

$$\langle f, \nu \rangle = \langle \Delta f, \mu \rangle = \left\langle \lim_{t \to 0} \frac{T_*(t)f - f}{t}, \mu \right\rangle = \lim_{t \to 0} \left\langle f, \frac{T(t)\mu - \mu}{t} \right\rangle.$$

From this we conclude that $\sigma(M(\mathbb{R}), C_b(\Omega)) - \lim_{t \to 0} \frac{T(t)\mu - \mu}{t}$ exists. Combining this with [43, Thm. 3.6.4] we obtain that $\mu \in D(A)$. This finishes the proof. $\square$

Now fix a positive, Lebesgue integrable (and unbounded) function $\psi : \mathbb{R} \to \mathbb{R}$ and define $B : D(A) \to M(\mathbb{R})$ by

$$B\mu := \psi \cdot \mu, \quad \mu \in D(A).$$

Observe that indeed $B\mu \in M(\mathbb{R})$ by an application of Remark 7.42. To show that the operator $(B, D(A))$ satisfies the conditions of Theorem 7.31 let $\lambda \in \rho(A)$, especially, since $(T(t))_{t \geq 0}$ is a bounded semigroup and hence $\mathbb{C}_+ \subseteq \rho(A)$ holds, we can choose $\lambda \in \mathbb{R}$ with $\lambda > 0$ and observe that by the duality between $C_b(\mathbb{R})$ and $M(\mathbb{R})$

$$p_f(BR(\lambda, A)\mu) = |\langle f, BR(\lambda, A)\mu \rangle| = |\langle f\psi, R(\lambda, A)\mu \rangle| = |\langle R(\lambda, A_*)(f\psi), \mu \rangle|$$

for all $f \in C_b(\mathbb{R})$ and $\mu \in D(A)$. Moreover, by an application of Fubini's theorem and an explicit calculation of an integral one obtains

$$
|R(\lambda, A_*)(f\psi)(x)| = \left| \int_0^\infty \int_{\mathbb{R}} e^{-\lambda t} \varphi_t(x - y) f(y)\psi(y) \, dy \, dt \right|
$$

$$
= \left| \int_{\mathbb{R}} \int_0^\infty e^{-\lambda t} \varphi_t(x - y) f(y)\psi(y) \, dt \, dy \right|
$$

$$
= \left| \int_{\mathbb{R}} \frac{1}{\sqrt{2\lambda}} e^{-\sqrt{2\lambda}|x-y|} f(y)\psi(y) \, dy \right|
$$

$$
= |(\xi_\lambda * f\psi)(x)|,
$$

where $\xi_\lambda(x) := \frac{1}{\sqrt{2\lambda}} e^{-\sqrt{2\lambda}|x|}$ is continuous and hence the convolution $\xi_\lambda * (f\psi)$ is continuous. Furthermore, by Young's convolution inequality and the assumption that $\psi$ is integrable we obtain

$$
\|\xi_\lambda * f\psi\|_\infty \le \|\xi_\lambda\|_\infty \|f\psi\|_1 \le \|\xi_\lambda\|_\infty \|f\|_\infty \|\psi\|_1 < \infty
$$

hence $h := \xi_\lambda * f\psi \in C_b(\mathbb{R})$ and $p_f(BR(\lambda, A)\mu) = p_h(\mu)$. Recall from Sect. 2.2.7 that $C_b(\mathbb{R})^\circ = M(\mathbb{R})$. Making use of this duality, we obtain the following norm estimate:

$$
\|BR(\lambda, A)\mu\| = \sup_{\substack{f \in C_b(\mathbb{R}) \\ \|f\|_\infty \le 1}} |\langle f, BR(\lambda, A)\mu \rangle| = \sup_{\substack{f \in C_b(\mathbb{R}) \\ \|f\|_\infty \le 1}} |\langle R(\lambda, A_*)f\psi, \mu \rangle| \le \sup_{\substack{f \in C_b(\mathbb{R}) \\ \|f\|_\infty \le 1}} \|R(\lambda, A_*)f\psi\|_\infty \|\mu\|.
$$

Moreover,

$$
|R(\lambda, A_*)(f\psi)(x)| = \left| \int_0^\infty e^{-\lambda t} T_*(t) f(x)\psi(x) \, dt \right|
$$

$$
= \left| \int_0^\infty e^{-\lambda t} \int_{\mathbb{R}} \varphi_t(x - y) f(y)\psi(y) \, dy dt \right|
$$

$$
\le \|f\|_\infty \int_0^\infty e^{-\lambda t} (\varphi_t * \psi)(x) \, dt
$$

$$
\le \|f\|_\infty \|\psi\|_1 \int_0^\infty e^{-\lambda t} \|\varphi_t\|_\infty \, dt
$$

$$
= \|f\|_\infty \|\psi\|_1 \int_0^\infty \frac{e^{-\lambda t}}{\sqrt{2\pi t}} \, dt
$$

$$
= \frac{\|f\|_\infty \|\psi\|_1}{\sqrt{2\lambda}}.
$$

Hence, for $\lambda$ large enough we obtain $\|R(\lambda, A_*)f\psi\| < 1$ and hence $\|BR(\lambda, A)\| < 1$. Now we can apply Theorem 7.31 and conclude that $A + B$ generates a positive bi-continuous semigroup on $M(\mathbb{R})$.

**Remark 7.44** Suppose that $f \in L^p(\mathbb{R})$ is unbounded. Observe that there exists $g \in L^1(\mathbb{R})$ and $h \in L^\infty(\mathbb{R})$ such that $f = g + h$. In particular, we consider the operator $B : D(A) \to M(\mathbb{R})$ defined by $B\mu := f \cdot \mu$ as above. Since the case for $f \in L^1(\mathbb{R})$ is treated before and $f \in L^\infty(\mathbb{R})$ gives rise to a bounded perturbation we conclude that we can extend the previous result to the whole scale of $L^p$-spaces.

## Notes on This Chapter

The perturbation theory of bi-continuous semigroups started with the work of B. Farkas, see, for example, [106–108] just to mention a few. For $C_0$-semigroups such a result was originally obtained by Miyadera [185] and later improved by Voigt [231].

For the fundamental concepts in the context of vector lattices, Banach lattices, and positive operators, we refer to the monographs by Schaefer [207] or Meyer-Nieberg [181]. An extensive discussion of positive operators and positive semigroups on Banach lattices can, for example, be found in the monographs by Arendt et al. [19], Arlotti and Banasiak [23] or Bátkai, Kramar-Fijavž and Rhandi [33]. There has been an enormous development in the perturbation theory of $C_0$-semigroups which is established, for example, by Engel and Nagel [101, Chap. III, Sect. 3], Voigt [231], Desch and Schappacher [83], Bombieri [47] and Adler, Bombieri and Engel in [1], just to mention a few. It is noteworthy that by the comments that have been made by Kruse and Schwenninger in [154], the theory presented in Sect. 7.4 has been corrected and refined in comparison with the original work [53].

# Desch–Schappacher Perturbations of Bi-continuous Semigroups

In this chapter, we discuss another class of perturbations, the so-called Desch–Schappacher perturbations. They provide a powerful framework for extending the perturbation theory of bi-continuous semigroups beyond the bounded and Miyadera–Voigt classes, discussed in Chap. 7. Desch–Schappacher perturbations demand a deeper understanding of the interplay between the generator and extrapolation spaces that were discussed in Chaps. 4 and 5. These spaces, constructed to accommodate unbounded perturbations, allow for the definition of perturbed operators that might not be well-defined on the original space. By employing this framework, one ensures the perturbed operator still generates a bi-continuous semigroup, maintaining the structural stability of the dynamics. In this chapter, we formulate and prove a Desch–Schappacher-type perturbation theorem for bi-continuous semigroups, showing that for a certain class of admissible operators the perturbation of a generator of a bi-continuous semigroup still gives rise to a semigroup generator.

## 8.1  An Abstract Desch–Schappacher Perturbation Result

Let us first fix some notation. Let $(T(t))_{t \geq 0}$ be a bi-continuous semigroup on a bi-admissible space $(X, \| \cdot \|, \tau)$ with generator $(A, D(A))$. Here, we assume that $\tau$ is generated by the family of seminorms $\mathcal{P}$. Furthermore, let $B \in \mathscr{L}(X, X_{-1})$ such that $B : (X, \tau) \to (X_{-1}, \tau_{-1})$ is a continuous linear operator and recall for $t_0 > 0$ the definition of $\mathfrak{X}_{t_0}$ from (7.1.1). We also recall that by Lemma 7.1 and Lemma 7.2 the space $\mathfrak{X}_{t_0}$ is a Banach space and in particular a Banach algebra.

For $F \in \mathfrak{X}_{t_0}$ and $t \in [0, t_0]$ we define the so-called *(abstract) Volterra operator* $V_B$ on $\mathfrak{X}_{t_0}$ by

$$(V_B F)(t)x := \int_0^t T_{-1}(t-r)BF(r)x \, dr. \tag{8.1}$$

The integral has to be understood in the sense of a $\tau_{-1}$-Riemann integral. Notice that in general for $x \in X$ we have $(V_B F)(t)x \in X_{-1}$. For the formulation of our main result we need the following definition.

**Definition 8.1** Let $B \in \mathscr{L}(X, X_{-1})$ such that also $B : (X, \tau) \to (X_{-1}, \tau_{-1})$ is continuous. The operator $B$ is said to be *admissible*, if there is $t_0 > 0$ such that the following conditions are satisfied:

1. $V_B F(t)x \in X$ for all $t \in [0, t_0]$ and $x \in X$.
2. $\mathrm{Ran}(V_B) \subseteq \mathfrak{X}_{t_0}$.
3. $\|V_B\| < 1$.

The set of all admissible operators $B : (X, \tau) \to (X_{-1}, \tau_{-1})$ will be denoted by $\mathcal{S}_{t_0}^{DS,\tau}$. We write $B \in \mathcal{S}_{t_0}^{DS,\tau}(T)$ whenever it is important to emphasize for which semigroup $(T(t))_{t \geq 0}$ the operator $B$ is admissible.

**Remark 8.2** By construction $A_{-1} : (X, \tau) \to (X_{-1}, \tau_{-1})$ is continuous, and actually an isomorphism. In particular, we have

$$\forall p \in \mathcal{P} \, \exists L > 0 \, \exists \gamma \in \mathcal{P}_{-1} \forall x \in X : \; p(x) \leq L(\gamma(x) + \gamma(A_{-1}x)).$$

This section contains the formulation of the Desch–Schappacher type perturbation result and its proof. Observe that the proof of the following theorem is at some points verbatim the same as the one of [101, Chap. III, Theorem 3.1].

**Theorem 8.3** *Let $(A, D(A))$ be the generator of a bi-continuous semigroup $(T(t))_{t \geq 0}$ on a bi-admissible space $(X, \| \cdot \|, \tau)$. Let $B : X \to X_{-1}$ such that $B \in \mathcal{S}_{t_0}^{DS,\tau}$ for some $t_0 > 0$. Then the operator $(A_{-1} + B)_{|X}$ with domain*

$$D((A_{-1} + B)_{|X}) := \{x \in X : \; A_{-1}x + Bx \in X\}$$

*generates a bi-continuous semigroup $(S(t))_{t \geq 0}$ on $X$ with respect to $\tau$. Moreover, the semigroup $(S(t))_{t \geq 0}$ satisfies the variation of parameters formula*

$$S(t)x = T(t)x + \int_0^t T_{-1}(t-r)BS(r)x \, dr, \tag{8.2}$$

*for every $t \geq 0$ and $x \in X$.*

***Proof*** Since $\|V_B\| < 1$ by hypothesis, we conclude that $1 \in \rho(V_B)$. Now let $t > 0$ be arbitrary and write $t = nt_0 + t_1$ for $n \in \mathbb{N}$ and $t_1 \in [0, t_0)$. Define

$$S(t) := ((R(1, V_B)T_{|[0,t_0]})^n(t_0) \cdot (R(1, V_B)T_{|[0,t_0]})(t_1).$$

We first show that $(S(t))_{t \geq 0}$ is a semigroup. For $0 \leq s, t \leq s + t \leq t_0$ and $n \in \mathbb{N}$ we prove the following identity (cf. [101, p. 184])

$$(V_B^n T)(t + s) = \sum_{k=0}^{n} (V_B^{n-k} T_{|[0,t_0]})(s) \cdot (V_B^k T)(t), \quad \forall n \in \mathbb{N} \tag{8.3}$$

by induction. We abbreviate $V := V_B$. Since $V^0 = I$, Eq. (8.3) is trivially satisfied for $n = 0$. Now assume that (8.3) is true for some $n \in \mathbb{N}$. Then we obtain by this hypothesis that

$$\sum_{k=0}^{n+1} (V^{n+1-k}T)(s) \cdot (V^k T)(t)$$

$$= \sum_{k=0}^{n} \left( \int_0^s T_{-1}(s - r)BV^{n-k}T(r)\, dr \right) \cdot V^k T(t) + T(s) \int_0^t T_{-1}(t - r)BV^n T(r)\, dr$$

$$= \int_0^s T_{-1}(s - r)B \sum_{k=0}^{n} V^{n-k}T(r) \cdot V^k T(t)\, dr + \int_0^t T_{-1}(s + t - r)BV^n T(r)\, dr$$

$$= \int_0^s T_{-1}(s - r)BV^n T(r + t)\, dr + \int_0^t T_{-1}(s + t - r)BV^n T(r)\, dr$$

$$= \int_t^{s+t} T_{-1}(s + t - r)BV^n T(r)\, dr + \int_0^t T_{-1}(s + t - r)BV^n T(r)\, dr$$

$$= V^{n+1} T(s + t).$$

By this we can conclude that $(S(t))_{t \geq 0}$ satisfies the semigroup law for $0 \leq s, t \leq s + t \leq t_0$. Indeed, for each $t \in [0, t_0]$ the point evaluation $\delta_t : \mathfrak{X}_{t_0} \to \mathcal{L}(X)$ is a contraction and since $\|V\| < 1$ by hypothesis the inverse of $I - V$ is given by the Neumann series. Therefore,

$$S(t) = \delta_t \left( \sum_{n=0}^{\infty} V^n T \right) = \sum_{n=0}^{\infty} (V^n T)(t), \quad t \in [0, t_0].$$

Moreover, we have

$$\left\| (V^n T)(t) \right\| \leq \left\| V^n \right\| \cdot \left\| T_{|[0,t_0]} \right\|,$$

and we conclude that the series above converges absolutely. Hence

$$S(s)S(t) = \sum_{n=0}^{\infty} (V^n T)(s) \cdot \sum_{n=0}^{\infty} (V^n T)(t)$$

$$= \sum_{n=0}^{\infty} \sum_{k=0}^{n} (V^{n-k} T)(s)(V^k T)(t)$$

$$= \sum_{n=0}^{\infty} (V^n T)(s+t) = S(s+t).$$

Now we show that $S(t)S(s) = S(t+s)$ for all $t, s > 0$. For that let $t, s > 0$ be arbitrary and $n, m \in \mathbb{N}$ and $t_1, t_2 \in [0, t_0)$ such that $t = nt_0 + t_1$ and $s = mt_0 + t_2$. Then we obtain the following

$$\begin{aligned}
S(t)S(s) &= S(t_0)^n S(t_1) S(t_0)^m S(t_2) \\
&= S(t_0)^n S(t_0)^m S(t_1) S(t_2) \\
&= \begin{cases} S(t_0)^{n+m} S(t_1 + t_2), & \text{if } t_1 + t_2 < t_0, \\ S(t_0)^{n+m+1} S(t_2 - (t_0 - t_1)), & \text{if } t_1 + t_2 \geq t_0. \end{cases}
\end{aligned}$$

But in both cases the right-hand side equals $S(t+s)$ by definition. Hence $(S(t))_{t \geq 0}$ satisfies the semigroup law. The next step is to show that it is a bi-continuous semigroup with respect to $\tau$. Notice that

$$S_{|[0,t_0]}(t) = R(1, V_B) T_{|[0,t_0]}(t)$$

and hence $(S(t))_{t \geq 0}$ is locally bounded and the set $\{S(t) : t \in [0, t_0]\}$ is bi-equicontinuous. For $t > 0$ let $m := \left\lfloor \frac{t}{t_0} \right\rfloor$ and notice that $\left\{ S(t_0)^k : 1 \leq k \leq m \right\}$ is bi-equicontinuous, hence we conclude that the set

$$\left\{ S(t_0)^k : 1 \leq k \leq m \right\} \cdot \{S(s) : s \in [0, t_0]\}$$

is also bi-equicontinuous. By definition of $(S(t))_{t \geq 0}$ we obtain $\tau$-strong continuity, and hence $(S(t))_{t \geq 0}$ is a bi-continuous semigroup with respect to $\tau$. We now prove

$$S(t)x = T(t)x + \int_0^t T_{-1}(t-r)BS(r)x \, dr$$

for each $t > 0$ and $x \in X$ by proceeding similarly to [101, Chap. III, Sect. 3]. For $t = nt_0 + t_1, n \in \mathbb{N}$ and $t_1 \in [0, t_0)$, we obtain:

$$\int_0^t T_{-1}(t-r)BS(r) \, dr$$

$$= \sum_{k=0}^{n-1} \int_{kt_0}^{(k+1)t_0} T_{-1}(t-r)BS(r) \, dr + \int_{nt_0}^{t} T_{-1}(t-r)BS(r) \, dr$$

$$= \sum_{k=0}^{n-1} T_{-1}(t - (k+1)t_0) \int_0^{t_0} T_{-1}(t_0 - r) B S(r) \, dr \cdot S(kt_0) + \int_0^{t_1} T_{-1}(t_1 - r) B S(r) \, dr \cdot S(nt_0)$$

$$= \sum_{k=0}^{n-1} T(t - (k+1)t_0)(S(t_0) - T(t_0)) S(kt_0) + (S(t_1) - T(t_1)) S(nt_0) = S(t) - T(t).$$

The next step is to show that the resolvent set of $(A_{-1} + B)_{|X}$ is non-empty. For this we claim that $R(\lambda, A_{-1})B$ is bounded with $\|R(\lambda, A_{-1})B\| < 1$ for $\lambda$ large enough. Choose $M \geq 0$ and $\omega \in \mathbb{R}$ such that $\|T(t)\| \leq Me^{\omega t}$ for all $t > 0$. Then for $\lambda > \omega$ we obtain:

$$R(\lambda, A_{-1})B = \int_0^\infty e^{\lambda r} T_{-1}(r) B \, dr = \sum_{n=0}^\infty e^{-\lambda n t_0} T(nt_0)(V_B F_\lambda)(t_0)$$

where $F_\lambda(r) := e^{-\lambda(t_0 - r)} I \in \mathfrak{X}_{t_0}$. From this we obtain the following estimate:

$$\|R(\lambda, A_{-1})B\| \leq \|V_B\| + \frac{Me^{(\omega - \lambda)t_0}}{1 - e^{(\omega - \lambda)t_0}} \|V_B\| .$$

Since $\|V_B\| < 1$ we conclude for sufficient large $\lambda$:

$$\|R(\lambda, A_{-1})B\| < 1.$$

This yields $1 \in \rho(R(\lambda, A_{-1})B)$ for large $\lambda$ and then invertibility of $\lambda - (A_{-1} + B)_{|X}$, since

$$\lambda - (A_{-1} + B)_{|X} = (\lambda - A)(I - R(\lambda, A_{-1})B).$$

Hence the resolvent set of $(A_{-1} + B)_{|X}$ contains each sufficiently large $\lambda$. In the last step we will show that $(A_{-1} + B)_{|X}$ is actually the generator of the bi-continuous semigroup $(S(t))_{t \geq 0}$ with respect to $\tau$. Denote by $(C, D(C))$ the generator of $(S(t))_{t \geq 0}$. Let $\lambda > \max(\omega_0(T), \omega_0(S))$, then by the variation of constants formula (8.2), the resolvent representation as Laplace transform (Sect. 3.3) and the fact that we may interchange the improper $\tau$-Riemann integral and the $\tau_{-1}$-Riemann integral by an application of Lemma 3.10 we obtain

$$R(\lambda, C) = R(\lambda, A) + R(\lambda, A_{-1}) B R(\lambda, C).$$

Whence we conclude

$$(I - R(\lambda, A_{-1})B) R(\lambda, C) = R(\lambda, A),$$

and therefore

$$I = (\lambda - A)(I - R(\lambda, A_{-1})B) R(\lambda, C) = (\lambda - (A_{-1} + B)_{|X}) R(\lambda, C).$$

It follows that $C \subseteq (A_{-1} + B)_{|X}$ and by the previous observations $C = (A_{-1} + B)_{|X}$. $\quad\square$

### 8.1.1  Abstract Favard Spaces and Comparison of Semigroups

In this subsection we want to combine the perturbation theory with the theory of abstract Favard spaces as described in Sect. 4.2. In the next proposition we show that Desch–Schappacher perturbations of bi-continuous semigroups, which satisfy a special range condition concerning the extrapolated Favard class, gives us semigroups which are close to each other in some sense.

**Proposition 8.4** *Let $(T(t))_{t\geq 0}$ be a bi-continuous semigroup on a bi-admissible space $(X, \|\cdot\|, \tau)$ generated by $(A, D(A))$. Suppose that $B \in \mathcal{S}_{t_0}^{DS,\tau}$ with $\mathrm{Ran}(B) \subseteq F_0(A)$ and let $(S(t))_{t\geq 0}$ be the perturbed semigroup. Then there exists $C \geq 0$ such that for each $t \in [0, 1]$ one has*

$$\|T(t) - S(t)\| \leq Ct.$$

***Proof*** We may assume $\omega_0(T) < 0$. We find $M \geq 0$ such that $\|T(t)\| \leq M$ and $\|S(t)\| \leq M$ for every $t \in [0, 1]$. Since $\mathrm{Ran}(B) \subseteq F_0(A)$ we conclude that $A_{-1}^{-1}B : X \to F_1(A)$. Hence $A_{-1}^{-1}B$ is bounded by the closed graph theorem and we find $K \geq 0$ such that $\left\|A_{-1}^{-1}Bx\right\|_{F_1(A)} \leq K\|x\|$ for each $x \in X$. Let $\mathcal{P}_{-1}$ the family of seminorms corresponding to the first extrapolation space (see Sect. 5.1.1). By using (8.2) we obtain

$$
\begin{aligned}
\|S(t)x - T(t)x\| &= \left\| A_{-1} \int_0^t T(t-r)A_{-1}^{-1}BS(r)x \, dr \right\| \\
&= \left\| \tau\lim_{h\to 0} \frac{T_{-1}(h) - I}{h} \int_0^t T(t-r)A_{-1}^{-1}BS(r)x \, dr \right\| \\
&= \left\| \tau\lim_{h\to 0} \int_0^t \frac{T(h) - I}{h} T(t-r)A_{-1}^{-1}BS(r)x \, dr \right\| \\
&= \sup_{p\in\mathcal{P}_{-1}} \lim_{h\to 0} p\left( \int_0^t \frac{T(h) - I}{h} T(t-r)A_{-1}^{-1}BS(r)x \, dr \right) \\
&\leq \sup_{p\in\mathcal{P}_{-1}} \lim_{h\to 0} \int_0^t p\left( \frac{T(h) - I}{h} T(t-r)A_{-1}^{-1}BS(r)x \right) dr \\
&\leq \limsup_{h\to 0} \int_0^t \left\| \frac{T(h) - I}{h} T(t-r)A_{-1}^{-1}BS(r)x \right\| dr \\
&\leq M \int_0^t \left\| A_{-1}^{-1}BS(r)x \right\|_{F_1(A)} dr \\
&\leq tKM^2 \cdot \|x\|
\end{aligned}
$$

for each $x \in X$ and $t \in [0, 1]$. $\qquad\square$

**Corollary 8.5** *Let $(T(t))_{t\geq 0}$ be a bi-continuous semigroup on a bi-admissible space $(X, \|\cdot\|, \tau)$ generated by $(A, D(A))$. If $B \in \mathcal{S}_{t_0}^{DS,\tau}$ and $\mathrm{Ran}(B) \subseteq F_0(A)$, then the perturbed semigroup $(S(t))_{t\geq 0}$ leaves the space of strong continuity $\underline{X} := \overline{D(A)}^{\|\cdot\|}$ invariant.*

## 8.2  Admissible Operators

Next we consider a sufficient condition for $B : (X, \tau) \to (X_{-1}, \tau_{-1})$ to be admissible. Throughout this section we denote the space of continuous functions $f : [0, t_0] \to (X, \tau)$ which are $\|\cdot\|$-bounded by $C_b ([0, t_0], (X, \tau))$. If equipped with the sup-norm, $C_b ([0, t_0], (X, \tau))$ becomes a Banach space. The proof of the following theorem is at points verbatim the same as the one of [101, Chap. III, Corollary 3.3].

**Theorem 8.6** *Let $(T(t))_{t\geq 0}$ be a bi-continuous semigroup with generator $(A, D(A))$ on a bi-admissible space $(X, \|\cdot\|, \tau)$. Let $\mathcal{P}$ be the set of generating continuous seminorms corresponding to $\tau$. Let $B \in \mathscr{L}(X, X_{-1})$ such that $B : (X, \tau) \to (X_{-1}, \tau_{-1})$ is a linear and continuous operator, and let $t_0 > 0$ be such that*

*(a)* $\displaystyle \int_0^{t_0} T_{-1}(t_0 - r)Bf(r)\, dr \in X$ *for each $f \in C_b ([0, t_0], (X, \tau))$.*

*(b) For every $\varepsilon > 0$ and every $p \in \mathcal{P}$ there exist $q \in \mathcal{P}$ and $K > 0$ such that for all $f \in C_b ([0, t_0], (X, \tau))$*

$$p \left( \int_0^{t_0} T_{-1}(t_0 - r)Bf(r)\, dr \right) \leq K \cdot \sup_{r \in [0, t_0]} |q(f(r))| + \varepsilon \|f\|_\infty . \tag{8.4}$$

*(c) There exists $M \in (0, 1)$ such that for all $f \in C_b ([0, t_0], (X, \tau))$*

$$\left\| \int_0^{t_0} T_{-1}(t_0 - r)Bf(r)\, dr \right\| \leq M \|f\|_\infty . \tag{8.5}$$

*Then $B \in \mathcal{S}_{t_0}^{DS,\tau}$, and as a consequence the operator $(A_{-1} + B)_{|X}$ defined on the domain*

$$D((A_{-1} + B)_{|X}) := \{x \in X : A_{-1}x + Bx \in X\}$$

*generates a bi-continuous semigroup with respect to $\tau$.*

***Proof*** We first show $\mathrm{Ran}(V_B) \subseteq \mathfrak{X}_{t_0}$. Let $f \in C_b ([0, t_0], (X, \tau))$ and define for $t \in [0, t_0]$ the auxiliary function $f_t : [0, t_0] \to X$ by

$$f_t(r) := \begin{cases} f(0), & r \in [0, t_0 - t], \\ f(r + t - t_0), & r \in [t_0 - t, t_0]. \end{cases}$$

Then $f_t \in C_b([0, t_0], (X, \tau))$ and

$$\int_0^t T_{-1}(t - r) B f(r) \, dr = \int_0^{t_0} T_{-1}(t_0 - r) B f_t(r) \, dr - \int_t^{t_0} T_{-1}(r) B f(0) \, dr. \quad (8.6)$$

By Theorem 3.3

$$\int_t^{t_0} T_{-1}(r) B f(0) \, dr = T(t) \int_0^{t_0 - t} T_{-1}(r) B f(0) \, dr \in D(A_{-1}) = X.$$

We conclude that the map $\psi : [0, t_0] \to X_{-1}$ defined by

$$\psi(t) := \int_0^t T_{-1}(t - r) B f(r) \, dr \quad (8.7)$$

has values in $X$. Moreover, for $\varepsilon > 0$ and $p \in \mathcal{P}$ we have the following estimate:

$$p(\psi(t) - \psi(s))$$
$$= p\left( \int_0^t T_{-1}(t - r) B f(r) \, dr - \int_0^s T_{-1}(s - r) B f(r) \, dr \right)$$
$$\leq p\left( \int_0^{t_0} T_{-1}(t_0 - r) B (f_t(r) - f_s(r)) \, dr \right) + p\left( \int_s^t T_{-1}(r) B f(0) \, dr \right)$$
$$\leq K \cdot \sup_{r \in [0, t_0]} q(f_t(r) - f_s(r)) + p\left( \int_s^t T_{-1}(r) B f(0) \, dr \right) + \varepsilon \| f_t - f_s \|_\infty$$
$$\leq K \cdot \sup_{r \in [0, t_0]} q(f_t(r) - f_s(r))$$
$$\quad + L \cdot \left( \gamma \left( \int_s^t T_{-1}(r) B f(0) \, dr \right) + \gamma \left( (T_{-1}(t) - T_{-1}(s)) B f(0) \right) \right) + \varepsilon \| f_t - f_s \|_\infty$$
$$\leq K \cdot \sup_{r \in [0, t_0]} |q(f_t(r) - f_s(r))|$$
$$\quad + L \cdot \left( \int_s^t \gamma(T_{-1}(r) B f(0)) \, dr + \gamma \left( (T_{-1}(t) - T_{-1}(s)) B f(0) \right) \right) + 2\varepsilon \| f \|_\infty$$

where the $\gamma \in \mathcal{P}_{-1}$ of the second to last inequality comes from Remark 8.2. The extrapolated semigroup $(T_{-1}(t))_{t \geq 0}$ is strongly $\tau_{-1}$-continuous and $\gamma \in \mathcal{P}_{-1}$, so that we can find $\delta_1 > 0$ such that

$$\gamma\left( (T_{-1}(t) - T_{-1}(s)) B f(0) \right) < \varepsilon \text{ whenever } |t - s| < \delta_1.$$

Moreover, $f$ is $\tau$-continuous and therefore uniformly $\tau$-continuous on compact sets, which gives us $\delta_2 > 0$ such that

$$\sup_{r\in[0,t_0]} |q(f_t(r) - f_s(r))| < \varepsilon \text{ if } |t - s| < \delta_2.$$

Last but not least, $\gamma(T_{-1}(r)Bf(0))$ is bounded by some constant $M > 0$, so for $\delta_3 = \frac{\varepsilon}{M}$ we have

$$|s - t| < \delta_3 \implies \int_s^t \gamma(T_{-1}(r)Bf(0))\, dr < \varepsilon.$$

Now, we take $\delta := \min\{\delta_1, \delta_2, \delta_3\}$ and obtain

$$p(\psi(t) - \psi(s)) < (K + 2\|f\|_\infty + 2L)\varepsilon,$$

showing that $\psi : [0, t_0] \to X$ is $\tau$-continuous.

Next, we prove the norm-boundedness using the same techniques and arguments as in [101, Chap. III, Sect. 3]. Let $f \in C_b([0, t_0], (X, \tau))$ and write

$$f = \widetilde{f}_\delta + h_\delta,$$

where

$$h_\delta(r) := \begin{cases} \left(1 - \frac{r}{\delta}\right) f(r), & 0 \le r < \delta, \\ 0, & \delta \le r \le t_0 \end{cases}$$

for some $\delta > 0$. Then $\widetilde{f}_\delta$ and $h_\delta$ are norm-bounded and continuous with respect to $\tau$, $\widetilde{f}_\delta(0) = 0$ and $\|\widetilde{f}_\delta\|_\infty \le \|f\|_\infty$. Now we obtain

$$\left\| \int_0^t T_{-1}(t - r)Bf(r)\, dr \right\| \le \left\| \int_0^t T_{-1}(t - r)B\widetilde{f}_\delta(r)\, dr \right\| + \left\| \int_0^t T_{-1}(t - r)Bh_\delta(r)\, dr \right\|$$

$$\le M \|\widetilde{f}_\delta\|_\infty + K \left( \left\| \int_0^t T_{-1}(t - r)Bh_\delta(r)\, dr \right\|_{-1} + \left\| A_{-1} \int_0^t T_{-1}(t - r)Bh_\delta(r)\, dr \right\|_{-1} \right)$$

$$\le M \|\widetilde{f}_\delta\|_\infty + K \left\| \int_0^\delta T_{-1}(t - r)\left(1 - \frac{r}{\delta}\right)Bf(0)\, dr \right\|_{-1}$$

$$+ K \left\| T_{-1}(t)Bf(0) - \frac{1}{\delta} \int_0^\delta T_{-1}(t - r)Bf(0)\, dr \right\|_{-1}.$$

By taking $\delta \searrow 0$ we obtain

$$\left\| \int_0^t T_{-1}(t - r)Bf(r)\, dr \right\| \le M \|f\|_\infty. \tag{8.8}$$

We proceed with showing local bi-equicontinuity. For that let $(x_n)_{n\in\mathbb{N}}$ be a norm-bounded $\tau$-null-sequence. Let $\varepsilon > 0$ and $p \in \mathscr{P}$, then by taking $f^n(r) = f(r)x_n$, we can find $q \in \mathcal{P}$ such that

$$p\left(V_B F(t)x_n\right) = p\left(\int_0^t T_{-1}(t-r)BF(r)x_n\,\mathrm{d}r\right)$$

$$\leq p\left(\int_0^{t_0} T_{-1}(t_0-r)Bf^n(r)\,\mathrm{d}r - \int_t^{t_0} T_{-1}(r)Bf^n(0)\,\mathrm{d}r\right)$$

$$\leq K\cdot\sup_{r\in[0,t_0]}\left|q(f^n(r))\right| + p\left(\int_t^{t_0} T_{-1}(r)Bf^n(0)\,\mathrm{d}r\right) + \varepsilon\left\|f_t^n\right\|_\infty$$

$$\leq K\cdot\sup_{r\in[0,t_0]}\left|q(f^n(r))\right| + \varepsilon\left\|f_t^n\right\|_\infty$$

$$+ L\cdot\left(\gamma\left(\int_t^{t_0} T_{-1}(r)Bf^n(0)\,\mathrm{d}r\right) + \gamma\left((T_{-1}(t_0)-T_{-1}(t))\,Bf^n(0)\right)\right).$$

Now we can argue by the local bi-equicontinuity of $(T(t))_{t\geq 0}$ and $(T_{-1}(t))_{t\geq 0}$ and with the arbitrarily small $\varepsilon > 0$ to conclude the local bi-equicontinuity of $V_B F$. Hence we see that $V_B$ maps $\mathfrak{X}_{t_0}$ to $\mathfrak{X}_{t_0}$ and by (8.8) that $\|V_B\| < 1$ since by assumption $M \in (0,1)$.  $\square$

---

## 8.3    Examples

### 8.3.1    The Translation Semigroup

In this section we want to give an application of the Desch–Schappacher perturbation result to an explicit example. Recall from Sect. 2.2.1 that on the space $X = C_b(\mathbb{R})$ the left-translation semigroup $(T(t))_{t\geq 0}$ is bi-continuous with respect to the compact-open topology $\tau_{co}$. Moreover, by Sect. 5.2.1 the resulting extrapolation spaces are given by:

$$\underline{X}_{-1} = \{F \in \mathscr{D}'(\mathbb{R}) : F = f - Df \text{ for some } f \in UC_b(\mathbb{R})\},$$
$$X_{-1} = \{F \in \mathscr{D}'(\mathbb{R}) : F = f - Df \text{ for some } f \in C_b(\mathbb{R})\},$$

where $UC_b(\mathbb{R})$ denotes the space of bounded uniformly continuous functions and $Df$ the distributional derivative of $f$. Recall that the generator of $(T(t))_{t\geq 0}$ is $A = \frac{\mathrm{d}}{\mathrm{d}x}$ with domain $D(A) := C_b^1(\mathbb{R})$, and $A_{-1} = D$ with domain $D(A_{-1}) = C_b(\mathbb{R})$, where $D$ denotes the distributional derivative. The extrapolated semigroup $(T_{-1}(t))_{t\geq 0}$ is the restriction to $X_{-1}$ of the left-translation semigroup on the space $\mathscr{D}'(\mathbb{R})$ of distributions.

Consider the function $g : \mathbb{R} \to \mathbb{R}$ defined by

$$g(x) = \begin{cases} 0, & x \leq -1,\ x > 1, \\ x, & -1 < x \leq 0, \\ 2-x, & 0 < x \leq 1. \end{cases} \tag{8.9}$$

The graph of this function is the following.

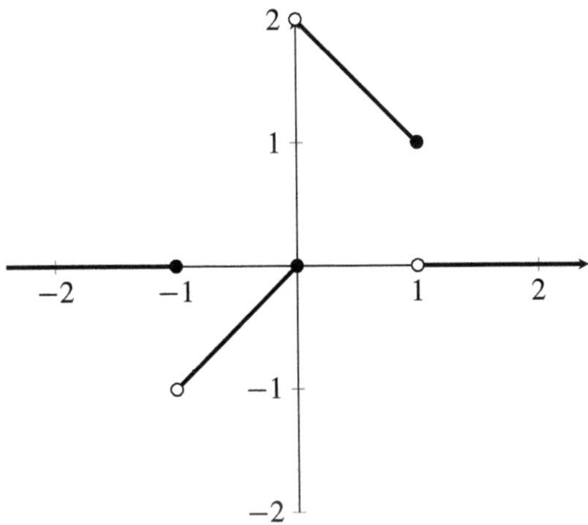

Notice that $g \in X_{-1}$, since $g = h - Dh$ where $h$ is the tent function on the real line defined by

$$h(x) = \begin{cases} 0, & x \leq -1, \ x > 1, \\ x + 1, & -1 < x \leq 0, \\ -x + 1, & 0 < x \leq 1. \end{cases}$$

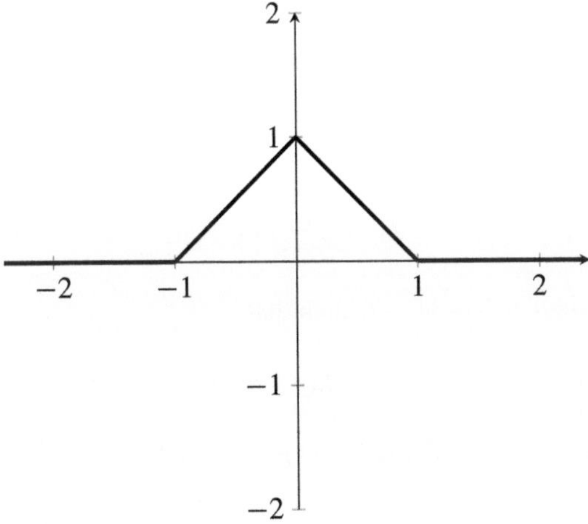

We now construct an operator $B \in \mathscr{L}(X, X_{-1})$ satisfying all conditions of Theorem 8.6, i.e., $((A_{-1} + B)_{|X}, D((A_{-1} + B)_{|X}))$ generates a bi-continuous semigroup. For this purpose let $\mu$ be a bounded regular Borel measure on $\mathbb{R}$ and define the continuous functional $\Phi : C_b(\mathbb{R}) \to \mathbb{R}$ by $\Phi(f) = \int_{\mathbb{R}} f \, d\mu$ and the operator $B : X \to X_{-1}$ by

$$Bf := \Phi(f)g.$$

This operator $B$ is by construction continuous with respect to the local convex topologies on the spaces $X$ and $X_{-1}$, and also for the norms. Moreover, $B$ has all properties required in Theorem 8.6. To see this let $f \in C_b([0, t_0], (X, \tau))$ be arbitrary. Define a map $\psi : \mathbb{R} \to \mathbb{R}$ by

$$\psi(\cdot) = \int_0^{t_0} T_{-1}(t_0 - r) Bf(r)(\cdot) \, dr.$$

Observe that

$$T_{-1}(t_0 - r)Bf(r)(x) = T_{-1}(t_0 - r)\Phi(f(r))g(x) = \Phi(f(r))g(x + t_0 - r).$$

We claim that $\psi$ is continuous. Indeed, let $\varepsilon > 0$ be arbitrary, and notice that by substitution for each $x \in \mathbb{R}$

$$\int_0^{t_0} \Phi(f(r))g(x + t_0 - r) \, dr = \int_x^{x+t_0} \Phi(f(x + t_0 - s))g(s) \, ds.$$

After this substitution we can make the following calculation for each $x, y \in \mathbb{R}$

$$\begin{aligned}
\psi(x) - \psi(y) &= \int_x^{x+t_0} \Phi(f(x + t_0 - s))g(s) \, ds - \int_y^{y+t_0} \Phi(f(x + t_0 - s))g(s) \, ds \\
&= \int_0^{x+t_0} \Phi(f(x + t_0 - s))g(s) \, ds - \int_0^{x} \Phi(f(x + t_0 - s))g(s) \, ds \\
&\quad - \int_0^{y+t_0} \Phi(f(y + t_0 - s))g(s) \, ds + \int_0^{y} \Phi(f(y + t_0 - s))g(s) \, ds \\
&= \int_{y+t_0}^{x+t_0} \left(\Phi(f(x + t_0 - s) - f(y + t_0 - s))\right) g(s) \, ds \\
&\quad + \int_x^{y} \left(\Phi(f(x + t_0 - s) - f(y + t_0 - s))\right) g(s) \, ds.
\end{aligned}$$

By the assumptions there exists $M > 0$ such that

$$\|(\Phi(f(x + t_0 - \cdot) - f(y + t_0 - \cdot))) g(\cdot)\|_\infty \leq M.$$

For $\delta := \frac{\varepsilon}{2M} > 0$ and for $x, y \in \mathbb{R}$ with $|x - y| < \delta$ we have

$$|\psi(x) - \psi(y)|$$

$$\leq \int_{y+t_0}^{x+t_0} |(\Phi(f(x+t_0-s) - f(y+t_0-s)))\,g(s)|\ ds$$

$$+ \int_{x}^{y} |(\Phi(f(x+t_0-s) - f(y+t_0-s)))\,g(s)|\ ds$$

$$\leq 2\,|x-y| \cdot \|(\Phi(f(x+t_0-\cdot) - f(y+t_0-\cdot)))\,g(\cdot)\|_{\infty}$$

$$\leq 2M \cdot |x-y| < \varepsilon.$$

This proves that $\psi \in C_b(\mathbb{R})$.

Observe that in general we only have

$$Q := \int_0^{t_0} T_{-1}(t_0-r)Bf(r)\ dr \in X_{-1},$$

so that point evaluation of this expression at $x \in \mathbb{R}$ does not make sense. We know, however, that $\psi \in C_b(\mathbb{R})$, and that the pointwise Riemann-sums $R_n(x)$ for the integral

$$\int_0^{t_0} \Phi(f(r))g(x+t_0-r)dr$$

converge for all $x \in \mathbb{R}$ to $\psi(x)$. If we can show that the sequence $(R_n)_{n \in \mathbb{N}}$ converges in the sense of distributions we can conclude that $Q := \int_0^{t_0} T_{-1}(t_0-r)Bf(r)\ dr = \psi \in X$. Let $\widetilde{\psi} \in \mathscr{D}(\mathbb{R})$ be a test function and define $\varphi := \widetilde{\psi} - \mathscr{D}\widetilde{\psi}$. Then

$$\langle (1-A_{-1})^{-1}R_n, \varphi \rangle \to \langle (1-A_{-1})^{-1}Q, \varphi \rangle.$$

By the meaning of this pairing we conclude that $\langle R_n, \widetilde{\psi} \rangle \to \langle Q, \widetilde{\psi} \rangle$. By the above we conclude that $Q \in X$.

The next step is to estimate the norm. Notice that

$$\left\| \int_0^{t_0} T_{-1}(t_0-r)Bf(r)\ dr \right\|_{\infty} = \sup_{x \in \mathbb{R}} \left| \int_0^{t_0} \Phi(f(r))g(x+t_0-r)\ dr \right|$$

$$\leq \sup_{x \in \mathbb{R}} \int_0^{t_0} |\Phi(f(r))| \cdot |g(x+t_0-r)|\ dr \leq 2 \int_0^{t_0} |\Phi(f(r))|\ dr$$

$$\leq 2 \int_0^{t_0} \int_{\mathbb{R}} |f(r)(x)|\ d\,|\mu|\,(x)\ dr \leq 2\,|\mu|\,(\mathbb{R}) \int_0^{t_0} \|f(r)\|_{\infty}\ dr$$

$$= 2\,|\mu|\,(\mathbb{R}) \int_0^{t_0} \|f(r)\|_{\infty}\ dr \leq 2\,|\mu|\,(\mathbb{R})t_0\,\|f\|_{\infty}.$$

In particular we can choose $t_0$ so small that $M := 2\,|\mu|\,(\mathbb{R})t_0 < 1$. Hence condition (c) of Theorem 8.6 is fulfilled. Condition (b) from Theorem 8.6 can be proven similarly. Let $K \subseteq \mathbb{R}$ be an arbitrary compact set and $\varepsilon > 0$. Then

$$p_K \left( \int_0^{t_0} T_{-1}(t_0 - r) B f(r)(x) \, dr \right) \leq \sup_{x \in K} \int_0^{t_0} |\Phi(f(r))| \cdot |g(x + t_0 - r)| \, dr$$

$$\leq 2 \sup_{x \in K} \int_0^{t_0} |\Phi(f(r))| \, dr$$

$$\leq 2 t_0 |\mu|(\mathbb{R}) \sup_{r \in [0, t_0]} \sup_{y \in K'} |f(r)(y)| + \varepsilon \|f\|_\infty ,$$

since by the regularity of the measure $\mu$ we choose $K' \subseteq \mathbb{R}$ such that $|\mu| (\mathbb{R} \setminus K') < \varepsilon$. By Theorem 8.6 we conclude that $(A_{-1} + B)_{|X}$ generates again a bi-continuous semigroup on $C_b(\mathbb{R})$ with respect to $\tau_{co}$. We now give an expression for the generator. Observe that $f \in D((A_{-1} + B)_{|X})$ if and only if $f \in C_b(\mathbb{R})$ and $f' + \Phi(f)g \in C_b(\mathbb{R})$ and this is precisely, when the following conditions are satisfied:

$$\begin{cases} \lim_{t \nearrow -1} \left( f'(t) + \Phi(f)g(t) \right) = \lim_{t \searrow -1} \left( f'(t) + \Phi(f)g(t) \right), \\ \lim_{t \nearrow 0} \left( f'(t) + \Phi(f)g(t) \right) = \lim_{t \searrow 0} \left( f'(t) + \Phi(f)g(t) \right), \\ \lim_{t \nearrow 1} \left( f'(t) + \Phi(f)g(t) \right) = \lim_{t \searrow 1} \left( f'(t) + \Phi(f)g(t) \right). \end{cases} \tag{8.10}$$

By the explicit expression for $g : \mathbb{R} \to \mathbb{R}$ we can rewrite Equation (8.10) as follows:

$$\begin{cases} \lim_{t \nearrow -1} f'(t) = \lim_{t \searrow -1} f'(t) - \Phi(f), \\ \lim_{t \nearrow 0} f'(t) = \lim_{t \searrow 0} f'(t) + 2\Phi(f), \\ \lim_{t \nearrow 1} f'(t) + \Phi(f) = \lim_{t \searrow 1} f'(t). \end{cases} \tag{8.11}$$

Or equivalently

$$\lim_{t \nearrow -1} f'(t) - \lim_{t \searrow -1} f'(t) = -\frac{1}{2} \left( \lim_{t \nearrow 0} f'(t) - \lim_{t \searrow 0} f'(t) \right) = \lim_{t \nearrow 1} f'(t) - \lim_{t \searrow 1} f'(t) = -\Phi(f).$$

$$\tag{8.12}$$

We see that the generator $(C, D(C))$ of the perturbed semigroup is given by

$$Cf = f' + \int_{\mathbb{R}} f \, d\mu \cdot g, \quad f \in D(C),$$

$$D(C) = \left\{ f \in C_b(\mathbb{R}) : f \in C_b^1(\mathbb{R} \setminus \{-1, 0, 1\}) \text{ and } (8.4) \text{holds} \right\} .$$

The previous example uses a function $g \in X_{-1}$ which has three points of discontinuity, with one sided limits at each of these points. We generalize this to a countable (discrete) set of jump discontinuities. For that assume that $g \in X_{-1}$ is a function such that $\|g\|_\infty < \infty$ and that the set of discontinuities of $g$ is discrete. One defines again an operator $B : X \to X_{-1}$

by

$$Bf := \Phi(f)g := \int_{\mathbb{R}} f \, d\mu \cdot g, \quad f \in C_b(\mathbb{R}).$$

Notice that none of previous calculations and arguments depend on the number of discontinuities (in fact, we only used that $g$ is bounded). So we can conclude that $(A_{-1} + B)_{|X}$ generates a bi-continuous semigroup on $X$ with respect to $\tau_{co}$. The only issue we have to care about are the conditions mentioned in (8.12), that is an "explicit" description of the domain. Let $Z := \{x_1, x_2, x_3, \ldots\}$ be the set of discontinuities of $g$ that is assumed to be discrete, and we suppose that all of these points are jump discontinuities. Let us define $a_n := \lim_{t \nearrow x_n} g(t)$ and $b_n := \lim_{t \searrow x_n} g(t)$. We observe that $f \in D((A_{-1} + B)_{|X})$ if and only if

$$\lim_{t \nearrow x_n} f'(t) + \Phi(f)a_n = \lim_{t \searrow x_n} f'(t) + \Phi(f)b_n, \quad \text{for each } n \in \mathbb{N},$$

or equivalently

$$\lim_{t \nearrow x_n} f'(t) - \lim_{t \searrow x_n} f'(t) = \Phi(f)(b_n - a_n), \quad \text{for each } n \in \mathbb{N}.$$

We conclude that the operator $(C, D(C))$ given by

$$Cf = f' + \int_{\mathbb{R}} f \, d\mu \cdot g,$$

$$D(C) = \left\{ f \in C_b(\mathbb{R}) : \ f \in C_b^1(\mathbb{R} \setminus Z), \lim_{t \nearrow x_n} f'(t) - \lim_{t \searrow x_n} f'(t) = \Phi(f)(b_n - a_n), \ n \in \mathbb{N} \right\}$$

generates a bi-continuous semigroup on $C_b(\mathbb{R})$ with respect to $\tau_{co}$.

### 8.3.2  The Implemented Semigroups

#### 8.3.2.1  Ideals in $\mathscr{L}(E)$ and Module Homomorphisms

Our purpose is to relate Desch–Schappacher perturbations of the $C_0$-semigroup $(T(t))_{t \geq 0}$ and Desch–Schappacher perturbations of the left implemented semigroup $(\mathcal{U}_L(t))_{t \geq 0}$. Before doing so we need some auxiliary results exploiting the algebraic structure of the domain of Hille–Yosida operators on $\mathscr{L}(E)$.

**Lemma 8.7** *Let $E$ be a Banach space and $(\mathcal{A}, D(\mathcal{A}))$ a Hille–Yosida operator on $\mathscr{L}(E)$, i.e., suppose there exists $\omega \in \mathbb{R}$ and $M \geq 1$ such that $(\omega, \infty) \subseteq \rho(\mathcal{A})$ and*

$$\left\| R(\lambda, \mathcal{A})^n \right\| \leq \frac{M}{(\lambda - \omega)^n},$$

*for each $\lambda > \omega$ and $n \in \mathbb{N}$. The following are equivalent:*

(a) $D(\mathcal{A})$ is a right ideal of the Banach algebra $\mathscr{L}(E)$, i.e., $CB \in D(\mathcal{A})$ whenever $C \in D(\mathcal{A})$, $B \in \mathscr{L}(E)$, and $\mathcal{A}$ is a right $\mathscr{L}(E)$-module homomorphism, i.e., $\mathcal{A}(CB) = \mathcal{A}(C)B$ for $C \in D(\mathcal{A})$ and $B \in \mathscr{L}(E)$.
(b) There exists a Hille–Yosida operator $(A, D(A))$ such that $\mathcal{A}(C) = AC$, where $D(\mathcal{A}) = \mathscr{L}(E, D(A))$.
(c) There exists a Hille–Yosida operator $(A, D(A))$ such that $\mathcal{A}(C) = A_{-1}C$, where $D(\mathcal{A}) = \{C \in \mathscr{L}(E) : A_{-1}C \in \mathscr{L}(E)\}$.

**Proof** The implication (c) $\Rightarrow$ (a) is just a checking of properties of an explicitly given operator. The equivalence (b) $\Leftrightarrow$ (c) follows from the fact that the operator $A$ and $A_{-1}$ coincide on the domain $D(A)$ of $A$.

(a) $\Rightarrow$ (b) By definition one has $R(\lambda, \mathcal{A}) \in \mathscr{L}(\mathscr{L}(E))$ whenever $\lambda \in \rho(\mathcal{A})$. Define for $\lambda \in \rho(\mathcal{A})$

$$R(\lambda) := R(\lambda, \mathcal{A})(I).$$

Since $R(\lambda, \mathcal{A})$, $\lambda \in \rho(\mathcal{A})$ satisfy the resolvent identity also $R(\lambda)$, $\lambda \in \rho(\mathcal{A})$ do:

$$R(\lambda) - R(\mu) = R(\lambda, \mathcal{A})(I) - R(\mu, \mathcal{A})(I) = (R(\lambda, \mathcal{A}) - R(\mu, \mathcal{A}))\,(I)$$
$$= ((\lambda - \mu)R(\lambda, \mathcal{A})R(\mu, \mathcal{A}))\,(I) = (\lambda - \mu)R(\lambda)R(\mu)$$

for each $\lambda, \mu \in \rho(\mathcal{A})$. Hence the family $(R(\lambda))_{\lambda \in \rho(\mathcal{A})}$ is a pseudoresolvent. If $R(\lambda)x = 0$ for some $x \in E$, then

$$0 = \lambda R(\lambda)x = \lambda R(\lambda, \mathcal{A})(I)x.$$

But since $\lambda R(\lambda, \mathcal{A})(I) \to I$ as $\lambda \to \infty$, it follow that $x = 0$. Therefore $R(\lambda)$ is injective, and hence there exists a closed operator $(A, D(A))$ such that $R(\lambda) = R(\lambda, A)$, i.e.,

$$R(\lambda, A) = R(\lambda, \mathcal{A})(I).$$

Let $C \in D(\mathcal{A})$, i.e., $C = R(\lambda, \mathcal{A})D$ for some $D \in \mathscr{L}(E)$. Then

$$\mathcal{A}(C) = \mathcal{A}(R(\lambda, \mathcal{A})D) = \lambda R(\lambda, \mathcal{A})D - D = (\lambda R(\lambda, \mathcal{A}) - I)D$$
$$= (\lambda R(\lambda, A) - (\lambda - A)R(\lambda, A))D = AR(\lambda, A)D = AC.$$

$\square$

**Lemma 8.8** Let $E$ be a Banach space and $(\mathcal{A}, D(\mathcal{A}))$ a generator of a bi-continuous semigroup $(\mathcal{T}(t))_{t \geq 0}$ on $\mathscr{L}(E)$ with respect to $\tau_{\mathrm{sot}}$. The following are equivalent:

(a) $D(\mathcal{A})$ is a right-ideal of $\mathscr{L}(E)$ and $\mathcal{A}$ is a right $\mathscr{L}(E)$-module homomorphism.
(b) The semigroup $(\mathcal{T}(t))_{t \geq 0}$ is left implemented, i.e., there exists a $C_0$-semigroup $(S(t))_{t \geq 0}$ such that $\mathcal{T}(t)C = S(t)C$ for each $t \geq 0$.

*Under these equivalent conditions, if $(B, D(B))$ is the generator of the $C_0$-semigroup $(S(t))_{t\geq 0}$, then $\mathcal{A}(C) = B_{-1}C$ for each $C \in \mathcal{L}(E, E_{-1}) = X_{-1}(\mathcal{A})$.*

**Proof** (b) $\Rightarrow$ (a) : If $C \in D(\mathcal{A})$, then the limit

$$(\mathcal{A}C)(x) := \lim_{t \searrow 0} \frac{T(t)Cx - Cx}{t},$$

exists for each $x \in X$. Since $(T(t))_{t\geq 0}$ is left implemented we obtain for $B \in \mathcal{L}(E)$

$$(\mathcal{A}(CB))(x) = \lim_{t \to 0} \frac{T(t)(CB)x - (CB)x}{t} = \lim_{t \to 0} \frac{(T(t)C)(Bx) - C(Bx)}{t},$$

and we conclude that $CB \in D(\mathcal{A})$ and $\mathcal{A}(CB) = \mathcal{A}(C)B$.

(a) $\Rightarrow$ (b) : For $\lambda \in \rho(\mathcal{A})$, $C \in D(\mathcal{A})$, $B \in \mathcal{L}(E)$ one has

$$(\lambda - \mathcal{A})(CB) = \lambda CB - \mathcal{A}(C)B = (\lambda C - \mathcal{A}(C))B.$$

Since $\lambda - \mathcal{A}$ is a bijective map we conclude that

$$R(\lambda, \mathcal{A})(DB) = (R(\lambda, \mathcal{A})D)B$$

for each $D \in \mathcal{L}(E)$. By the Euler-Formula (see [59, Theorem 4.6] and [101, Chap. II, Sect. 3]) we obtain

$$T(t)C = \tau_{\mathrm{sot}} \lim_{n \to \infty} \left( \frac{n}{t} R\left( \frac{n}{t}, \mathcal{A} \right) \right)^n C.$$

From this we deduce the equality

$$T(t)(CB)(x) = \left( \tau \lim_{n \to \infty} \left( \frac{n}{t} R\left( \frac{n}{t}, \mathcal{A} \right) \right)^n C \right) B = (T(t)C)B.$$

Set $S(t) := T(t)I$, and we are done

$$T(t)C = T(t)(I \cdot C) = (T(t)I)C = S(t)C.$$

Finally, $\mathcal{A}$ is multiplication operator by the generator $(B, D(B))$ of the semigroup $(S(t))_{t\geq 0}$ by Lemma 8.7. $\qquad\square$

**Proposition 8.9** *Let $(T(t))_{t\geq 0}$ and $(S(t))_{t\geq 0}$ be $C_0$-semigroups on the Banach space $E$, and let $(A, D(A))$ denote the generator of $(T(t))_{t\geq 0}$. Let $(\mathcal{U}(t))_{t\geq 0}$ and $(\mathcal{V}(t))_{t\geq 0}$ be the semigroups left implemented by $(T(t))_{t\geq 0}$ and $(S(t))_{t\geq 0}$, respectively. Let $(\mathcal{G}, D(\mathcal{G}))$ be the generator of $(\mathcal{U}(t))_{t\geq 0}$ and let $\mathcal{K} : \mathcal{L}(E) \to \mathcal{L}(E, E_{-1}(A))$ be such that $\mathcal{K} \in \mathcal{S}_{t_0}^{DS, \tau_{\mathrm{sot}}}(\mathcal{U})$ and such that $C := (\mathcal{G}_{-1} + \mathcal{K})_{|\mathcal{L}(E)}$ (with maximal domain) is the generator of $(\mathcal{V}(t))_{t\geq 0}$. Then $\mathcal{K}$ has the property that*

$$\mathcal{K}(CD) = \mathcal{K}(C)D,$$

*for each $C, D \in \mathscr{L}(E)$.*

**Proof** Since by assumption $\mathcal{G}$ and $\mathcal{C} = (\mathcal{G}_{-1} + \mathcal{K})_{|\mathscr{L}(E)}$ both generate implemented semigroups we conclude by Lemma 8.8 that $\mathcal{G}$, and hence $\mathcal{G}_{-1}$, and $\mathcal{C}$ are all multiplication operators. One has $\mathcal{G}_{-1}(C) = A_{-1}C$ for each $C \in \mathscr{L}(E)$ and there exists an operator $M : E \to E_{-1}(L)$ such that $\mathcal{C}(C) = MC$ for each $C \in D(\mathcal{C})$. We conclude that

$$\mathcal{K}(C) = MC - A_{-1}C$$

for each $C \in D(\mathcal{C})$. Since $(\mathcal{C}, D(\mathcal{C}))$ is bi-dense in $\mathscr{L}(E)$, for each $C \in \mathscr{L}(E)$, there exists a sequence of operators $(C_n)_{n\in\mathbb{N}}$ in $D(\mathcal{C})$ such that $\sup_{n\in\mathbb{N}} \|C_n\| < \infty$ and

$$C_n x \to C x,$$

for each $x \in E$. The continuity of $\mathcal{K}$ and $\mathcal{G}_{-1}$ yields

$$\mathcal{K}(C_n) \overset{\tau_{\mathrm{sot}}}{\to} \mathcal{K}(C),$$

$$\mathcal{G}(C_n) \overset{\tau_{\mathrm{sot}}}{\to} \mathcal{G}(C)$$

with convergence in $\mathscr{L}_{\mathrm{sot}}(E, E_{-1}(A))$. Therefore, for each $x \in E$ the sequence $(MC_n x)_{n\in\mathbb{N}}$ is Cauchy in $E_{-1}(A)$ and we can define

$$Lx := \lim_{n\to\infty} MC_n x, \quad x \in E.$$

By construction we obtain $L \in \mathscr{L}(E, E_{-1}(A))$ and $\mathcal{C}(C) = LC$ for each $C \in \mathscr{L}(E)$ and therefore

$$\mathcal{K}(C) = A_{-1}C + LC,$$

for $C \in D(\mathcal{C})$. Now we define $B := L + A_{-1}$ as an operator in $\mathscr{L}(E, E_{-1}(A))$ and conclude that

$$\mathcal{K}(C) = BC$$

for each $C \in \mathscr{L}(E)$ and that was to be proven. $\qquad\square$

### 8.3.2.2 A One-to-One Correspondence

We are now prepared to relate Desch–Schappacher perturbations of the implemented semigroup with the perturbations of the underlying $C_0$-semigroup. To do so we have to use the class of Desch–Schappacher admissible operators $\mathcal{S}_{t_0}^{DS}$ for $C_0$-semigroups. Recall from [101, Chap. III, Sect. 3a] the following definitions for a strongly continuous semigroup $(T(t))_{t\geq 0}$ on a Banach space $E$. We define

$$\mathcal{S}_{t_0}^{DS}(T) := \{B \in \mathscr{L}(E, E_{-1}) : V_B \in \mathscr{L}\left(C\left([0, t_0], \mathscr{L}_{\mathrm{sot}}(E)\right)\right), \ \|V_B\| < 1\},$$

where $V_B$ denotes the corresponding Volterra operator on $E$ defined by

$$(V_B F)(t) := \int_0^t T_{-1}(t - r) B F(r)\, dr, \quad F \in C\left([0, t_0]\,, E\right),\ t \in [0, t_0]\,.$$

The following result shows that Desch–Schappacher perturbations of a $C_0$-semigroup always give us Desch–Schappacher perturbations of the corresponding implemented semigroup.

**Theorem 8.10** *Let $(\mathcal{U}(t))_{t\geq 0}$ be the semigroup on $\mathscr{L}(E)$ left implemented by the $C_0$-semigroup $(T(t))_{t\geq 0}$. Suppose that $B \in \mathcal{S}_{t_0}^{DS}$ and let $(S(t))_{t\geq 0}$ be the perturbed $C_0$-semigroup. Define the operator $\mathcal{K} : \mathscr{L}(E) \to \mathscr{L}(E, E_{-1})$ by*

$$\mathcal{K}S := BS, \quad S \in \mathscr{L}(E).$$

*Then $\mathcal{K} \in \mathcal{S}_{t_0}^{DS, \tau_{\mathrm{sot}}}$ and the perturbed semigroup $(\mathcal{V}(t))_{t\geq 0}$ is left implemented by $(S(t))_{t\geq 0}$.*

**Proof** First of all we show that $V_{\mathcal{K}} F(t)C \in \mathscr{L}(E)$ for $F \in \mathfrak{X}_{t_0}, t \in [0, t_0]$ and $C \in \mathscr{L}(E)$. Define $f \in C\left([0, t_0]\,, \mathscr{L}_{\mathrm{sot}}(E)\right)$ by $f(r) := F(r)C$ and observe

$$(V_{\mathcal{K}} F)(t)Cx = \int_0^t \mathcal{U}_{-1}(t - r)\mathcal{K} F(r)Cx\, dr = \int_0^t T_{-1}(t - r)B f(r)x\, dr.$$

Since by assumption $B \in \mathcal{S}_{t_0}^{DS}$, we obtain $(V_{\mathcal{K}} F)(t)Cx \in E$. The following estimate will be crucial for what follows

$$\|(V_{\mathcal{K}} F)(t)Cx\| = \left\| \int_0^t \mathcal{U}_{-1}(t - r)\mathcal{K} F(r)Cx\, dr \right\| = \left\| \int_0^t T_{-1}(t - r)B F(r)Cx \right\|$$

$$= \|(V_B f)(t)x\| \leq \|V_B\| \cdot \|f\| \cdot \|Cx\| \leq \|V_B\| \cdot \|f\| \cdot \|C\| \cdot \|x\|\,.$$

This estimate shows that $(V_{\mathcal{K}} F)(t)C \in \mathscr{L}(E)$. Moreover, we directly see that $\mathrm{Ran}(V_{\mathcal{K}}) \subseteq \mathfrak{X}_{t_0}$, since $\tau_{\mathrm{sot}}$-strong continuity, norm boundedness and bi-equicontinuity of $V_{\mathcal{K}} F$ follow also from the previous estimate. Also the fact that $\|V_{\mathcal{K}}\| < 1$ is immediate, due to the assumption that $B \in \mathcal{S}_{t_0}^{DS}$. Finally, we show that $(\mathcal{G}_{-1} + \mathcal{K})_{|\mathscr{L}(E)}$ generates the semigroup left implemented by $(S(t))_{t\geq 0}$. For this notice that for sufficiently large $\lambda > 0$ we have

$$R(\lambda, (A_{-1} + B)_E)Cx = \int_0^\infty e^{-\lambda t} S(t)Cx\, dt = \int_0^\infty e^{-\lambda t} \mathcal{V}(t)Cx\, dt$$

$$= R(\lambda, (\mathcal{G}_{-1} + \mathcal{K})_{|\mathscr{L}(E)})Cx,$$

for all $x \in E$ and $C \in \mathscr{L}(E)$. Whence we conclude that $(\mathcal{G}_{-1} + \mathcal{K})_{|\mathscr{L}(E)}$ generates the semigroup left implemented by $(S(t))_{t\geq 0}$. $\qquad\square$

The converse of Theorem 8.10 is also true.

**Theorem 8.11** *Let $(\mathcal{U}(t))_{t\geq 0}$ and $(\mathcal{V}(t))_{t\geq 0}$ be two semigroups on $\mathscr{L}(E)$, left implemented by the $C_0$-semigroups $(T(t))_{t\geq 0}$ and $(S(t))_{t\geq 0}$, respectively. Let $(A, D(A))$ be the generator of $(T(t))_{t\geq 0}$ and let $\mathcal{K} \in \mathcal{S}_{t_0}^{DS,\tau_{\mathrm{sot}}}(\mathcal{U})$ be such that $(\mathcal{V}(t))_{t\geq 0}$ is the corresponding perturbed semigroup. Define $B \in \mathscr{L}(E, E_{-1})$ by*

$$Bx := (\mathcal{K}I)x, \quad x \in E.$$

*Then $B \in \mathcal{S}_{t_0}^{DS}(T)$ and $(A_{-1} + B)_{|E}$ generates $(S(t))_{t\geq 0}$.*

**Proof** Let $f \in C([0, t_0], \mathscr{L}_{\mathrm{sot}}(E))$ and $x \in E$. We observe that by Lemma 8.7 one has that $(\mathcal{K}I)f(r) = \mathcal{K}(If(r)) = \mathcal{K}f(r)$ for each $r \in [0, t_0]$. For $f \in C([0, t_0], \mathscr{L}_{\mathrm{sot}}(E))$ we define $F \in \mathfrak{X}_{t_0}$ by $F(r) := M_{f(r)}$, the multiplication with $f(r)$, i.e., $F(r)C = f(r)C$ for each $C \in \mathscr{L}(E)$. The following computation is crucial for the proof:

$$\begin{aligned}
V_B f(t)x &= \int_0^t T_{-1}(t - r)Bf(r)x \, dr = \int_0^t T_{-1}(t - r)(\mathcal{K}I)f(r)x \, dr \\
&= \int_0^t \mathcal{U}_{-1}(t - r)\mathcal{K}f(r)x \, dr = \int_0^t \mathcal{U}_{-1}(t - r)\mathcal{K}F(r)Ix \, dr \\
&= (V_{\mathcal{K}}F)(t)Ix.
\end{aligned}$$

From this and from the assumption that $\mathcal{K} \in \mathcal{S}_{t_0}^{DS,\tau_{\mathrm{sot}}}$, we conclude that $B \in \mathcal{S}_{t_0}^{DS}$. Moreover we have

$$S(t)x = \mathcal{V}(t)Ix = \mathcal{U}(t)Ix + \int_0^t \mathcal{U}_{-1}(t - r)\mathcal{K}\mathcal{V}(r)Ix \, dr = T(t)x + \int_0^t T_{-1}(t - r)BS(r)x \, dr,$$

for each $x \in E$. This yields that $(A_{-1} + B)_{|E}$ generates $(S(t))_{t\geq 0}$. $\qquad\square$

Summarizing Theorems 8.11 and 8.10 we can state the following.

**Corollary 8.12** *Let $(\mathcal{U}(t))_{t\geq 0}$ and $(\mathcal{V}(t))_{t\geq 0}$ be two semigroups on $\mathscr{L}(E)$ left implemented by the $C_0$-semigroups $(T(t))_{t\geq 0}$ and $(S(t))_{t\geq 0}$ on $E$, respectively. Let us denote the generators of $(\mathcal{U}(t))_{t\geq 0}$ and $(T(t))_{t\geq 0}$ by $(\mathcal{G}, D(\mathcal{G}))$ and $(A, D(A))$, respectively. The following are equivalent:*

(i) *There exists $\mathcal{K} \in \mathcal{S}_{t_0}^{DS,\tau_{\mathrm{sot}}}(\mathcal{U})$ such that $(\mathcal{V}(t))_{t\geq 0}$ is generated by $(\mathcal{G}_{-1} + \mathcal{K})_{|\mathscr{L}(E)}$.*
(ii) *There exists $B \in \mathcal{S}_{t_0}^{DS}(T)$ such that $(S(t))_{t\geq 0}$ is generated by $(A_{-1} + B)_{|E}$.*

**Remark 8.13** Notice that not every Desch–Schappacher perturbation of an implemented semigroup gives again an implemented semigroup. To see this let $(\mathcal{G}, D(\mathcal{G}))$ be the generator of the left implemented semigroup $(\mathcal{U}(t))_{t\geq 0}$ and $\Phi \in (\mathscr{L}(E), \tau_{\mathrm{sot}})'$. Define, as above, an operator $\mathcal{K} : \mathscr{L}(E) \to \mathscr{L}(E, E_{-1})$ by

$$\mathcal{K}(C) := \Phi(C)\mathcal{G}_{-1}(I), \quad C \in \mathscr{L}(E).$$

Such an operator $\mathcal{K}$ is not multiplicative if $\Phi \neq 0$.

### 8.3.2.3 Comparisons

Now we relate comparison properties of the implemented semigroup and properties of the underlying $C_0$-semigroup. First of all, for $B \in \mathscr{L}(E)$ we define the multiplication operator $M_B \in \mathscr{L}(\mathscr{L}(E), \mathscr{L}(E))$ by $M_B S := BS$. Then one has $\|M_B\| = \|B\|$. By taking $B := T(t) - S(t)$ for $t > 0$ we directly obtain the following result.

**Lemma 8.14** *Let $(\mathcal{U}(t))_{t \geq 0}$ and $(\mathcal{V}(t))_{t \geq 0}$ be two semigroups on $\mathscr{L}(E)$ left implemented by the $C_0$-semigroups $(T(t))_{t \geq 0}$ and $(S(t))_{t \geq 0}$, respectively. Then the following are equivalent:*

*(a) There exists $M \geq 0$ such that $\|\mathcal{U}(t) - \mathcal{V}(t)\| \leq Mt$ for each $t \in [0, 1]$.*
*(b) There exists $M \geq 0$ such that $\|T(t) - S(t)\| \leq Mt$ for each $t \in [0, 1]$.*

Recall from [59, Prop. 6.1] that the Favard spaces of the implemented semigroup and of the underlying semigroup for $\alpha \in [0, 1]$ are connected by

$$F_\alpha(\mathcal{U}) = \mathscr{L}(E, F_\alpha(T)). \tag{8.13}$$

This yields to the following result.

**Lemma 8.15** *For $B \in \mathscr{L}(E, E_{-1})$ we define $\mathcal{K} : \mathscr{L}(E) \to \mathscr{L}(E, E_{-1})$ by $\mathcal{K}S := BS$. Then $\operatorname{Ran}(\mathcal{K}) \subseteq F_\alpha(\mathcal{U})$ if and only if $\operatorname{Ran}(B) \subseteq F_\alpha(T)$.*

By [9] and [59] the extrapolated implemented semigroup is defined by

$$\mathcal{U}_{-1}(t)S = T_{-1}(t)S, \quad S \in \overline{\mathscr{L}(E)}^{\mathscr{L}_{\mathrm{sot}}(E,E_{-1})} = \mathscr{L}(E, E_{-1}).$$

This gives

$$F_0(\mathcal{U}) = F_1(\mathcal{U}_{-1}) = \mathscr{L}(E, F_1(T_{-1})) = \mathscr{L}(E, F_0(T)).$$

**Proposition 8.16** *Let $(\mathcal{U}(t))_{t \geq 0}$ and $(\mathcal{V}(t))_{t \geq 0}$ be two semigroups on $\mathscr{L}(E)$ left implemented by $(T(t))_{t \geq 0}$ and $(S(t))_{t \geq 0}$, respectively. Furthermore, let $(\mathcal{G}, D(\mathcal{G}))$ denote the generator of $(\mathcal{U}(t))_{t \geq 0}$. Suppose that there exists $M \geq 0$ such that*

$$\|\mathcal{U}(t) - \mathcal{V}(t)\| \leq Mt$$

*for each $t \in [0, 1]$. Then there exists $\mathcal{K} \in \mathcal{S}_{t_0}^{DS, \tau_{\mathrm{sot}}}$ with $\operatorname{Ran}(\mathcal{K}) \subseteq F_0(\mathcal{G})$.*

***Proof*** Since $\|\mathcal{U}(t) - \mathcal{V}(t)\| \leq Mt$ for each $t \in [0, 1]$ we can use Lemma 8.14 to conclude that $\|T(t) - S(t)\| \leq Mt$ for each $t \in [0, 1]$. If $(A, D(A))$ denotes the generator of $(T(t))_{t \geq 0}$, then by [101, Chap. III, Theorem 3.9] we find $B \in \mathcal{L}(E, E_{-1})$ such that $B \in \mathcal{S}_{t_0}^{DS}$ and $\mathrm{Ran}(B) \subseteq F_0(A)$. As in Theorem 8.10 this gives rise to an multiplication operator $\mathcal{K} : \mathcal{L}(E) \to \mathcal{L}(E, E_{-1})$ defined by

$$\mathcal{K}S := BS, \quad S \in \mathcal{L}(E).$$

By Lemma 8.15 we conclude that $\mathrm{Ran}(\mathcal{K}) \subseteq F_0(\mathcal{G})$. It remains to show that $(\mathcal{G}_{-1} + \mathcal{K})_{|\mathcal{L}(E)}$ generates $(\mathcal{V}(t))_{t \geq 0}$. But, by [101, Chap. III, Theorem 3.9], $(A_{-1} + B)_{|E}$ generates $(S(t))_{t \geq 0}$. $\qquad\square$

Combining Propositions 8.16, 8.4 and [101, Chap. III, Theorem 3.9] we obtain the following theorem.

**Theorem 8.17** *Let $(\mathcal{U}(t))_{t \geq 0}$ and $(\mathcal{V}(t))_{t \geq 0}$ be two semigroups on $\mathcal{L}(E)$ left implemented by $(T(t))_{t \geq 0}$ and $(S(t))_{t \geq 0}$, respectively. Denote by $(\mathcal{G}, D(\mathcal{G}))$ the generator of $(\mathcal{U}(t))_{t \geq 0}$ and by $(A, D(A))$ the generator of $(T(t))_{t \geq 0}$. If $\mathcal{K} \in \mathcal{S}_{t_0}^{DS, \tau_{\mathrm{sot}}}(\mathcal{U})$ such that $\mathrm{Ran}(\mathcal{K}) \subseteq F_0(\mathcal{G})$, then there exists $B \in \mathcal{S}_{t_0}^{DS}(T)$ with $\mathrm{Ran}(B) \subseteq F_0(A)$ such that $\mathcal{K}S = BS$ for each $S \in \mathcal{L}(E)$.*

***Proof*** By Proposition 8.4 we find $M \geq 0$ such that $\|\mathcal{U}(t) - \mathcal{V}(t)\| \leq Mt$ for each $t \in [0, 1]$. Following the proof of Proposition 8.16 there exists $B \in \mathcal{S}_{t_0}^{DS}$ such that $\mathrm{Ran}(B) \subseteq F_0(A)$. $\qquad\square$

From this we can deduce the following equivalence.

**Theorem 8.18** *Let $(\mathcal{U}(t))_{t \geq 0}$ and $(\mathcal{V}(t))_{t \geq 0}$ be two semigroups on $\mathcal{L}(E)$ left implemented by the $C_0$-semigroups $(T(t))_{t \geq 0}$ and $(S(t))_{t \geq 0}$ on $E$, respectively. Let us denote the generators of $(\mathcal{U}(t))_{t \geq 0}$ and $(T(t))_{t \geq 0}$ by $(\mathcal{G}, D(\mathcal{G}))$ and $(A, D(A))$, respectively. The following are equivalent:*

(a) *There exists $\mathcal{K} \in \mathcal{S}_{t_0}^{DS, \tau_{\mathrm{sot}}}(\mathcal{U})$ such that $\mathrm{Ran}(\mathcal{K}) \subseteq F_0(\mathcal{G})$ and such that $(\mathcal{V}(t))_{t \geq 0}$ is generated by $(\mathcal{G}_{-1} + \mathcal{K})_{|\mathcal{L}(E)}$.*

(b) *There exists $B \in \mathcal{S}_{t_0}^{DS}(T)$ such that $\mathrm{Ran}(B) \subseteq F_0(A)$ and such that $(S(t))_{t \geq 0}$ is generated by $(A_{-1} + B)_{|E}$.*

## Notes on This Chapter

As suggested by G. Greiner in [123] abstract perturbation theory of one-parameter semigroups provides good means to change the domain of a semigroup generator. For this an enlargement of the underlying Banach space may be necessary and extrapolation spaces become important. One of the well-known results in this direction goes back to the papers of W. Desch and W. Schappacher, see [83] and [84]. Another prominent example of such general perturbation techniques is due to Staffans and Weiss [222, 223], and an elegant abstract operator theoretic/algebraic approach has been developed by Adler, Bombieri and Engel in [1]. A general theory of unbounded domain perturbations is given by Hadd, Manzo and Rhandi [127]. A more recent paper by Bátkai, Jacob, Voigt and Wintermayr [32] extends the notion of positivity to extrapolation spaces, and studies positive perturbations for positive semigroups on AM-spaces. Hence, the study of abstract Desch–Schappacher type perturbations is a lively research field. The reason for such an active interest in this area is that the range of application is vast. We mention here only a selection from the most recent ones: boundary perturbations by Nickel [195], boundary feedback by Casarino, Engel, Nagel and Nickel [66], boundary control by Engel, Kramar Fijavž, Klöss, Nagel and Sikolya [99] and Engel and Kramar Fijavž [98], port-Hamiltonian systems by Baroun and Jacob [28], control theory by Jacob, Nabiullin, Partington and Schwenninger [136, 137] and Jacob, Schwenninger and Zwart [138] and vertex control in networks by Engel and Kramar Fijavž [97, 100]. This chapter is mainly based on the work by Budde and Farkas [60]. There is also work on positive Desch–Schappacher perturbations for bi-continuous semigroups that is not included here [52].

# Part III
# Applications and Theoretical Advances

# Bi-continuous Cosine Families

**9**

In this chapter, we want to study the second-order autonomous abstract Cauchy problem of the form

$$\begin{cases} \ddot{u}(t) = Au(t), & t \geq 0, \\ u(0) = x \in X, \\ \dot{u}(0) = y \in X, \end{cases}$$

for some linear operator $(A, D(A))$ on a Banach space $X$. In the classical Banach space setting, the so-called cosine families serve as solutions to this problem.

In this chapter, we want to use the idea of bi-continuous semigroups to introduce the concept of bi-continuous cosine families. We strongly believe that a systematic study of cosine families via the bi-continuous approach can be as fruitful as the theory of bi-continuous semigroups.

## 9.1  Bi-continuous Cosine Families

The following definition of our main objects is inspired by a combination of Definition 2.4 and [221, Definition 2.2].

**Definition 9.1** Let $(X, \|\cdot\|, \tau)$ be a bi-admissible space. We call a family of bounded linear operators $(C(t))_{t \geq 0}$ a *bi-continuous cosine family* if the following holds:

(i)  $2C(t)C(s) = C(t+s) + C(t-s)$ and $C(0) = I$ for all $t \geq s \geq 0$.

(ii)  $(C(t))_{t \geq 0}$ is strongly $\tau$-continuous.

(iii)  There exist $M \geq 1$ and $\omega \in \mathbb{R}$ such that $\|C(t)\| \leq M e^{\omega t}$ for each $t \geq 0$.

© The Author(s), under exclusive license to Springer Nature Switzerland AG 2026
C. Budde, *Bi-Continuous Operator Semigroups*, Frontiers in Mathematics,
https://doi.org/10.1007/978-3-032-12948-2_9

(iv)  $(C(t))_{t\geq 0}$ is *locally-bi-equicontinuous*, i.e., if $(x_n)_{n\in\mathbb{N}}$ is a norm-bounded sequence in $X$ which is $\tau$-convergent to 0, then also $(C(s)x_n)_{n\in\mathbb{N}}$ is $\tau$-convergent to 0 uniformly for $s \in [0, t_0]$ for each fixed $t_0 \geq 0$.

**Remark 9.2** Similarly to the definition of bi-continuous semigroups, cf. Definition 2.4, we also require exponential boundedness for bi-continuous cosine families, see Definition 9.1(iii). This assumption is reasonable in view of [221, Proposition 2.4].

For the following result, we follow the lines of the proof of [221, Theorem 2.7].

**Lemma 9.3** *Let $(C(t))_{t\geq 0}$ be a bi-continuous cosine family on a bi-admissible space $(X, \|\cdot\|, \tau)$. Then $(C(t))_{t\geq 0}$ is strongly $\tau$-continuous $\Longleftrightarrow$ $(C(t))_{t\geq 0}$ is strongly $\tau$-continuous in $t = 0$.*

***Proof*** The implication "$\Longrightarrow$" is of course trivial. For the other implication, we argue by contradiction. Assume that there exists $x_0 \in X$ and $t_0 > 0$ such that $t \mapsto C(t)x_0$ is not continuous at $t_0 > 0$ with respect to $\tau$. For $n \in \mathbb{N}$ and $p \in \mathcal{P}$, we define

$$K_{n,p} := \sup\left\{ p(C(t)x_0 - C(s)x_0) : |t - t_0| < \frac{t_0}{8n}, \; |s - t_0| < \frac{t_0}{8n}, \; t \geq s \geq 0\right\}.$$

By construction, one has $K_{n,p} \geq 0$ and $K_{n+1,p} \leq K_{n,p}$ for all $n \in \mathbb{N}$. Hence, there exists $K \in \mathbb{R}$ such that $K_{n,p} \to K$. Due to the discontinuity assumption, there exists $p \in \mathcal{P}$ such that $K > 0$. Let us choose such a $p \in \mathcal{P}$. The previous construction also yields $K_{n,p} \geq K$ for all $n \in \mathbb{N}$. The latter fact will eventually lead to the desired contradiction.

The definition of $K_{n,p}$ by means of suprema yields that there exist sequences $(\tau_n)_{n\in\mathbb{N}}$ and $(\sigma_n)_{n\in\mathbb{N}}$ with $\tau_n < \sigma_n$ for all $n \in \mathbb{N}$ such that

$$|\tau_n - t_0| < \frac{t_0}{8n}, \tag{9.1}$$

$$|\sigma_n - t_0| < \frac{t_0}{8n} \tag{9.2}$$

$$p(C(\tau_n)x_0 - C(\sigma_n)x_0) \geq K_{n,p} - \frac{1}{n}. \tag{9.3}$$

Let us show that $2\tau_n - \sigma_n > 0$ for all $n \in \mathbb{N}$. For this, it suffices to show that $\tau_n > \sigma_n - \tau_n$ for all $n \in \mathbb{N}$. From (9.1)–(9.2) it actually follows that $\sigma_n - \tau_n \leq \frac{t_0}{4n}$ and

$$\tau_n \geq t_0 - \frac{t_0}{8n} \geq \frac{4t_0}{8n} > \frac{t_0}{2n}.$$

We also deduce that

$$p(C(\sigma_{4n})x_0 - C(2\tau_{4n} - \sigma_{4n})x_0) < K_{n,p}, \tag{9.4}$$

for all $n \in \mathbb{N}$, as $|\sigma_{4n} - t_0| < \frac{t_0}{8n}$ and $|2\tau_{4n} - \sigma_{4n} - t_0| < \frac{t_0}{8n}$ by (9.1)–(9.2). We will now finally derive the desired contradiction by showing that $K_{n,p} \to 0$ for $n \to \infty$. Indeed, by Definition 9.1(i) we obtain that

$$C(t - s) - C(t + s) + 2(C(t + s) - C(t)) = 2C(t)(C(s) - I),$$

for all $t \geq s \geq 0$, so that

$$2p(C(t + s)x - C(t)x) \leq 2p(C(t)(C(s) - I)x) + p(C(t + s)x - C(t - s)x),$$

for all $x \in X$ and $t \geq s \geq 0$. By choosing $t = \tau_{4n}$ and $s = \sigma_{4n} - \tau_{4n}$ as well as $x = x_0$ we obtain

$$2p(C(\sigma_{4n})x_0 - C(\tau_{4n})x_0) \leq 2p(C(\tau_{4n})(C(\sigma_{4n} - \tau_{4n}) - I)x_0) + p(C(\sigma_{4n})x_0 - C(2\tau_{4n} - \sigma_{4n})x), \quad n \in \mathbb{N}.$$

Now Eqs. (9.3)–(9.4) together yield

$$2\left(K_{n,p} - \frac{1}{4n}\right) \leq 2p(C(\tau_{4n})(C(\sigma_{4n} - \tau_{4n}) - I)x_0) + K_{n,p}, \quad n \in \mathbb{N}$$

or, equivalently,

$$K_{n,p} \leq \frac{1}{2n} + 2p(C(\tau_{4n})(C(\sigma_{4n} - \tau_{4n}) - I)x_0), \quad n \in \mathbb{N}.$$

We notice that by assumption $(C(t))_{t \geq 0}$ is strongly $\tau$-continuous at $t = 0$. This implies that $(C(\sigma_{4n} - \tau_{4n}) - I)x_0 \to 0$ for $n \to \infty$ with respect to $\tau$. Moreover, by (9.1)–(9.2) and Definition 9.1(iii) one obtains

$$\|(C(\sigma_{4n} - \tau_{4n}) - I)x_0\| \leq (Me^{\frac{\omega t_0}{16n}} + 1)\,\|x_0\|,$$

so that the sequence $((C(\sigma_{4n} - \tau_{4n}) - I)x_0)_{n \in \mathbb{N}}$ is also $\|\cdot\|$-bounded. Now, by Definition 9.1, the family $(C(t))_{t \geq 0}$ is also locally bi-equicontinuous, i.e., for the sequence $y_n := (C(\sigma_{4n} - \tau_{4n}) - I)x_0$ it holds that $C(s)y_n \to 0$ for $n \to \infty$ with respect to $\tau$ uniformly for $s \in [0, t_0]$. We observe, that certainly $(\sigma_{4n})_{n \in \mathbb{N}}$ is a bounded sequence and hence $p(C(\tau_{4n})(C(\sigma_{4n} - \tau_{4n}) - I)x_0) \to 0$ for $n \to \infty$, which indeed shows that $K_{n,p} \to 0$ for $n \to \infty$. Contradiction. $\qquad\square$

From Lemma 9.3 we can deduce that the operators in the cosine family $(C(t))_{t \geq 0}$ commute. We could also follow the proof of [17, Lemma 3.14.3(b)], however, we want to stay self-contained and stick to a proof that is similar to [221, Lemma 2.8 & Theorem 2.9].

**Proposition 9.4** *Let $(C(t))_{t \geq 0}$ be a bi-continuous cosine family on a bi-admissible space $(X, \|\cdot\|, \tau)$. Then $C(t)C(s) = C(s)C(t)$ for all $t \geq s \geq 0$.*

***Proof*** From Definition 9.1(i) we obtain by induction that for each $t \geq 0$ and $n \in \mathbb{N}$ there exist constants $\alpha_0, \ldots, \alpha_n$ such that $C(nt) = \alpha_0 I + \alpha_1 C(t) + \alpha_2 C(t)^2 + \cdots + \alpha_n C(t)^n$. This implies that for all $t \geq 0$ one has that $C(mt)C(nt) = C(nt)C(mt)$ for all $n, m \in \mathbb{N}$. For $t_1 = \frac{r}{2^k}$ and $t_2 = \frac{s}{2^l}$, $r, k, s, l \in \mathbb{N}$, one has $C(t_1)C(t_2) = C(t_2)C(t_1)$. However, it is a common fact that the dyadic rationals are dense. Finally, Lemma 9.3 yields the desired conclusion. $\qquad\square$

**Example 9.5** We follow the lines of [221, Example 2.27]. Let us consider the bi-admissible space $(\mathrm{C_b}(\mathbb{R}), \| \cdot \|_\infty, \tau_{\mathrm{co}})$. On $X = \mathrm{C_b}(\mathbb{R})$ we define a family $(C(t))_{t \geq 0}$ of operators by

$$(C(t)f)(x) := \frac{1}{2}\left( f(x+t) + f(x-t) \right), \quad t \geq 0, \ f \in \mathrm{C_b}(\mathbb{R}), \ x \in \mathbb{R}.$$

The first observation is that $C(t) \in \mathscr{L}(X)$ for all $t \geq 0$ since

$$\|C(t)f\|_\infty = \sup_{x \in \mathbb{R}} |(C(t)f)(x)| = \frac{1}{2} \sup_{x \in \mathbb{R}} |f(x+t) - f(x-t)| \leq \|f\|_\infty,$$

for all $f \in \mathrm{C_b}(\mathbb{R})$ (therefore, $(C(t))_{t \geq 0}$ is also exponentially bounded). Indeed, $(C(t))_{t \geq 0}$ also satisfies the functional equation from Definition 9.1(i), see also [221, Example 2.27(b)].

Let us first show that $(C(t))_{t \geq 0}$ is not strongly continuous with respect to $\|\cdot\|_\infty$. To do so, we take $f \in \mathrm{C_b}(\mathbb{R})$ defined by $f(x) = e^{ix^2}$, $x \in \mathbb{R}$. Then $\|C(t)f - f\|_\infty \not\to 0$ due to the fact that $f$ is indeed bounded and continuous but lacks uniform continuity. In fact, $\left|\frac{1}{2}\left( f(x+t) + f(x-t) \right) - f(x)\right|$ attains for sufficiently small $t > 0$ its maximum for $x = \frac{(2n-1)\pi}{2t}$ with maximum $2\left|\cos(\frac{t^2}{2})\right|$. For exactly this reason, one obtains that $(C(t))_{t \geq 0}$ is strongly continuous with respect to the compact-open topology $\tau_{\mathrm{co}}$. Later on, in Example 9.16, we will further investigate the bi-continuous cosine family $(C(t))_{t \geq 0}$.

## 9.2     The Generator

For bi-continuous cosine families, we are able to define the (infinitesimal) generator. In order to do so, we combine the definition of strongly continuous cosine families, cf. [221, Definition 2.12], as well as the definition of the generator of bi-continuous semigroups, cf. Chap. 3.

**Definition 9.6** Let $(C(t))_{t \geq 0}$ be a bi-continuous cosine family on a bi-admissible space $(X, \|\cdot\|, \tau)$. The *generator* $(A, D(A))$ of $(C(t))_{t \geq 0}$ is the linear operator on $X$ defined by

$$Ax := \tau\!\!\lim_{t \to 0} \frac{2}{t^2}(C(t)x - x),$$

$$D(A) := \left\{ x \in X : \tau\!\!\lim_{t \to 0} \frac{2}{t^2}(C(t)x - x) \text{ exists in } X, \ \sup_{(0,1]} \left\| \frac{2}{t^2}(C(t)x - x) \right\| < \infty \right\}.$$

We start with investigating some basic properties of bi-continuous cosine families and their generators.

**Lemma 9.7** *Let* $(C(t))_{t\geq 0}$ *be a bi-continuous cosine family on a bi-admissible space* $(X, \|\cdot\|, \tau)$ *and let* $(A, D(A))$ *denote its generator. If* $x \in D(A)$, *then* $C(t)x \in D(A)$ *and* $AC(t)x = C(t)Ax$ *for all* $t \geq 0$.

*Proof* Let $x \in D(A)$. Then for arbitrary $t_0 \geq 0$, we see that by Proposition 9.4 one has

$$AC(t_0)x = \tau\!\!\lim_{t \to 0} \frac{2}{t^2}(C(t)C(t_0)x - C(t_0)x) = \tau\!\!\lim_{t \to 0} \frac{2}{t^2}(C(t_0)(C(t)x - x)) = C(t_0)Ax.$$

$\square$

**Proposition 9.8** *Let* $(C(t))_{t\geq 0}$ *be a bi-continuous cosine family on a bi-admissible space* $(X, \|\cdot\|, \tau)$ *and let* $(A, D(A))$ *denote its generator. For all* $t \geq 0$ *and* $x \in X$ *one has*

$$\int_0^t (t - s)C(s)x \, ds \in D(A) \text{ and } A \int_0^t (t - s)C(s)x \, ds = C(t)x - x.$$

*Here, the integral has to be understood as a* $\tau$-*Riemann integral.*

**Remark 9.9** Similarly to the argumentation in Proposition 3.2, the existence of the $\tau$-Riemann integral in Proposition 9.8 follows from Assumption 2.1, Definition 9.1 and [146, Proposition 1.1]. In what follows, we will use $\tau$-Riemann integrals without mentioning it explicitly anymore.

**Proof of Proposition** 9.8. We subdivide the proof of this result into several parts. Recall that, by Lemma 9.3, the strong $\tau$-continuity and the strong $\tau$-continuity in $t = 0$ for $(C(t))_{t\geq 0}$ are equivalent. We will make use of this throughout the proof without referring to Lemma 9.3.

**Claim 1:** $\tau\!\!\lim_{\substack{h \to 0 \\ 0 < h < t}} \frac{1}{h^2} \left( \int_t^{t+h} (t - s)C(s)x \, ds - \int_{t-h}^t (t - s)C(s)x \, ds \right) = -C(t)x$ *for all* $t \geq$ $0, x \in X$.

Certainly, the identity

$$\frac{1}{h^2}\left(\int_t^{t+h}(s-t)\,\mathrm{d}s + \int_{t-h}^t (t-s)\,\mathrm{d}s\right) = 1,$$

holds. Let $\varepsilon > 0$ be arbitrary. Then for $p \in \mathcal{P}$ there exists $\delta > 0$ such that $p(C(s)x - C(t)x) < \varepsilon$ whenever $|s-t| < \delta$. Now choose $0 < h < t$ such that $|h| < \delta$. Then

$$p\left(\frac{1}{h^2}\left(\int_t^{t+h}(t-s)C(s)x\,\mathrm{d}s - \int_{t-h}^t (t-s)C(s)x\,\mathrm{d}s\right) + C(t)x\right)$$

$$\leq p\left(\frac{1}{h^2}\left(\int_t^{t+h}(s-t)(C(t)x - C(s)x)\,\mathrm{d}s - \int_{t-h}^t (t-s)(C(s)x - C(t)x)\,\mathrm{d}s\right)\right)$$

$$\leq \frac{1}{h^2}\left(\int_t^{t+h}(s-t)p(C(t)x - C(s)x)\,\mathrm{d}s - \int_{t-h}^t (t-s)p(C(s)x - C(t)x)\,\mathrm{d}s\right)$$

$$\leq \frac{\varepsilon}{h^2}\left(\int_t^{t+h}(s-t)\,\mathrm{d}s - \int_{t-h}^t (t-s)\,\mathrm{d}s\right) = \varepsilon.$$

**Claim 2:** $\tau\lim\limits_{h\to 0}\dfrac{2}{h^2}\displaystyle\int_0^h sC(s)x\,\mathrm{d}s = x$ for all $x \in X$.

Let $\varepsilon > 0$ and $x \in X$ be arbitrary. For $p \in \mathcal{P}$ we find $\delta > 0$ such that $p(C(s)x - x) < \varepsilon$ whenever $|s| < \delta$. For $|h| < \delta$ we then obtain

$$p\left(\frac{2}{h^2}\int_0^h sC(s)x\,\mathrm{d}s - x\right) = p\left(\frac{2}{h^2}\int_0^h s(C(s)x - x)\,\mathrm{d}s\right)$$

$$\leq \frac{2}{h^2}\int_0^h s\cdot p(C(s)x - x)\,\mathrm{d}s$$

$$< \frac{2\varepsilon}{h^2}\int_0^h s\,\mathrm{d}s = \varepsilon.$$

**Claim 3:** $\tau\lim\limits_{\substack{h\to 0\\ 0<h<t}}\dfrac{1}{h}\displaystyle\int_{t-h}^{t+h}C(s)x\,\mathrm{d}s = 2C(t)x$ for all $t \geq 0$ and $x \in X$.

Let $\varepsilon > 0$ and $x \in X$ be arbitrary. For $p \in \mathcal{P}$ we find $\delta > 0$ such that $p(C(s)x - C(t)x) < \varepsilon$ whenever $|s-t| < \delta$. For $|h| < \delta$ we obtain

$$p\left(\frac{1}{h}\int_{t-h}^{t+h}C(s)x\,\mathrm{d}s - 2C(t)x\right) = p\left(\frac{1}{h}\int_{t-h}^{t+h}C(s)x - C(t)x\,\mathrm{d}s\right)$$

$$\leq \frac{1}{h}\int_{t-h}^{t+h}p(C(s)x - C(t)x)\,\mathrm{d}s$$

$$< \frac{\varepsilon}{h}\int_{t-h}^{t+h}1\,\mathrm{d}s = \varepsilon$$

**Claim 4:** $\tau\lim\limits_{h\to 0}\dfrac{1}{h}\displaystyle\int_0^h C(s)x\,\mathrm{d}s = x$ for all $x \in X$.

Let $\varepsilon > 0$ and $x \in X$ be arbitrary. For $p \in \mathcal{P}$ we find $\delta > 0$ such that $p(C(s)x - x) < \varepsilon$ whenever $|s| < \delta$. For $|h| < \delta$ we obtain

$$
\begin{aligned}
p\left(\frac{1}{h}\int_0^h C(s)x\,\mathrm{d}s - x\right) &= p\left(\frac{1}{h}\int_0^h C(s)x - x\,\mathrm{d}s\right)\\
&\le \frac{1}{h}\int_0^h p(C(s)x - x)\,\mathrm{d}s\\
&< \frac{\varepsilon}{h}\int_0^h 1\,\mathrm{d}s = \varepsilon.
\end{aligned}
$$

We are now able to prove the desired result. Let $x \in X$ and $t > 0$ be arbitrary and define

$$
y = \int_0^t (t - s)C(s)x\,\mathrm{d}s.
$$

We will show that

$$
\tau\lim_{h\to 0}\frac{2}{h^2}(C(h)y - y) - C(t)x - x.
$$

For this assume that $0 < h < t$. By Definition 9.1 and Proposition 9.4, we can get in the same way as in the proof of [221, Lemma 2.14] that

$$
\begin{aligned}
&\frac{2}{h^2}(C(h)y - y)\\
={}&\frac{1}{h^2}\left(\int_t^{t+h}(t - s)C(s)x\,\mathrm{d}s - \int_{t-h}^t (t - s)C(s)x\,\mathrm{d}s\right) + \frac{2}{h^2}\int_0^h sC(s)x\,\mathrm{d}s\\
&+ \frac{1}{h}\int_{t-h}^{t+h} C(s)x\,\mathrm{d}s - \frac{1}{h}\int_0^h C(s)x\,\mathrm{d}s.
\end{aligned}
$$

By using Claim 1-4 we therefore obtain the desired result. $\qquad\square$

**Corollary 9.10** *Let $(C(t))_{t\ge 0}$ be a bi-continuous cosine family on a bi-admissible space $(X, \|\cdot\|, \tau)$ and let $(A, D(A))$ denote its generator. For all $t \ge 0$ and $x \in D(A)$ one has*

$$
\int_0^t (t - s)C(s)Ax\,\mathrm{d}s = C(t)x - x.
$$

*Proof* By Lemma 9.7 we have that $AC(t)x = C(t)Ax$ for all $t \ge 0$ and $x \in D(A)$. Proposition 9.8 yields that

$$
y := \int_0^t (t - s)C(s)x\,\mathrm{d}s \in D(A),
$$

where $t \geq 0$. By combining these two results we obtain

$$C(t)x - x = A \int_0^t (t - s)C(s)x \, ds = \tau\text{-}\lim_{h \to 0} \frac{2}{h^2}(C(h)y - y)$$

$$= \int_0^t (t - s)C(s) \cdot \tau\text{-}\lim_{h \to 0} \frac{2}{h^2}(C(h)x - x) \, ds = \int_0^t (t - s)C(s)Ax \, ds.$$

$\square$

**Lemma 9.11** *Let $(C(t))_{t \geq 0}$ be a bi-continuous cosine family on a bi-admissible space $(X, \|\cdot\|, \tau)$. For all $x \in X$ one has*

$$\frac{2}{t^2} \int_0^t (t - s)C(s)x \, ds \to x,$$

*whenever $t \to 0$ with respect to the locally convex topology $\tau$.*

**Proof** Let $\varepsilon > 0$ and $x \in X$ be arbitrary and take $p \in \mathcal{P}$. By the $\tau$-strong continuity of $(C(t))_{t \geq 0}$ there exists $\delta > 0$ such that $p(C(s)x - x) < \varepsilon$ whenever $|s| < \delta$. By taking $|t| < \delta$ we observe that

$$p\left(\frac{2}{t^2} \int_0^t (t - s)C(s)x \, ds - x\right) = p\left(\frac{2}{t^2} \int_0^t (t - s)(C(s)x - x) \, ds\right)$$

$$\leq \frac{2}{t^2} \int_0^t (t - s)p(C(s)x - x) \, ds$$

$$\leq \frac{2}{t^2} \int_0^t (t - s)\varepsilon \, ds = \varepsilon.$$

$\square$

**Corollary 9.12** *Let $(C(t))_{t \geq 0}$ be a bi-continuous cosine family on a bi-admissible space $(X, \|\cdot\|, \tau)$ and let $(A, D(A))$ denote its generator. Then $D(A)$ is bi-dense, i.e., for every $x \in X$ there exists a $\|\cdot\|$-bounded sequence $(x_n)_{n \in \mathbb{N}}$ in $D(A)$ such that $x_n \to x$ for $n \to \infty$ with respect to $\tau$.*

**Proof** This is an immediate consequence of Lemma 9.11 and Proposition 9.8. In fact, we set

$$x_n := 2n^2 \int_0^{\frac{1}{n}} \left(\frac{1}{n} - s\right)C(s)x \, ds, \quad n \in \mathbb{N}.$$

Then

$$\|x_n\| = \left\|2n^2 \int_0^{\frac{1}{n}} \left(\frac{1}{n} - s\right)C(s)x \, ds\right\| \leq 2n^2 \int_0^{\frac{1}{n}} \left(\frac{1}{n} - s\right)Me^{\omega s} \|x\| \, ds = \frac{2Mn(n(e^{\frac{\omega}{n}} - 1) - \omega)\|x\|}{\omega^2},$$

which stays uniformly bounded for all $n \in \mathbb{N}$. By Proposition 9.8 we have $x_n \in D(A)$ for all $n \in \mathbb{N}$. Finally, Lemma 9.11 yields that $x_n \to x$ for $n \to \infty$ with respect to $\tau$. $\qquad\square$

**Proposition 9.13** *Let $(C(t))_{t\geq 0}$ be a bi-continuous cosine family on a bi-admissible space $(X, \|\cdot\|, \tau)$ and let $(A, D(A))$ denote its generator. Then $(A, D(A))$ is bi-closed, i.e., whenever $(x_n)_{n\in\mathbb{N}}$ is a sequence in $D(A)$ such that $x_n \to x$ and $Ax_n \to y$ for $n \to \infty$ for some $y \in X$ with respect to $\tau$ and both sequences $(x_n)_{n\in\mathbb{N}}$ and $(Ax_n)_{n\in\mathbb{N}}$ are $\|\cdot\|$-bounded, then $x \in D(A)$ and $Ax = y$.*

**Proof** Let $(x_n)_{n\in\mathbb{N}}$ be a sequence in $D(A)$ such that $x_n \to x$ and $Ax_n \to y$ for $n \to \infty$ for some $y \in X$ with respect to $\tau$ and both sequences $(x_n)_{n\in\mathbb{N}}$ and $(Ax_n)_{n\in\mathbb{N}}$ are $\|\cdot\|$-bounded. We have to show that $x \in D(A)$ and $Ax = y$. By Corollary 9.10 we have that

$$C(t)x_n - x_n = \int_0^t (t-s)C(s)Ax_n \, ds, \quad t \geq 0.$$

As $(C(t))_{t\geq 0}$ is locally bi-equicontinuous by Definition 9.1, we can conclude that

$$C(t)x - x = \int_0^t (t-s)C(s)y \, ds, \quad t \geq 0.$$

By Proposition 9.8 and Lemma 9.11 the proof is complete. $\qquad\square$

**Remark 9.14** The generator $(A, D(A))$ of a bi-continuous cosine family can be interpreted as the second $\tau$-derivative in $t = 0$. Let us also investigate the first $\tau$-derivative. We observe that by Corollary 9.10 for $x \in D(A)$ one has

$$\frac{C(t)x - x}{t} = \int_0^t C(s)Ax \, ds - \frac{1}{t}\int_0^t sC(s)Ax \, ds, \quad t > 0.$$

The claim is that the limit for $t \to 0$ yields 0. To see this, we observe that for $p \in \mathcal{P}$ one has

$$p\left(\int_0^t C(s)Ax \, ds\right) \leq \int_0^t p(C(s)Ax) \, ds \leq \int_0^t \|C(s)Ax\| \, ds$$

$$\leq \int_0^t Me^{\omega s} \|Ax\| \, ds = \frac{M(e^{\omega t} - 1)}{\omega} \|Ax\| \to 0,$$

for $t \to 0$. Similarly, for the second term, we obtain

$$p\left(\frac{1}{t}\int_0^t sC(s)Ax \, ds\right) \leq \frac{1}{t}\int_0^t s \cdot p(C(s)Ax) \, ds \to 0,$$

as $t \to 0$ by the continuity of the integrand. This is of course due to the continuity of the seminorms as well as the bi-continuous cosine family $(C(t))_{t\geq 0}$. Hence, we conclude that $\tau\lim_{t\to 0} \frac{C(t)x - x}{t} = 0$ for all $x \in D(A)$.

**Proposition 9.15** *Let $(C(t))_{t\geq 0}$ be a bi-continuous cosine family on a bi-admissible space $(X, \|\cdot\|, \tau)$ and let $(A, D(A))$ denote its generator. For $x \in D(A)$ the following assertions hold:*

(i) *The map $\mathbb{R}_+ \ni t \mapsto C(t)x \in X$ is twice $\tau$-differentiable.*

(ii) $\frac{d}{dt}C(t)x \to 0$ *for* $t \to 0$.

(iii) $\frac{d^2}{dt^2}C(t)x = C(t)Ax$ *for all* $t \geq 0$.

***Proof*** We observe that Corollary 9.10 yields

$$C(t)x = x + \int_0^t (t-s)C(s)Ax\,ds = x + t\int_0^t C(s)Ax\,ds - \int_0^t sC(s)Ax\,ds.$$

Differentiating yields

$$\frac{d}{dt}C(t)x = \int_0^t C(s)Ax\,ds.$$

The same approach as in Remark 9.14 shows that $\frac{d}{dt}C(t) \to 0$ for $t \to 0$. By taking again the derivative then gives

$$\frac{d^2}{dt^2}C(t)x = C(t)Ax, \quad t \geq 0.$$

$\square$

**Example 9.16** Let us continue the investigation of the bi-continuous cosine family considered in Example 9.5. In particular, we want to determine its generator $(A, D(A))$. In fact, one has that $Af = f''$ with $D(A) = C_b^2(\mathbb{R})$. This follows from the fact that for twice continuously differentiable functions one has that

$$\lim_{t\to 0} \frac{2((C(t)f)(x) - f(x))}{t^2} = \lim_{t\to 0} \frac{f(x+t) + f(x-t) - 2f(x)}{t^2} = f''(x), \quad x \in \mathbb{R}.$$

From the previous observation, we obtain that $C_b^2(\mathbb{R}) \subseteq D(A)$ and $Af = f''$. Conversely, assume that $f \in D(A)$ and $Af = g$. By Proposition 9.15, the function $C(\cdot)f$ is twice $\tau$-differentiable showing that $f$ is twice differentiable. As $Af = g$, we have $g \in C_b(\mathbb{R})$ so that $f \in C_b^2(\mathbb{R})$.

We are also able to characterize the space of strong continuity.

**Lemma 9.17** *Let $(C(t))_{t\geq 0}$ be a bi-continuous cosine family on a bi-admissible space $(X, \|\cdot\|, \tau)$ and let $(A, D(A))$ denote its generator. Then*

$$X_{\mathrm{cont}} := \{x \in X : t \mapsto C(t)x \text{ is } \|\cdot\|\text{-continuous}\}$$

*is a closed subspace of $X$ which is invariant under $(C(t))_{t\geq 0}$. Moreover, one has that*
$\overline{D(A)} = X_{\mathrm{cont}}$.

**Proof** Both the closedness and the invariance of $X_{\mathrm{cont}}$ follow from the definition of the space. Let $x \in D(A)$. As $\|C(t)\| \leq Me^{\omega t}$ for some $M \geq 1$ and $\omega \in \mathbb{R}$ we obtain in combination with Corollary 9.10 that

$$\|C(t)x - x\| \leq \sup_{\|y\|\leq 1} \int_0^t (t-s)\,|\langle C(s)Ax, y\rangle|\;\mathrm{d}s \leq M \cdot \|Ax\| \cdot \frac{e^{\omega t} - t\omega - 1}{\omega^2} \to 0$$

for $t \to 0$. This yields $D(A) \subseteq X_{\mathrm{cont}}$ so that by the closedness of $X_{\mathrm{cont}}$ one has $\overline{D(A)} \subseteq X_{\mathrm{cont}}$. For the converse inclusion let $x \in X_{\mathrm{cont}}$ and define

$$x_n := 2n^2 \int_0^{\frac{1}{n}} \left(\frac{1}{n} - s\right) C(s)x\;\mathrm{d}s, \quad n \in \mathbb{N}.$$

Then $(x_n)_{n\in\mathbb{N}}$ is a sequence in $D(A)$ by means of Proposition 9.8. We observe that

$$\|x_n - x\| \leq \sup_{\|y\|\leq 1} 2n^2 \int_0^{\frac{1}{n}} \left|\left\langle \left(\frac{1}{n} - s\right)(C(s)x - x), y\right\rangle\right|\;\mathrm{d}s \leq 2n^2 \int_0^{\frac{1}{n}} \left(\frac{1}{n} - s\right) \|C(s)x - x\|\;\mathrm{d}s.$$

Because of the strong continuity of $s \mapsto C(s)$ with respect to the norm, we see that $x_n \to x$ so that $X_{\mathrm{cont}} \subseteq \overline{D(A)}$. $\qquad\square$

**Corollary 9.18** *Let $(C(t))_{t\geq 0}$ be a bi-continuous cosine family on a bi-admissible space $(X, \|\cdot\|, \tau)$ and let $(A, D(A))$ denote its generator. The restriction of $(C(t))_{t\geq 0}$ to $\overline{D(A)}$ defines a strongly continuous cosine family which is generated by the part of $A$ in $\overline{D(A)}$.*

## 9.3  Generators and Resolvents

There is another way to introduce generators of bi-continuous cosine families by means of resolvent operators. Generators of bi-continuous semigroups have also been introduced in this way, see Definition 3.13. This approach has also been used for strongly continuous cosine families, see for example [17, Proposition 3.14.4]. Our approach will be a mix of both methodologies.

Before we start the investigation of generators of bi-continuous cosine families and their resolvents, we consider for $a \geq 0$ and $\lambda \in \mathbb{C}$ the operator

$$Q_a(\lambda)x := \int_0^a e^{-\lambda t} C(t)x\;\mathrm{d}t, \quad x \in X,$$

where the integral has to be understood as a $\tau$-Riemann integral. We observe that such an operator indeed exists as a linear map $Q_a(\lambda) : X \to X$ in view of Proposition 3.13. In fact, the proofs of those results do not make use of the semigroup property at all but only use strong continuity. For the same reason, the following result will be stated without a proof.

**Lemma 9.19** *Let* $(C(t))_{t\geq 0}$ *be a bi-continuous cosine family on a bi-admissible space* $(X, \|\cdot\|, \tau)$ *such that* $\|C(t)\| \leq Me^{\omega t}$ *for all* $t \geq 0$*. Then the following properties hold:*

(i) *Let* $\lambda \in \mathbb{C}$ *and* $a \geq 0$*. Then* $Q_a(\lambda) \in \mathscr{L}(X)$ *and*

$$Q(\lambda) := \lim_{a \to \infty} Q_a(\lambda)$$

  *exists for all* $\lambda \in \mathbb{C}$ *with* $\mathrm{Re}(\lambda) > \omega$ *with respect to the operator norm.*
(ii) *For every* $x \in X$ *one has*

$$\tau\lim_{\lambda \to \infty} \lambda Q(\lambda)x = x.$$

The following result shows that the Laplace transform of a bi-continuous cosine family $(C(t))_{t\geq 0}$ gives rise to a resolvent of a certain operator. For bi-continuous semigroups, this result was forumlated in Proposition 3.11. The idea here is the same: one needs to show that a certain family of operators is a pseudo-resolvent.

**Proposition 9.20** *Let* $(C(t))_{t\geq 0}$ *be a bi-continuous cosine family on a bi-admissible space* $(X, \|\cdot\|, \tau)$ *such that* $\|C(t)\| \leq Me^{\omega t}$ *for all* $t \geq 0$*. Then there exists a uniquely determined linear operator* $(B, D(B))$ *such that* $(\omega^2, \infty) \subseteq \rho(B)$ *and*

$$\lambda R(\lambda^2, B)x = \int_0^\infty e^{-\lambda t} C(t)x \, dt, \quad \lambda > \omega.$$

***Proof*** We first observe that by Proposition 9.4 the equation $2C(t)C(s) = C(t+s) + C(t-s)$ holds for all $t \geq s \geq 0$. Moreover, according to Definition 9.1, there exist $M \geq 1$ and $\omega \in \mathbb{R}$ such that $\|C(t)\| \leq Me^{\omega t}$ for all $t \geq 0$. We consider the operators $Q(\lambda)$ from Lemma 9.19, i.e.,

$$Q(\lambda)x := \int_0^\infty e^{-\lambda t} C(t)x \, dt, \quad \lambda > \omega.$$

By following the same calculations as in the proof of [17, Proposition 3.14.4] one obtains that for $\lambda \neq \mu$ one has

$$\frac{\mu Q(\lambda)x - \lambda Q(\mu)x}{\mu^2 - \lambda^2} = Q(\lambda)Q(\mu)x, \quad x \in X,$$

see also [17, pp. 205–206]. By taking $R(\lambda) := \frac{1}{\sqrt{\lambda}} Q(\sqrt{\lambda})$ for $\lambda > \omega^2$ we obtain

$$R(\lambda)R(\mu)x = \frac{R(\lambda)x - R(\mu)x}{\mu - \lambda},$$

see also [17, p. 206]. This shows that the family $\{R(\lambda) : \lambda > \omega\}$ is a pseudo-resolvent. Suppose that $R(\lambda)x = 0$ for all $\lambda > \omega^2$, i.e., $x \in \mathrm{Ker}(R(\lambda))$. Then by construction $x \in \mathrm{Ker}(Q(\sqrt{\lambda}))$. Let $(\lambda_n)_{n\in\mathbb{N}}$ be an unbounded sequence in $(\omega^2, \infty)$. Then $(\sqrt{\lambda_n})_{n\in\mathbb{N}}$ is also unbounded in $(|\omega|, \infty)$. We have that $x \in \mathrm{Ker}(Q(\sqrt{\lambda_n}))$ for all $n \in \mathbb{N}$ and by Lemma 9.19 we obtain $0 = \sqrt{\lambda_n}Q(\sqrt{\lambda_n})x \to x$ for $n \to \infty$ with respect to $\tau$. As $\tau$ is supposed to be Hausdorff, limits are unique which yields $x = 0$. The rest follows by [101, Chap. III, Proposition 4.6].

We observe that for a given bi-continuous cosine family, Definition 9.6 and Proposition 9.20 yield two a priori different operators $(A, D(A))$ and $(B, D(B))$, respectively. However, we will show that both operators coincide. For strongly continuous cosine families this is [17, Proposition 3.14.5(d)]. First we will show that the operator $(B, D(B))$ behaves similarly to $(A, D(A))$.

**Proposition 9.21** *Let $(C(t))_{t\geq 0}$ be a bi-continuous cosine family on a bi-admissible space $(X, \|\cdot\|, \tau)$. Furthermore, let $(B, D(B))$ be the operator from Proposition 9.20. Then the following assertions hold:*

(a) $\displaystyle\int_0^t (t - s)C(s)x\,ds \in D(B)$ *for all $t \geq 0$ and $x \in D(B)$.*

(b) *For $x \in D(B)$ it holds that*

$$\tau\lim_{t\to 0} \frac{2}{t^2}(C(t)x - x) = Bx.$$

(c) $\displaystyle C(t)x - x = B\int_0^t (t - s)C(s)x\,ds = \int_0^t (t - s)C(s)Bx\,ds$ *for all $x \in D(B)$ and $t \geq 0$.*

(d) *Let $x, y \in X$. Then $x \in D(B)$ and $Bx = y \iff C(t)x - x = \displaystyle\int_0^t (t - s)C(s)y\,ds$ for all $t \geq 0$.*

**Proof** (a) Let $t \geq 0$ and $x \in D(B)$ be arbitrary. We notice that $D(B) = \mathrm{Ran}(R(\lambda^2, B))$. In particular, $x = R(\lambda^2, B)y$ for some $y \in X$. Together with Proposition 9.4 this yields

$$\int_0^t (t - s)C(s)x\,ds = \int_0^t (t - s)C(s)R(\lambda^2, B)y\,ds$$

$$= \frac{1}{\lambda}\int_0^t (t - s)C(s)\int_0^\infty e^{-\lambda\xi}C(\xi)y\,d\xi\,ds$$

$$= \frac{1}{\lambda}\int_0^t \int_0^\infty (t - s)e^{-\lambda\xi}C(s)C(\xi)y\,d\xi\,ds$$

$$= \frac{1}{\lambda} \int_0^\infty e^{-\lambda \xi} C(\xi) \int_0^t (t-s) C(s) y \, ds \, d\xi$$

$$= R(\lambda^2, B) \int_0^t (t-s) C(s) y \, ds \in \mathrm{Ran}(R(\lambda^2, B)) = D(B).$$

(b) If $x \in D(B)$ then $x = R(\lambda^2, B)y$ for some $y \in X$. Then

$$C(t)x - x = C(t)R(\lambda^2, B)y - R(\lambda^2, B)y = \frac{1}{\lambda} \int_0^\infty e^{-\lambda \xi} C(t) C(\xi) y \, d\xi - \frac{1}{\lambda} \int_0^\infty e^{-\lambda \xi} C(\xi) y \, d\xi.$$

The desired equality follows from a calculation similar to [176, Satz 14].

(c) Put $y = \int_0^t (t-s) C(s) x \, ds$. By (b) we have that $\tau\lim_{t\to 0} \frac{2}{t^2}(C(t)y - y) = Ay$. The assertion now follows from Proposition 9.8 or Corollary 9.10.

(d) The implication "$\Longrightarrow$" follows from both parts (a) and (c). For the converse implication "$\Longleftarrow$" assume that $C(t)x - x = \int_0^t (t-s) C(s) y \, ds$ for all $t \geq 0$. Taking the Laplace transform on both sides of the equation yields that $\frac{1}{\lambda}R(\lambda^2, B)y = \lambda R(\lambda^2, B)x - \frac{1}{\lambda}x$ showing that $x \in D(B)$ and $Bx = y$.

□

With Proposition 9.21 in hand, we are now able to show that the operators $(A, D(A))$ and $(B, D(B))$ coincide.

**Proposition 9.22** *Let $(C(t))_{t\geq 0}$ be a bi-continuous cosine family on a bi-admissible space $(X, \|\cdot\|, \tau)$. Then the two operators $(A, D(A))$ and $(B, D(B))$ from Definition 9.6 and Proposition 9.20, respectively, coincide.*

**Proof** Let $x \in D(B)$ such that $Bx = y$. It follows from Proposition 9.21(b) that

$$\frac{2}{t^2}(C(t)x - x) - y = \frac{2}{t^2} \int_0^t (t-s)(C(s)y - y) \, ds \to 0,$$

as $t \to 0$ with respect to $\tau$. Conversely, let $x \in D(A)$. By Corollary 9.10 we know that

$$C(t)x - x = \int_0^t (t-s) C(s) Ax \, ds, \quad t \geq 0.$$

Hence, by Proposition 9.21(d) we conclude that $x \in D(B)$ and $Bx = Ax$.  □

**Remark 9.23** We saw in Proposition 9.13 that $(A, D(A))$ is a bi-closed operator. However, due to Proposition 9.22 the operator $(A, D(A))$ coincides with the operator $(B, D(B))$ which by Proposition 9.20 is a closed operator. Hence, the generator of a bi-continuous cosine family is in addition also closed.

## 9.4   A Hille–Yosida Type Generation Theorem for Bi-continuous Cosine Families

In this section, we will show that the resolvent of the generator of a bi-continuous cosine family satisfies certain Hille–Yosida type estimates as we are used to from $C_0$-semigroups [101, Corollary 1.11] or bi-continuous semigroups (Proposition 3.15). In particular, we will show that norm estimates similarly to [221, Theorem 3.1] hold. The highlight is a Hille–Yosida type generation theorem for bi-continuous cosine families, similar to [221, Theorem 3.2].

**Proposition 9.24**  *Let $(C(t))_{t\geq0}$ be a bi-continuous cosine family on a bi-admissible space $(X, \|\cdot\|, \tau)$ and let $(A, D(A))$ denote its generator. Assume that $\|C(t)\| \leq Me^{\omega t}$ for all $t \geq 0$. Then $\lambda^2 \in \rho(A)$ whenever $\mathrm{Re}(\lambda) > \omega$ and*

$$\left\| \frac{\mathrm{d}^n}{\mathrm{d}\lambda^n} \lambda R(\lambda^2, A) \right\| \leq \frac{Mn!}{(\mathrm{Re}(\lambda) - \omega)^{n+1}}, \quad n \in \mathbb{N}.$$

***Proof***  First of all, we show that

$$\frac{\mathrm{d}}{\mathrm{d}\lambda} \lambda R(\lambda^2, A)x = -\int_0^\infty te^{-\lambda t} C(t)x \, \mathrm{d}t, \quad x \in X.$$

Indeed, by Proposition 9.20 and the norming property, cf. Assumption 2.1, we obtain

$$\left\| \frac{\lambda R(\lambda^2, A)x - \mu R(\mu^2, A)}{\lambda - \mu} + \int_0^\infty te^{-\lambda t} C(t)x \, \mathrm{d}t \right\|$$

$$\leq \sup_{\substack{\varphi \in (X,\tau)' \\ \|\varphi\| \leq 1}} \int_0^\infty \left| \frac{e^{-\lambda t} - e^{-\mu t}}{\lambda - \mu} + te^{-\lambda t} \right| \cdot |\varphi(C(t)x)| \, \mathrm{d}t$$

$$\leq M \|x\| \int_0^\infty \left| \frac{e^{-\lambda t} - e^{-\mu t}}{\lambda - \mu} + te^{-\lambda t} \right| e^{\omega t} \, \mathrm{d}t.$$

Of course, this calculation runs parallel to the one in Proposition 3.15. The latter expression converges to 0 whenever $\mu$ tends to $\lambda$ by means of dominated convergence. Induction then yields that for $n \in \mathbb{N}$ one has

$$\frac{\mathrm{d}^n}{\mathrm{d}\lambda^n} \lambda R(\lambda^2, A)x = (-1)^n \int_0^\infty t^n e^{-\lambda t} C(t)x \, \mathrm{d}t, \quad x \in X. \tag{9.5}$$

From (9.5) we therefore obtain for $x \in X$ that

$$\left\| \frac{\mathrm{d}^n}{\mathrm{d}\lambda^n} \lambda R(\lambda^2, A)x \right\| = \left\| \int_0^\infty t^n \mathrm{e}^{-\lambda t} C(t)x \, \mathrm{d}t \right\|$$

$$\leq \sup_{\substack{\varphi \in (X,\tau)' \\ \|\varphi\| \leq 1}} \int_0^\infty t^n \mathrm{e}^{-\mathrm{Re}(\lambda)t} |\varphi(C(t)x)| \, \mathrm{d}t$$

$$\leq M \|x\| \int_0^\infty t^n \mathrm{e}^{-\mathrm{Re}(\lambda)t} \mathrm{e}^{\omega t} \, \mathrm{d}t$$

$$= \frac{Mn!}{(\mathrm{Re}(\lambda) - \omega)^{n+1}} \|x\|.$$

$\square$

**Remark 9.25** We observe that in Proposition 9.24 all derivatives of the operator $\lambda R(\lambda^2, A)$ are needed. However, for bi-continuous semigroups, one "only" needs all natural powers of the resolvent, cf. Proposition 3.15. This is due to the fact that the equality $\frac{\mathrm{d}^n}{\mathrm{d}\lambda^n} R(\lambda, A) = (-1)^n n! R(\lambda, A)^{n+1}$ holds, whenever $(A, D(A))$ is the generator of a bi-continuous semigroup, see also (3.4.2).

Similarly to Lemma 3.10 we have the following result.

**Lemma 9.26** *Let* $(C(t))_{t \geq 0}$ *be a bi-continuous cosine family on a bi-admissible space* $(X, \|\cdot\|, \tau)$ *such that* $\|C(t)\| \leq M\mathrm{e}^{\omega t}$ *for all* $t \geq 0$. *By* $(A, D(A))$ *we denote its generator. Then*

$$\tau\text{-}\lim_{\lambda \to \infty} \lambda^2 R(\lambda^2, A)x = x, \quad x \in X.$$

***Proof*** Let $\varepsilon > 0$ and $p \in \mathcal{P}$ be arbitrary. Due to the strong continuity of $(C(t))_{t \geq 0}$ with respect to $\tau$ there exists $\delta > 0$ such that $p(C(t)x - x) < \varepsilon$ for $t \in (0, \delta)$. Let $\lambda \in \mathbb{R}$ such that $\lambda > \omega$. By Proposition 9.20 we obtain for $x \in X$ that

$$p(\lambda^2 R(\lambda^2, A)x - x) \leq \lambda \int_0^\infty \mathrm{e}^{-\lambda t} p(C(t)x - x) \, \mathrm{d}t$$

$$= \lambda \int_0^\delta \mathrm{e}^{-\lambda t} p(C(t)x - x) \, \mathrm{d}t + \lambda \int_\delta^\infty \mathrm{e}^{-\lambda t} p(C(t)x - x) \, \mathrm{d}t$$

$$\leq \lambda \int_0^\delta \mathrm{e}^{-\lambda t} p(C(t)x - x) \, \mathrm{d}t + \lambda \int_\delta^\infty \mathrm{e}^{-\lambda t} \|C(t)x - x\| \, \mathrm{d}t$$

$$< \lambda\varepsilon \int_0^\delta \mathrm{e}^{-\lambda t} \, \mathrm{d}t + \lambda \|x\| \int_\delta^\infty \mathrm{e}^{-\lambda t} (M\mathrm{e}^{\omega t} + 1) \, \mathrm{d}t$$

$$\leq \lambda\varepsilon \int_0^\infty \mathrm{e}^{-\lambda t} \, \mathrm{d}t + \|x\| \, \mathrm{e}^{-\delta\lambda} (M\mathrm{e}^\omega + 1)$$

$$= \varepsilon + \|x\| \, \mathrm{e}^{-\delta\lambda} (M\mathrm{e}^\omega + 1).$$

The second part of the latter sum vanishes for $\lambda \to \infty$. This proves the claim. $\qquad\square$

**Lemma 9.27** *Let* $(C(t))_{t\geq0}$ *be a bi-continuous cosine family on a bi-admissible space* $(X, \|\cdot\|, \tau)$ *such that* $\|C(t)\| \leq Me^{\omega t}$ *for all* $t \geq 0$. *By* $(A, D(A))$ *we denote its generator. Then*

$$\tau\lim_{\lambda\to\infty} \frac{\lambda^{n+1}}{n!} \frac{d^n}{d\lambda^n} \lambda R(\lambda^2, A)x = x, \quad x \in X, \ n \in \mathbb{N}.$$

***Proof*** Let $\varepsilon > 0$ and $p \in \mathcal{P}$ be arbitrary. Due to the strong continuity of $(C(t))_{t\geq0}$ with respect to $\tau$ there exists $\delta > 0$ such that $p(C(t)x - x) < \varepsilon$ for $t \in (0, \delta)$. Let $\lambda \in \mathbb{R}$ such that $\lambda > \omega$, then by using Proposition 9.20 we obtain for all $\lambda > \omega$ and for $x \in X$ that

$$p\left(\frac{\lambda^{n+1}}{n!} \frac{d^n}{d\lambda^n} \lambda R(\lambda^2, A)x - x\right)$$

$$\leq \frac{\lambda^{n+1}}{n!} \int_0^\infty t^n e^{-\lambda t} p(C(t)x - x)\, dt$$

$$= \frac{\lambda^{n+1}}{n!} \int_0^\delta t^n e^{-\lambda t} p(C(t)x - x)\, dt + \frac{\lambda^{n+1}}{n!} \int_\delta^\infty t^n e^{-\lambda t} p(C(t)x - x)\, dt$$

$$< \varepsilon + M\,\|x\| \cdot \frac{\lambda^{n+1}}{n!} \int_\delta^\infty t^n e^{(\omega-\lambda)t}\, dt + \|x\| \cdot \frac{\lambda^{n+1}}{n!} \int_\delta^\infty t^n e^{-\lambda t}\, dt$$

$$= \varepsilon + M\,\|x\| \cdot \frac{\lambda^{n+1}}{n!(\lambda - \omega)^n} \Gamma(n + 1, (\lambda - \omega)\delta) + \|x\| \cdot \frac{\lambda}{n!}\Gamma(n + 1, \lambda\delta),$$

where $\Gamma(s, x)$ denotes the upper incomplete Gamma function defined by

$$\Gamma(s, x) = \int_x^\infty t^{s-1} e^{-t}\, dt, \quad s \geq 0, \ x \geq 0.$$

It is known that

$$\lim_{\lambda\to\infty} \frac{\lambda}{n!}\Gamma(n + 1, \lambda) = 0, \tag{9.6}$$

uniformly for $n \in \mathbb{N}$. This proves the claim. $\qquad\square$

The following results make use of Lemma 9.27 and follow the proofs of Theorem 3.15. Therefore, we skip the proof here.

**Proposition 9.28** *Let* $(C(t))_{t\geq0}$ *be a bi-continuous cosine family on a bi-admissible space* $(X, \|\cdot\|, \tau)$ *such that* $\|C(t)\| \leq Me^{\omega t}$ *for all* $t \geq 0$. *By* $(A, D(A))$ *we denote its generator. Then the following assertions hold true:*

(a) $\{e^{-\alpha t} C(t) : t > 0\}$ *is bi-equicontinuous for all* $\alpha > \omega$.

(b) *Let $(x_n)_{n\in\mathbb{N}}$ be a $\|\cdot\|$-bounded sequence which is $\tau$-convergent to $x \in X$. Then for $\alpha > \omega$
one has*

$$\tau\lim_{n\to\infty} \frac{(\lambda - \alpha)^{k+1}}{k!} \frac{d^k}{d\lambda^k} \lambda R(\lambda^2, A)(x_n - x) = 0,$$

*uniformly for $k \in \mathbb{N}$ and $\lambda \geq \alpha$.*

We are now ready to prove a Hille–Yosida type generation theorem for bi-continuous cosine families. This result reads similar to [221, Theorem 3.2] and follows the lines of Theorem 3.18 and [17, Theorem 3.15.3].

**Theorem 9.29** *Let $(A, D(A))$ be a linear operator on a Banach space $X$. The following assertions are equivalent:*

(a) *$(A, D(A))$ is the generator of a bi-continuous cosine family $(C(t))_{t\geq 0}$ such that
$\|C(t)\| \leq Me^{\omega t}$ for all $t \geq 0$.*

(b) *$(A, D(A))$ is a bi-densely defined operator satisfying*

$$\left\| \frac{d^n}{d\lambda^n} \lambda R(\lambda^2, A) \right\| \leq \frac{Mn!}{(\operatorname{Re}(\lambda) - \omega)^{n+1}}, \tag{9.7}$$

*for all $n \in \mathbb{N}$ and $\lambda > \omega$ and the family*

$$\left\{ \frac{(\lambda - \alpha)^{k+1}}{k!} \frac{d^k}{d\lambda^k} \lambda R(\lambda^2, A) : k \in \mathbb{N}, \ \lambda > \alpha \right\}, \tag{9.8}$$

*is bi-equicontinuous for each $\alpha > \omega$.*

***Proof*** The implication (a)$\Longrightarrow$(b) follows directly from Corollary 9.12, Propositions 9.24 and 9.28.

For the converse implication let us first assume that $\alpha = 0$ and that assertion (b) holds true. By [17, Theorem 2.4.1] there exists a function $F : [0, \infty) \to \mathscr{L}(X)$ such that $F(0) = 0$ and

$$\|F(t + h) - F(t)\| \leq M \int_t^{t+h} e^{\omega r} \, dr, \tag{9.9}$$

for all $t > 0$ and $h > 0$. Moreover, one has that

$$\lambda R(\lambda^2, A) = \int_0^\infty e^{-\lambda t} \, dF(t). \tag{9.10}$$

We will show that for $x \in D(A)$ one has

$$F(t)x - tx = \int_0^t (t - s)F(s)Ax \, ds, \quad t \geq 0.$$

We observe that the function $F$ is locally Lipschitz continuous and, following the notations of [159, Appendix A] and [17, Chaps. I.1 and I.2], one has $F \in \mathrm{Lip}_\omega(\mathbb{R}_+, \mathscr{L}(X))$. In particular, the function $F$ is of bounded semivariation, see also [17, Chap. I, Sect. 1.9]. In particular, one can use integration by parts, cf. [17, p. 54]. Hence, by applying [17, Theorem 2.4.1] to the function $(\omega, \infty) \ni \lambda \to \lambda R(\lambda^2, A) \in \mathscr{L}(X)$ and by using Eqs. (9.10) and [159, Theorem A.1] as well as the previous observations, we obtain that

$$\lambda R(\lambda^2, A) = \int_0^\infty \lambda e^{-\lambda t} F(t)\, dt \quad \text{or} \quad R(\lambda^2, A) = \int_0^\infty e^{-\lambda t} F(t)\, dt$$

where the integral converges with respect to the norm. We obtain that for $x \in D(A)$ one has

$$\lambda^2 \int_0^\infty e^{-\lambda t} t x \, dt = x = \lambda^2 R(\lambda^2, A)x - R(\lambda^2, A)Ax$$

$$= \lambda^2 \int_0^\infty e^{-\lambda t} F(t)x \, dt - \int_0^\infty e^{-\lambda t} F(t)Ax \, dt$$

$$= \lambda^2 \int_0^\infty e^{-\lambda t} \left[ F(t)x - \int_0^t (t - s)F(s)Ax \, ds \right] dt.$$

By the uniqueness of the Laplace transform we obtain that

$$F(t)x - tx = \int_0^t (t - s)F(s)Ax \, ds, \quad x \in D(A).$$

The previous observation yields that $F(\cdot)x \in C^1(\mathbb{R}_{\geq 0}, (X, \|\cdot\|))$ for all $x \in D(A)$. Integration by parts also yields that

$$\lambda R(\lambda^2, A)x = \int_0^\infty e^{-\lambda t} F'(t)x \, dt.$$

It also follows that on $D(A)$ one has that $F(0) = 0$ and

$$F(t) = \lim_{k \to \infty} \frac{(-1)^k}{k!} \int_{\frac{k}{t}}^\infty s^k \frac{d^{k+1}}{ds^{k+1}} s R(s^2, A)\, ds, \quad t > 0.$$

From the previous observations and Corollary 9.18, we obtain that $F'(t)$ exists on $X :=\overline{D(A)}$ and is a strongly continuous cosine family. Let us now show that $F(\cdot)x \in C^1(\mathbb{R}_{\geq 0}, (X, \tau))$ for all $x \in X$. To do so, we define

$$D_m(t, x) := \frac{F(t + 1/m)x - F(t)x}{1/m}, \quad x \in X, \ t \geq 0.$$

We want to use Assumption 2.1(ii) to show that $D_m(t, x)$ is $\tau$-convergent. By (9.9) this sequence is $\|\cdot\|$-bounded. As $D(A)$ is assumed to be bi-dense in $X$ we can find a $\|\cdot\|$-

bounded sequence $(x_n)_{n\in\mathbb{N}}$ in $D(A)$ such that $x_n \to x$ with respect to $\tau$. Moreover, we have that

$$p\left(\frac{F(t+1/m)-F(t)}{1/m}(x_n-x)\right)$$

$$= p\left(m\cdot\lim_{k\to\infty}\frac{(-1)^k}{k!}\int_{\frac{k}{t}}^{\infty} s^k\frac{\mathrm{d}^{k+1}}{\mathrm{d}s^{k+1}}sR(s^2,A)(x_n-x)\,\mathrm{d}s - \frac{(-1)^k}{k!}\int_{\frac{k}{t+1/m}}^{\infty} s^k\frac{\mathrm{d}^{k+1}}{\mathrm{d}s^{k+1}}sR(s^2,A)(x_n-x)\,\mathrm{d}s\right)$$

$$= p\left(m\cdot\lim_{k\to\infty}\frac{(-1)^k}{k!}\int_{\frac{k}{t+1/m}}^{\frac{k}{t}} s^k\frac{\mathrm{d}^{k+1}}{\mathrm{d}s^{k+1}}sR(s^2,A)(x_n-x)\,\mathrm{d}s\right)$$

$$\leq m\cdot\lim_{k\to\infty}\frac{1}{k!}\int_{\frac{k}{t+1/m}}^{\frac{k}{t}} p\left(s^k\frac{\mathrm{d}^{k+1}}{\mathrm{d}s^{k+1}}sR(s^2,A)(x_n-x)\right)\,\mathrm{d}s$$

$$= m\cdot\lim_{k\to\infty}\frac{(k+1)!}{k!}\int_{\frac{k}{t+1/m}}^{\frac{k}{t}}\frac{1}{s^2}\cdot p\left(\frac{s^{k+2}}{(k+1)!}\frac{\mathrm{d}^{k+1}}{\mathrm{d}s^{k+1}}sR(s^2,A)(x_n-x)\right)\,\mathrm{d}s$$

$$\leq \frac{\varepsilon}{3}\cdot m\cdot\lim_{k\to\infty}(k+1)\cdot\frac{1}{km} = \frac{\varepsilon}{3}$$

showing that $(D_m(t,x_n))_{m\in\mathbb{N}}$ is a $\tau$-Cauchy sequence for all $n \in \mathbb{N}$. Because $X$ is sequentially $\tau$-complete on $\|\cdot\|$-bounded sets we conclude that $F(\cdot)x$ is $\tau$-differentiable for all $x \in X$. By following the same arguments as in the proof of Theorem 3.18 one obtains in addition that $\{F'(t) : t \geq 0\}$ is globally bi-equicontinuous, that $F(\cdot)x \in C^1(\mathbb{R}_{\geq 0},(X,\tau))$ for all $x \in X$, that $F'(\cdot)$ is exponentially bounded and that for all $x \in X$ and $\lambda > \omega$ one has

$$\lambda R(\lambda^2,A)x = \int_0^\infty e^{-\lambda t}F'(t)x\,\mathrm{d}t.$$

By rescaling one obtains the above-mentioned properties for arbitrary $\alpha > \omega$. In particular, for arbitrary $\alpha \in \mathbb{R}$, we make use of the operator $(A-\alpha, D(A))$ instead of $(A,D(A))$ in the previous steps. Lastly, we define

$$C(t)x := F'(t)x = \tau\lim_{h\to 0}\frac{F(t+h)x - F(t)x}{h}, \quad x \in X,\ t \geq 0.$$

We already know that the functional equation from Definition 9.1(i) holds for all $x \in X = \overline{D(A)}$. Similar to Lemma 9.26 one shows that $\tau\lim_{\lambda\to\infty}\lambda^2 R(\lambda^2,A)x = x$ for all $x \in X$. By following the same approximation arguments as in the proof of Theorem 3.18 we obtain the desired result.

## Notes on This Chapter

The investigation of strongly continuous cosine families, or strongly continuous cosine operator functions, started with the work of M. Sova [221]. Cosine families serve as solutions to second-order autonomous abstract Cauchy problems which are of the form

$$\begin{cases} \ddot{u}(t) = Au(t), & t \geq 0, \\ u(0) = x \in X, \\ \dot{u}(0) = y \in X. \end{cases}$$

Since then, strongly continuous cosine functions have been studied extensively, see for example [15, 130, 145, 191, 201, 210, 215, 224]. They also turned out to work well in applications, cf. [30, 31, 41, 42, 69]. Monographs containing theory on strongly continuous cosine families have been written for example by H.O Fattorini [112] or Arendt et al. [17]. It is worth mentioning that cosine families have already been investigated previously in the general framework of locally convex spaces by H.O. Fattorini [110, 111], Y. Konishi [148] and others [128, 214]. This work can also be found here [57]. This chapter could be used for forthcoming research in the direction of bi-continuous cosine families.

# Stability of Bi-continuous Semigroups

10

The study of asymptotics of operator semigroups yields powerful tools for the exploration of the convergence of solutions of evolution equations.

The known theory has been developed for $C_0$-semigroups. In this chapter, we want to suggest how to study stability of bi-continuous semigroups. The problem that arises immediately is that one is not longer dealing with a norm but also with a locally convex topology. Stability of operator semigroups on locally convex spaces has only be treated, according to the best knowledge of the author, in the work of Jacob and Wegner [139]. It turns out that in the situation of locally convex spaces there are more concepts of stability in comparison to the classical $C_0$-semigroup case. As a matter of fact, we make use of the so-called mixed topology, see Appendix A, in order to deal with the norm topology and the additional locally convex topology at the same time.

## 10.1 Stability of Bi-continuous Semigroups

To study the stability of bi-continuous semigroups, one has to take both the norm topology and the locally convex topology $\tau$ into account. Stability of $C_0$-semigroups on Banach spaces has been extensively studied and can, for example, be found in the monograph by Eisner [93]. A few years back, Jacob and Wegner investigated the asymptotics of $C_0$-semigroups on locally convex spaces where the situation actually changes entirely, cf. [139]. We want to make use of those results as one can consider bi-continuous semigroups as operator semigroups on spaces equipped with the mixed topology. We refer the reader to recall the definition of $C_0$-semigroups on locally convex spaces from Sect. 6.3. For the convenience of the reader and as we are dealing with different locally convex topologies in this section, we write $\mathcal{P}_\tau$ for the family of seminorms associated to a locally convex topology $\tau$.

© The Author(s), under exclusive license to Springer Nature Switzerland AG 2026

C. Budde, *Bi-Continuous Operator Semigroups*, Frontiers in Mathematics,

https://doi.org/10.1007/978-3-032-12948-2_10

Notice that we call a $C_0$-semigroup on a locally convex space *bounded* if for each bounded subset $B \subseteq X$ and each $p \in \mathcal{P}$ there exists a constant $C \geq 0$ such that

$$\sup_{x \in B} p(T(t)x) \leq C$$

for all $t \geq 0$.

We recall the following stability notions for semigroups on locally convex spaces, cf. [139, Definition 2.2].

**Definition 10.1** Let $(X, \tau)$ be a locally convex space and $(T(t))_{t \geq 0}$ a $C_0$-semigroup on $X$. We say that $(T(t))_{t \geq 0}$ is

   (i) *strongly exponentially stable* if $\forall x \in X \, \exists \omega > 0 : \tau\text{-}\lim_{t \to \infty} e^{\omega t} T(t)x = 0$,

  (ii) *strongly stable* if $\forall x \in X : \tau\text{-}\lim_{t \to \infty} T(t)x = 0$,

 (iii) *uniformly stable* if for each bounded subset $B \subseteq X$ and for each $p \in \mathcal{P}$ one has that $\lim_{t \to \infty} \sup_{x \in B} p(T(t)x) = 0$.

Following the line of the previous definition, we now introduce asymptotic behavior for bi-continuous semigroups (which actually will be operator semigroups with asymptotic behavior with respect to the mixed topology).

**Definition 10.2** Let $(T(t))_{t \geq 0}$ be a bi-continuous semigroup on a Banach space $X$ with respect to the locally convex topology $\tau$. By $\gamma$ we denote the mixed topology on $X$. We say that $(T(t))_{t \geq 0}$ is

   (i) a *strongly exponentially stable bi-continuous semigroup* if

$$\forall x \in X \, \exists \omega > 0 : \gamma\text{-}\lim_{t \to \infty} e^{\omega t} T(t)x = 0,$$

  (ii) a *strongly stable bi-continuous semigroup* if

$$\forall x \in X : \gamma\text{-}\lim_{t \to \infty} T(t)x = 0,$$

 (iii) a *uniformly stable bi-continuous semigroup* if for each $\gamma$-bounded subset $B \subseteq X$ and every $p \in \mathcal{P}_\gamma$ one has $\lim_{t \to \infty} \sup_{x \in B} p(T(t)x) = 0$.

We observe that we can substitute the $\gamma$-boundedness in Definition 10.2(iii) by $\|\cdot\|$-boundedness by [73, Chap. I, Proposition 1.11].

**Remark 10.3** It is noteworthy that the concept of strong continuity of a semigroup on a locally convex space (Sect. 6.3) and the concept of strong continuity of a bi-continuous semigroup (Definition 2.4) differ. In fact, the definition of $C_0$-semigroups on locally convex spaces, cf. Definitions 6.11 and 6.12, requires that each operator $T(t)$, $t \geq 0$, is both linear and continuous. In general, the continuity condition is not fulfilled for a bi-continuous semigroup $(T(t))_{t \geq 0}$ when considering $(T(t))_{t \geq 0}$ as an operator semigroup with respect to the mixed topology. Effectively, Definition 2.4 only requires each operator $T(t)$, $t \geq 0$, to be continuous with respect to the Banach space norm. Nonetheless, the stability concepts of Definition 10.2 still make sense.

Recall from [220, p. 273] that a locally convex space $X$ is called *C-sequential*, if every convex sequentially open subset of $X$ is already open. Moreover, a locally convex space is called *semi-Montel* if its bounded sets are relatively compact. These concepts are important for the following result which in fact transfers [139, Theorem 2.7] to the setting of bi-continuous semigroups. As a matter of fact, we consider $C$-sequential semi-Montel spaces in the result below as Montel spaces, used in [139, Theorem 2.7], are barrelled semi-Montel spaces, which is a too strong assumption as in this case $\tau = \|\cdot\|$ by [73, Chap. I, Proposition 1.15].

**Proposition 10.4** *Let $(X, \|\cdot\|, \tau)$ be a bi-admissible space. Let $\gamma$ denote the mixed topology and assume that $(X, \gamma)$ is a semi-Montel space. Furthermore, let $(T(t))_{t \geq 0}$ be bi-continuous semigroup which is $\gamma$-equicontinuous on $X$. Then, $(T(t))_{t \geq 0}$ is a uniformly stable bi-continuous semigroup if and only if it is a strongly stable bi-continuous semigroup.*

***Proof.*** The implication "$\Longrightarrow$" is obvious. For the implication "$\Longleftarrow$" let $\mathcal{P}_\gamma$ denote the family of continuous seminorms generating the mixed topology $\gamma$. By assumption $(T(t))_{t \geq 0}$ is $\gamma$-equicontinuous, i.e., for given $q \in \mathcal{P}_\gamma$ there exists $p \in \mathcal{P}_\gamma$ and $C \geq 0$ such that $q(T(t)x) \leq Cp(x)$ for all $x \in X$ and $t \geq 0$. Let $\varepsilon > 0$ and $B \in \mathcal{B}_X$ be arbitrary but fixed. Then we notice that for $\varepsilon_0 := \frac{\varepsilon}{1+C}$ the sets $B_p(x, \varepsilon_0) := \{y \in X : p(x - y) < \varepsilon_0\}$, where $x \in \overline{B}$ forms an open cover for $\overline{B}$. Due to the assumption that $X$ is semi-Montel, there exist $x_1, \ldots, x_n \in \overline{B}$ such that $\overline{B} \subseteq \bigcup_{m=1}^{n} B_p(x_m, \varepsilon)$. For fixed $1 \leq m \leq n$ there exists $t_m \geq 0$ such that $q(T(t)x_m) < \varepsilon_0$ for $t \geq t_m$. Define $t_* := \max_{1 \leq m \leq n} t_m$ and let $x \in B \subseteq \overline{B}$ be arbitrary. Then there exists $1 \leq m \leq n$ such that $x \in B_p(x_m, \varepsilon_0)$. For $t \geq t_*$ we see that

$$q(T(t)x) \leq q(T(t)(x_n - x)) + q(T(t)x_n) \leq Cp(x - x_n) + q(T(t)x_n) < C\varepsilon_0 + \varepsilon_0 = \varepsilon.$$

As $x \in B$ was arbitrarily chosen, obtain that $\sup_{x \in B} q(T(t)x) < \varepsilon$ for all $t \geq t_*$. $\quad\square$

**Remark 10.5** One is able to formulate a weaker version of Proposition 10.4. Under the assumption that $(X, \gamma)$ is both $C$-sequential and semi-Montel, by [154, Theorem 3.17] or [150, Theorem 7.4] the semigroup $(T(t))_{t \geq 0}$ is quasi-equicontinuous with respect to the mixed topology $\gamma$. If one assumes that the semigroup is exponential stable in the sense that there exist $M \geq 0$ and $\omega < 0$ such that $\|T(t)\| \leq Me^{\omega t}$ for all $t \geq 0$, then the $\gamma$-

equicontinuity of $(T(t))_{\geq 0}$ follows automatically. However, this is in general already a strong assumption one has to make on the operator semigroup itself.

## 10.2  Examples

### 10.2.1 The Multiplication Semigroup

Consider the bi-admissible space $(C_b(\mathbb{R}), \|\cdot\|_\infty, \tau_{co})$, cf. Sect. 2.1.1. It was shown by Wiweger [239, Example D)] that $C_b(\mathbb{R})$ satisfies Proposition B.6(ii) so that the mixed and the submixed topologies coincide on $C_b(\mathbb{R})$. This fact was also used by Goldys et al. when they studied diffusion semigroups on spaces of continuous functions or Markov processes, cf. [121, 122].

Let $q : \mathbb{R} \to \mathbb{C}$ be a continuous function satisfying $\sup_{x \in \mathbb{R}} \mathrm{Re}(q(x)) < \infty$ and consider on $C_b(\mathbb{R})$ the multiplication semigroup given by

$$(T(t)f)(x) := e^{tq(x)} f(x), \quad t \geq 0, \ f \in C_b(\mathbb{R}), \ x \in \mathbb{R}.$$

Then, $(T(t))_{t \geq 0}$ is a bi-continuous semigroup on $C_b(\mathbb{R})$ with respect to the compact-open topology, see Sect. 5.2.2. We now will investigate the stability of this bi-continuous semigroup according to Definition 10.2. Firstly, assume that $\sup_{x \in \mathbb{R}} \mathrm{Re}(q(x)) < 0$, then

$$\widetilde{p}_{(p_{K_n}, a_n)_{n \in \mathbb{N}}}(T(t)f) = \sup_{n \in \mathbb{N}} \sup_{x \in K_n} a_n e^{t\mathrm{Re}(q(x))} |f(x)| \leq \widetilde{p}_{p_{K_n}, a_n}(f) \cdot \sup_{n \in \mathbb{N}} \sup_{x \in K_n} e^{t\mathrm{Re}(q(x))} \overset{t \to \infty}{\longrightarrow} 0,$$

so that $(T(t))_{t \geq 0}$ is a strongly stable bi-continuous semigroup according to Definition 10.2. The previous estimate also shows that under the assumption that $\sup_{x \in \mathbb{R}} \mathrm{Re}(q(x)) < 0$ the semigroup $(T(t))_{t \geq 0}$ is equicontinuous. We also obtain for any $\|\cdot\|$-bounded (or equivalently $\gamma$-bounded) subset $B \subseteq C_b(\mathbb{R})$ that

$$\sup_{f \in B} \widetilde{p}_{(p_{K_n}, a_n)_{n \in \mathbb{N}}}(T(t)f) \leq \sup_{f \in B} \widetilde{p}_{p_{K_n}, a_n}(f) \cdot \sup_{n \in \mathbb{N}} \sup_{x \in K_n} e^{t\mathrm{Re}(q(x))} \leq \sup_{f \in B} \sup_{n \in \mathbb{N}} a_n \|f\|_\infty \cdot \sup_{n \in \mathbb{N}} \sup_{x \in K_n} e^{t\mathrm{Re}(q(x))}$$

which also tends to $0$ for $t \to \infty$, showing that $(T(t))_{t \geq 0}$ is a uniformly stable bi-continuous semigroup according to Definition 10.2. Hence, the strong stability that we observed previously is in fact just a conclusion for that. However, we want to emphasize that the other implication, i.e., Proposition 10.4 is not applicable here as $C_b(\mathbb{R})$ fails to be semi-Montel due to the fact that the closed $\|\cdot\|_\infty$-unit ball of $C_b(\mathbb{R})$ is not $\gamma$-compact, see also [152, Remark 2.5(a)].

If we just assume that $\sup_{x \in \mathbb{R}} \mathrm{Re}(q(x)) < \infty$, then for $\omega > 0$ we have

$$e^{\omega t} \cdot \widetilde{p}_{(p_{K_n}, a_n)_{n \in \mathbb{N}}}(T(t)f) = e^{\omega t} \cdot \sup_{n \in \mathbb{N}} \sup_{x \in K_n} a_n e^{t\mathrm{Re}(q(x))} |f(x)| \leq \widetilde{p}_{p_{K_n}, a_n}(f) \cdot \sup_{n \in \mathbb{N}} \sup_{x \in K_n} e^{t(\omega + \mathrm{Re}(q(x)))},$$

showing that $(T(t))_{t \geq 0}$ is not a strongly exponentially stable bi-continuous semigroup excepted for the case when $\sup_{x \in \mathbb{R}} \mathrm{Re}(q(x)) < 0$ which implies strongly stability for this bi-continuous semigroup.

### 10.2.2 The Adjoint Semigroup

Let $(T(t))_{t \geq 0}$ be a $C_0$-semigroup on a Banach space $X$. Then we can consider the adjoint semigroup $(T'(t))_{t \geq 0}$ on the bi-admissible space $(X', \|\cdot\|_{X'}, \tau_{w*})$, see also Sect. 2.1.2. The dual semigroup of a $C_0$-semigroup is bi-continuous semigroups with respect to the weak*-topology on $X'$, cf. Sect. 2.2.3. We will now apply our theory to dual semigroups.

**Remark 10.6** Let us comment on the mixed topology $\gamma(\|\cdot\|_{X'}, \tau_{w*})$ which is needed for our stability concepts:

(a) From [154, Remark 3.19(b)] we obtain that $\gamma(\|\cdot\|_{X'}, \tau_{w*})$ is $C$-sequential if $X$ is separable.
(b) Moreover, by [152, Remark 2.5(a)] the space $(X', \gamma)$ is semi-Montel as Proposition B.6(i) is fulfilled, see also [154, Example 3.11(b)]. In particular, the mixed and submixed topology coincide in this case.

Let us consider the nilpotent left-translation semigroup $(T(t))_{t \geq 0}$ on the Banach space $X = \mathrm{L}^1([0, 1])$. This semigroup is a strongly stable $C_0$-semigroup according to [93, Chap. III, Example 3.2]. As $\mathrm{L}^1([0, 1])$ is known to be separable, we conclude by Remark 10.6 that the dual space $\mathrm{L}^\infty([0, 1])$ equipped with the mixed topology $\gamma$ is $C$-sequential and semi-Montel. The adjoint semigroup $(T'(t))_{t \geq 0}$ is the nilpotent right-translation semigroup on $\mathrm{L}^\infty([0, 1])$. This semigroup is a strongly stable bi-continuous semigroup. Indeed, let $f \in \mathrm{L}^\infty([0, 1])$ and $(g_n)_{n \in \mathbb{N}}$ be a sequence in $\mathrm{L}^1([0, 1])$, then

$$\widetilde{p}_{(p_{g_n}, a_n)_{n \in \mathbb{N}}}(T'(t)f) = \sup_{n \in \mathbb{N}} a_n \left| \int_0^1 (T'(t)f)(x) g_n(x)\, \mathrm{d}x \right| = \sup_{n \in \mathbb{N}} a_n \left| \int_0^1 f(x)(T(t)g_n)(x)\, \mathrm{d}x \right| \to 0$$

uniformly in $n \in \mathbb{N}$ as $T(t) = 0$ for all for all $t > 1$ and $(a_n)_{n \in \mathbb{N}}$ is bounded. One can also show strong stability when considering the Banach space valued spaces $\mathrm{L}^1([0, 1], Y)$ as this also has been done in the framework of flows on networks [61]. However, one has to assume that $Y'$ has the Radon–Nikodym property as in this case the dual space of $\mathrm{L}^1([0, 1], Y)$ will be $\mathrm{L}^\infty([0, 1], X')$, see also [85, Chap. IV, Sect. 1]. However, also in this case Proposition 10.4 and Remark 10.5 fail as $(T'(t))_{t \geq 0}$ is neither $\gamma$-equicontinuous nor exponentially stable.

**Remark 10.7** The above example is somehow a toy example, especially as the semigroup is nilpotent and therefore also exponentially stable in the classical sense. It is noteworthy that if the $C_0$-semigroup $(T(t))_{t \geq 0}$ is weakly stable, then the adjoint semigroup $(T'(t))_{t \geq 0}$

is weak*-stable. However, the whole point of the stability discussion for bi-continuous semigroups in this chapter is to take both the norm and the locally convex topology into account by means of the mixed topology.

## Notes on This Chapter

This chapter is based on [55] and provides an opportunity to explore the stability and long-term behavior of bi-continuous semigroups. While the results presented here may not constitute a major breakthrough, they offer a solid foundation and a promising starting point for future research on the asymptotic behavior of bi-continuous semigroups.

In the last decades, a lot of research has been dedicated to asymptotic behavior of operator semigroups and especially to the study of the relations between their asymptotic and spectral properties, see, for example, the works of Batty [34–36], van Neerven [37, 227, 228], Eisner [29, 93, 95], or Glück [22, 79, 118].

In this chapter, we want to investigate mean ergodic bi-continuous semigroups by means of
Cesáro means. We will also see that those semigroups have applications to Feller semigroups
generated by autonomous and non-autonomous second-order differential operators with
unbounded coefficients in $C_b(\mathbb{R}^N)$. The basic idea of *mean ergodicity* is that the time mean
equals the space mean within a dynamical system. This so-called ergodic hypothesis goes
back to Boltzmann [44] which has been mathematically accentuated later on by Birkhoff
[40] and von Neumann [235]. The most common technique is the linearization of dynamical
systems by means of Koopman operators which actually gives the chance to pass from
dynamics to linear operators, and back. The fundamental concepts of this operator theoretical
approach can, for example, be found in [94]. The interest in such operators has its origins
appearing in the context of statistical mechanics and probability theory highly motivates
these kind of operators.

## 11.1 Cesáro Means and Mean Ergodicity

Let us start with the definition of Cesáro means.

**Definition 11.1** Let $(T(t))_{t \geq 0}$ be a bi-continuous contraction semigroup on a Banach space
$X$ with respect to a locally convex topology $\tau$. For $r > 0$ we define $C(r) : X \to X$ by

$$C(r)x := \frac{1}{r} \int_0^r T(t)x \, dt, \quad x \in X.$$

Here, we understand the integrals as a $\tau$-Riemann integral. The family $(C(r))_{r>0}$ is called
the *Cesáro means* of $(T(t))_{t \geq 0}$.

We start by summarizing some main properties of the *Cesáro means.*

**Proposition 11.2** *Let $(T(t))_{t\geq 0}$ be a bi-continuous contraction semigroup on X with respect to $\tau$. For every $r > 0$ the Cesáro mean $C(r)$ is a bounded linear operator on X. Moreover, for every $x \in X$ and every $\|\cdot\|$-bounded sequence $(x_n)_{n\in\mathbb{N}}$ in X with $x_n \to x$ for $n \to \infty$ with respect to $\tau$ one has $C(r)x_n \to C(r)x$ as $n \to \infty$ with respect to $\tau$ uniformly on compact sets.*

**Proof** The linearity is a consequence of the linearity of the semigroup. As our standing assumption is that the Banach space $X$ together with the locally convex topology $\tau$ satisfies Assumption 2.1 the norming property yields

$$\|C(r)x\| = \sup_{\substack{\varphi\in(X,\tau)' \\ \|\varphi\|\leq 1}} |\varphi(C(r)x)| = \sup_{\substack{\varphi\in(X,\tau)' \\ \|\varphi\|\leq 1}} \left| \frac{1}{r} \int_0^r \varphi(T(t)x)\,\mathrm{d}t \right| \leq \|x\|,$$

for all $x \in X$. For the second part of the assertion, fix $x \in X$ and let $(x_n)_{n\in\mathbb{N}}$ be a $\|\cdot\|$-bounded sequence in $X$ with $x_n \to x$ for $n \to \infty$ with respect to $\tau$. As $(T(t))_{t\geq 0}$ is assumed to be bi-continuous, the semigroup is locally bi-equicontinuous by Definition 2.4. Hence, we know that $T(t)x_n \to T(t)x$ as $n \to \infty$ with respect to $\tau$ uniformly on compact sets. Therefore, for every $t_0 > 0$, $p \in \mathcal{P}$ and $\varepsilon > 0$ there exists $N \in \mathbb{N}$ such that

$$\sup_{t\in[0,t_0]} p(T(t)(x_n - x)) < \varepsilon, \quad n \geq N.$$

This yields

$$p(C(r)(x_n - x)) = p\left( \frac{1}{r} \int_0^r T(t)(x_n - x)\,\mathrm{d}t \right) \leq \frac{1}{r} \int_0^r p(T(t)(x_n - x))\,\mathrm{d}t < \varepsilon,$$

for every $r \in [0, t_0]$ and $n \geq N$.                                                      $\square$

The following result connects the kernel of the generator of a bi-continuous semigroup and the space of fixed points of the semigroup.

**Lemma 11.3** *Let $(T(t))_{t\geq 0}$ be a bi-continuous contraction semigroup on X with respect to $\tau$ and let $(A, D(A))$ be the corresponding generator. Then*

$$\mathrm{Ker}(A) = \mathrm{Fix}(T(t))_{t\geq 0} = \{x \in X : T(t)x = x \ \forall t \geq 0\}.$$

**Proof** The inclusion "$\subseteq$" takes $x \in \mathrm{Ker}(A)$, i.e., $x \in D(A)$ and $Ax = 0$. By Theorem 3.3(b) one obtains

$$T(t)x - x = \int_0^t T(s)Ax\,\mathrm{d}s = 0, \quad t \geq 0,$$

which yields $T(t)x = x$ and therefore $x \in \text{Fix}(T(t))_{t\geq 0}$. For the converse inclusion "$\supseteq$," let $x \in \text{Fix}(T(t))_{t\geq 0}$, i.e., $T(t)x = x$ for all $t \geq 0$. Then the quotient $\frac{1}{t}(T(t)x - x)$ clearly converges to 0 with respect to $\tau$. Hence, $x \in D(A)$ and $Ax = 0$. $\qquad\square$

**Proposition 11.4** *Let $(T(t))_{t\geq 0}$ be a bi-continuous contraction semigroup on $X$ with respect to $\tau$ and let $(A, D(A))$ be the corresponding generator. The Cesáro means $(C(r))_{r>0}$ of $(T(t))_{t\geq 0}$ have the following properties:*

(i)  $C(r)x \in \overline{\text{co}}^\tau \{T(t)x : t \geq 0\}$ *for every* $x \in X$.
(ii)  *For every $t > 0$ and $r > 0$ and any $x \in X$ one has*

$$(I - T(t))C(r)x = C(r)(I - T(t)x) = \frac{1}{r}(I - T(r)) \int_0^t T(s)x \, ds. \qquad (11.1)$$

(iii)  *If $C(r)x \to y$ for $r \to \infty$ for some $y \in X$ with respect to $\tau$, then $y \in \text{Ker}(A) = \text{Fix}(T(t))_{t\geq 0}$.*

*Proof* The first two statements follow directly from the definition of the $\tau$-Riemann integral and the strong $\tau$-continuity of the semigroup and the semigroup law. For the third assertion, assume that there exists $x \in X$ such that $C(r)x \to y$ for $r \to \infty$ for some $y \in X$ with respect to $\tau$. Let $(r_n)_{n\in\mathbb{N}}$ be a positive and increasing sequence in $\mathbb{R}$ such that $r_n \to \infty$ as $n \to \infty$. We saw in the proof of Proposition 11.2 that $\|C(r_n)x\| \leq \|x\|$ for each $n \in \mathbb{N}$ so that the sequence $(C(r_n)x)_{n\in\mathbb{N}}$ is $\|\cdot\|$-bounded and $\tau$-convergent. By (11.1) and the fact that $(T(t))_{t\geq 0}$ is locally bi-equicontinuous we obtain

$$(I - T(t))y = \tau\lim_{n\to\infty} (I - T(t))C(r_n)x = \tau\lim_{n\to\infty} \frac{1}{r_n}(I - T(r_n)) \int_0^t T(s)x \, ds = 0,$$

for every $t > 0$. This shows indeed that $y \in \text{Fix}(T(t))_{t\geq 0}$. $\qquad\square$

We now introduce the main definition.

**Definition 11.5** Let $(T(t))_{t\geq 0}$ be a bi-continuous contraction semigroup on $X$ with respect to $\tau$. We call $(T(t))_{t\geq 0}$ to be $\tau$-*mean ergodic* if $C(r)x$ $\tau$-converges for $r \to \infty$ in $X$ for every $x \in X$.

If a bi-continuous semigroup $(T(t))_{t\geq 0}$ is mean ergodic we are able to define an operator $P : X \to X$ by

$$Px := \tau\lim_{r\to\infty} C(r)x, \quad x \in X. \qquad (11.2)$$

By the norming property of the space, cf. Assumption 2.1, we obtain

$$\|Px\| = \sup_{\substack{\varphi \in (X,\tau)' \\ \|\varphi\| \le 1}} |\varphi(Px)| = \sup_{\substack{\varphi \in (X,\tau)' \\ \|\varphi\| \le 1}} \lim_{r \to \infty} |\varphi(C(r)x)| \le \|x\|,$$

for all $x \in X$, i.e., $P \in \mathscr{L}(X)$. As a matter of fact, $P$ is not only a bounded operator on $X$ but also a projection.

**Proposition 11.6** *Let $(T(t))_{t \ge 0}$ be a bi-continuous contraction semigroup on $X$ with respect to $\tau$ and let $(A, D(A))$ be the corresponding generator. If $(T(t))_{t \ge 0}$ is $\tau$-mean ergodic, then the operator $P$ defined by (11.2) is a projection on $X$ such that*

$$P = T(t)P = PT(t) = P^2, \quad t \ge 0. \tag{11.3}$$

*Therefore, one has $X = \mathrm{Ran}(P) \oplus \mathrm{Ker}(P)$ with*

(i) $\mathrm{Ran}(P) = \mathrm{Fix}(T(t))_{t \ge 0} = \mathrm{Ker}(A)$,
(ii) $\overline{\mathrm{Span}}^{\|\cdot\|} \{x - T(t)x : x \in X, \ t \ge 0\} \subseteq \mathrm{Ker}(P) \subseteq \overline{\mathrm{Span}}^{\tau} \{x - T(t)x : x \in X, \ t \ge 0\}$,
(iii) $\overline{\mathrm{Span}}^{\tau} \{x - T(t)x : x \in X, \ t \ge 0\} = \overline{\mathrm{Ran}(A)}^{\tau}$.

*Proof* By Proposition 11.4(iii) and Lemma 11.3 we conclude that $Px \in \mathrm{Fix}(T(t))_{t \ge 0}$ showing that $Px = T(t)Px$ for all $x \in X$ and $t > 0$. We will now show that $P$ also commutes with $T(t)$ for all $t > 0$. For this it suffices to show that $C(n)$ commutes with $T(t)$ for all $n \in \mathbb{N}$ and $t > 0$, cf. (11.1). As $(T(t))_{t \ge 0}$ is $\tau$-mean ergodic, we know that for all $x \in X$ the sequence $(C(n)x)_{n \in \mathbb{N}}$ is $\|\cdot\|$-bounded and $\tau$-convergent to $Px$ whenever $n \to \infty$. Due to the bi-continuity of $(T(t))_{t \ge 0}$ we also have that $T(t)C(n)x \to T(t)Px$ with respect to $\tau$ if $n \to \infty$. As obviously $T(t)C(n)x = C(n)T(t)x$ by the semigroup law, we obtain by taking the limit $n \to \infty$ that $T(t)Px = PT(t)x$ for all $x \in X$. Furthermore, we obtain

$$Px = \frac{1}{r} \int_0^r Px \, \mathrm{d}s = \frac{1}{r} \int_0^r T(s)Px \, \mathrm{d}s = C(r)Px,$$

for all $r > 0$ and $x \in X$. By taking the limit $r \to \infty$ one obtains $Px = P^2 x$ for all $x \in X$. This proves (11.3). We will now subsequently prove the properties (i)–(iii).

(i) As $P$ is a $\|\cdot\|$-bounded projection, we obtain that $X = \mathrm{Ran}(P) \oplus \mathrm{Ker}(P)$. Previously, we showed already that for all $x \in X$ one has that $Px \in \mathrm{Fix}(T(t))_{t \ge 0}$. Conversely, assume that $x \in \mathrm{Fix}(T(t))_{t \ge 0}$, then

$$C(r)x = \frac{1}{r} \int_0^r T(s)x \, \mathrm{d}s = \frac{1}{r} \int_0^r x \, \mathrm{d}s = x,$$

for all $r > 0$. Taking the limit $r \to \infty$ yields $Px = x$ showing that $x \in \mathrm{Ran}(P)$.

(ii) We have to prove two inclusions. In what follows let $Y := \mathrm{Span}$ $\{x - T(t)x : x \in X, \ t \ge 0\}$. For the first inclusion, let $y = x - T(t)x$ for some $t > 0$

and $x \in X$. By Proposition 11.4(ii) we have

$$C(r)y = \frac{1}{r}(I - T(r)) \int_0^t T(s)x \, ds.$$

Due to the fact that $(T(t))_{t \geq 0}$ is a contraction semigroup and by taking the limit $r \to \infty$ we deduce $Py = 0$ or $y \in \mathrm{Ker}(P)$. Hence, we conclude that $\overline{Y}^{\|\cdot\|} \subseteq \mathrm{Ker}(P)$. For the second inclusion, we first choose $\varphi \in (X, \tau)'$ that vanishes on $Y$. For all $t \geq 0$ and $x \in X$ we have $0 = \varphi(x - T(t)x) = \varphi(x) - \varphi(T(t)x)$ so that

$$\varphi(C(r)x) = \frac{1}{r} \int_0^r \varphi(T(s)x) \, ds = \varphi(x).$$

Now, if $x \in \mathrm{Ker}(P)$ we have

$$\varphi(x) = \tau\lim_{r \to \infty} \varphi(C(r)x) = \varphi(P(x)) = 0.$$

In summary, if $\varphi \in (X, \tau)'$ vanished on $Y$, then it also vanished on $\mathrm{Ker}(P)$. For the sake of a contradiction assume that $\mathrm{Ker}(P)$ is not contained in $\overline{Y}^\iota$. In particular, there exists $x_0 \in \mathrm{Ker}(P)$ such that $x_0 \notin \overline{Y}^\tau$. The Hahn–Banach theorem therefore yields $\varphi \in (X, \tau)'$ such that $\varphi(x_0) = 1$ and $\varphi(x - T(t)x) = 0$ for all $x \in X$ and $t \geq 0$. However, this is a contradiction as by the previous argument we have $\varphi(x_0) = 0$.

(iii) Again, two inclusions need to be justified. Firstly, assume that $y = x - T(t)x \in Y$. We know by Theorem 3.15(a) that $D(A)$ is bi-dense in $X$. Therefore, there exists a sequence $(x_n)_{n \in \mathbb{N}}$ in $D(A)$ which is $\|\cdot\|$-bounded and $\tau$-convergent to $x$. Hence,

$$y = \tau\lim_{n \to \infty} x_n - T(t)x_n,$$

due to the assumption that $(T(t))_{t \geq 0}$ is bi-continuous. On the other hand, by Theorem 3.3(b) we have

$$x_n - T(t)x_n = \int_0^t AT(s)x_n \, ds \in \overline{\mathrm{Ran}(A)}^\tau.$$

Combining the previous observations yields $\overline{Y}^\tau \subseteq \overline{\mathrm{Ran}(A)}^\tau$. For the converse, let $y \in \mathrm{Ran}(A)$ and $x \in D(A)$ such that $y = Ax$. The operator $(A, D(A))$ is the generator of the bi-continuous semigroup $(T(t))_{t \geq 0}$ so that

$$t = \tau\lim_{t \to 0} \frac{T(t)x - x}{t}.$$

This shows that $y \in \overline{Y}^\tau$ and also finally $\overline{\mathrm{Ran}(A)}^\tau \subseteq \overline{Y}^\tau$.

$\square$

The next result shows that for a local bi-continuous semigroup $(T(t))_{t\geq0}$ (recall Definition 7.4) the $\tau$-mean ergodicity is implied by the mean ergodicity of the strongly continuous semigroup $(\underline{T}(t))_{t\geq0}$ on $\underline{X}$. For the notation, we refer to Chap. 5. For mean ergodic $C_0$-semigroups we refer, for example, to [101, Chap. V, Sect. 4].

**Proposition 11.7** *Let* $(T(t))_{t\geq0}$ *be a local bi-continuous contraction semigroup and let* $(A, D(A))$ *denote the generator. If* $(\underline{T}(t))_{t\geq0}$ *is mean ergodic (as $C_0$-semigroup), then* $(T(t))_{t\geq0}$ *is $\tau$-mean ergodic.*

***Proof*** Let $x \in X$. The goal is to show that $\tau\lim_{r\to\infty} C(r)x$ exists. By Theorem 3.15(a) the domain $D(A)$ is bi-dense in $X$, i.e., there exists a sequence $(x_n)_{n\in\mathbb{N}}$ in $D(A)$ which is $\|\cdot\|$-bounded and $\tau$-convergent to $x$. As $(T(t))_{t\geq0}$ is local, for $\varepsilon > 0$ and $p \in \mathcal{P}$ there exists $K > 0$ $q \in \mathcal{P}$ such that $p(T(t)x) \leq Kq(x) + \varepsilon\|x\|$. Moreover, due to the $\tau$-convergence of $(x_n)_{n\in\mathbb{N}}$ to $x$ there exists $N \in \mathbb{N}$ such that $q(x_n - x_m) \leq \frac{\varepsilon}{K}$. Therefore,

$$p(C(r)x_n - C(r)x_m) \leq Kq(x_n - x_m) + \varepsilon\|x_n - x_m\| < \varepsilon(1 + M),$$

for all $r > 0$, $n, m \geq N$, and $M := 2\sup_{n\in\mathbb{N}} \|x_n\| < \infty$. By choosing $m = N$ and by taking the limit $n \to \infty$ we obtain by Proposition 11.2 that for all $r > 0$ one has

$$p(C(r)x - C(r)x_N) \leq \varepsilon(1 + M).$$

This leads to the observation that for all $r, s > 0$ one has

$$
\begin{aligned}
&p(C(r)x - C(s)x)\\
&\leq p(C(r)x - C(r)x_N) + p(C(r)x_N - C(s)x_N) + p(C(s)x_N - C(s)x) \qquad (11.4)\\
&\leq 2\varepsilon(1 + M) + \|C(r)x_N - C(s)x_N\|.
\end{aligned}
$$

Because $x_N \in D(A) \subseteq \underline{X}$, we have by the assumption of mean ergodicity of $(\underline{T}(t))_{t\geq0}$ on $\underline{X}$ that $C(r)x_N \to y$ for some $y \in X$ with respect to $\|\cdot\|$ as $r \to \infty$. Hence, $(C(r)x_N)_{r>0}$ is also a Cauchy net with respect to $\|\cdot\|$ and there exists $r_0 > 0$ such that $\|C(r)x_N - C(s)x_N\| < \varepsilon$ for all $r, s \geq r_0$. Therefore, for $r, s \geq r_0$ we obtain from (11.4) that $p(C(r)x - C(s)x) < 2\varepsilon(1 + M) + \varepsilon$. This shows that $(C(r)x)_{r>0}$ is a $\tau$-Cauchy net. Now, let $(r_n)_{n\in\mathbb{N}}$ be any increasing sequence in $(0, \infty)$ such that $r_n \to \infty$ for $n \to \infty$. Then $(C(r_n)x)_{n\in\mathbb{N}}$ $\tau$-converges to some $y \in X$ as $r \to \infty$ showing that $\tau\lim_{r\to\infty} C(r)x = y$ exists as desired. $\qquad\square$

The next result is a bi-continuous version of [101, Chap. V, Theorem 4.10] and [81, Theorem 15].

**Proposition 11.8** *Let $(T(t))_{t\geq 0}$ be a bi-continuous contraction semigroup with generator $(A, D(A))$. The following statements are equivalent:*

(i) *$(T(t))_{t\geq 0}$ is $\tau$-mean ergodic.*
(ii) *$\lambda R(\lambda, A)x$ converges with respect to $\tau$ for every $x \in X$ as $\lambda \to \infty$.*

*Moreover, if one of the above is satisfied, then*

$$\tau\lim_{r\to\infty} C(r)x = \tau\lim_{\lambda\to\infty} \lambda R(\lambda, A)x = Px.$$

In order to prove Proposition 11.8 we will need the following two results that are related to Wiener's theorem.

**Theorem 11.9** *Let $f \in L^1(\mathbb{R}_{>0})$ be such that*

$$\int_0^\infty f(x)x^{-i\xi}\,dx \neq 0, \tag{11.5}$$

*for every $\xi \in \mathbb{R}$. Then, the linear span for the set of $(f_\alpha)_{\alpha>0}$ of the functions defined by $f_\alpha(x) := f(\alpha x)$ are dense in $L^1(\mathbb{R}_{>0})$.*

*Proof* We consider the map $F : L^1(\mathbb{R}) \to L^1(\mathbb{R}_{>0})$ defined by

$$(Fg)(x) := \frac{1}{x}g(\log(x)), \quad g \in L^1(\mathbb{R}), \ x > 0.$$

This clearly defines an isometry with inverse $(F^{-1}h)(y) = e^y h(e^y)$ for $h \in L^1(\mathbb{R}_{>0})$ and $y \in \mathbb{R}$. By assumption, the function $f \in L^1(\mathbb{R}_{>0})$ satisfies (11.5), thus we have

$$\int_\mathbb{R} (F^{-1}f)(y)e^{-i\xi y}\,dy = \int_\mathbb{R} e^y f(e^y)e^{-i\xi y}\,dy = \int_0^\infty f(x)x^{-i\xi}\,dx \neq 0, \tag{11.6}$$

for all $\xi \in \mathbb{R}$. Now, fix $h \in L^1(\mathbb{R}_{>0})$ and set $g := F^{-1}h \in L^1(\mathbb{R})$. Since $F^{-1}f$ satisfies (11.6), we are able to apply Wiener's theorem. In particular, for every $\varepsilon > 0$ there exist $n \in \mathbb{N}$ and $\beta_1, \ldots, \beta_n \in \mathbb{C}$ as well as $\sigma_1, \ldots, \sigma_n \in \mathbb{R}$ such that

$$\int_\mathbb{R} \left| g(y) - \sum_{i=1}^n \beta_i (F^{-1}f)(y + \sigma_i) \right| dy < \varepsilon. \tag{11.7}$$

By choosing $\alpha_i = e^{\sigma_i}$ and $\gamma_i = \beta_i \alpha_i$ for all $1 \leq i \leq n$, we obtain by using (11.7) that

$$\int_0^\infty \left| h(x) - \sum_{i=1}^n \gamma_i f(\alpha_i x) \right| dx < \varepsilon.$$

This proves the result.                                                                  □

**Lemma 11.10** *Let* $K \in L^1(\mathbb{R}_{>0})$ *be such that*

$$\int_0^\infty K(y) y^{-i\xi} \, dy \neq 0,$$

*for all* $\xi \in \mathbb{R}$. *If* $\Phi : \mathbb{R}_{\geq 0} \to X$ *is a* $\|\cdot\|$*-bounded and* $\tau$*-continuous function satisfying*

$$\tau \lim_{\lambda \to 0} \lambda \int_0^\infty K(\lambda t) \Phi(t) \, dt = a \int_0^\infty K(t) \, dt, \tag{11.8}$$

*for some* $a \in X$, *then*

$$\tau \lim_{\lambda \to 0} \lambda \int_0^\infty f(\lambda t) \Phi(t) \, dt = a \int_0^\infty f(t) \, dt,$$

*for all* $f \in L^1(\mathbb{R}_{>0})$.

***Proof*** We split the proof into two parts in order to enable the reader to show what exactly happens to prove the desired result.

**Step 1:** Formula (11.8) holds for all $f \in \mathrm{Span}\{K_\alpha : \alpha > 0\} \subseteq L^1(\mathbb{R}_{>0})$, where $K_\alpha(x) = K(\alpha x)$ for all $\alpha, x > 0$.

Fix $\alpha > 0$ and $p \in \mathcal{P}$ and observe that

$$p\left(\lambda \int_0^\infty K_\alpha(\lambda t) \Phi(t) \, dt - a \int_0^\infty K_\alpha(t) \, dt\right)$$
$$= \frac{1}{\alpha} p\left(\mu \int_0^\infty K(\mu t) \Phi(t) \, dt - a \int_0^\infty K(s) \, ds\right),$$

where $\mu = \alpha\lambda$ and $s = \alpha t$. By taking the limit $\lambda \to 0$, and therefore $\mu \to 0$, we obtain

$$p\left(\lambda \int_0^\infty K_\alpha(\lambda t) \Phi(t) \, dt - a \int_0^\infty K_\alpha(t) \, dt\right) \to 0.$$

As $p \in \mathcal{P}$ was arbitrarily chosen and by linearity, we conclude that (11.8) holds for every $f \in \mathrm{Span}\{K_\alpha : \alpha > 0\}$.

After we have shown that (11.8) holds for the functions $K_\alpha, \alpha > 0$, the final and second step is to show that (11.8) also holds for all $f \in L^1(\mathbb{R}_{>0})$. For this, we use the denseness of the span of the functions $K_\alpha, \alpha > 0$, in $L^1(\mathbb{R}_{>0})$, cf. Theorem 11.9.

**Step 2:** Formula (11.8) holds for all $f \in L^1(\mathbb{R}_{>0})$.

By Theorem 11.9 there exists a sequence $(f_n)_{n \in \mathbb{N}}$ in $\mathrm{Span}\{K_\alpha : \alpha > 0\}$ such that $f_n \to f$ in $L^1(\mathbb{R}_{>0})$. In particular, for every $p \in \mathcal{P}$ one has

$$p\left(\lambda \int_0^\infty f(\lambda t)\Phi(t)\,dt - a\int_0^\infty f(t)\,dt\right) \tag{11.9}$$

$$\le p\left(\lambda \int_0^\infty (f(\lambda t) - f_n(\lambda t))\Phi(t)\,dt\right) \tag{11.10}$$

$$+ p\left(\lambda \int_0^\infty f_n(\lambda t)\Phi(t)\,dt - a\int_0^\infty f_n(t)\,dt\right) \tag{11.11}$$

$$+ p\left(a\int_0^\infty (f_n(t) - f(t))\,dt\right). \tag{11.12}$$

Now, by Step 1 of the proof, the second summand (11.11) vanishes for $\lambda \to 0$ as $f_n \in$ Span$\{K_\alpha : \alpha > 0\}$. Moreover, as $f_n \to f$ in $L^1(\mathbb{R}_{>0})$ and $\Phi$ is $\|\cdot\|$-bounded the dominated convergence theorem yields that also (11.10) and (11.12) vanish for $n \to \infty$. This proves the result. $\qquad\square$

**Proof of Proposition** 11.8. For the implication (i)$\Rightarrow$(ii) we assume that $(T(t))_{t\ge 0}$ is $\tau$-mean ergodic. For $x \in X$ we obtain by definition that $\tau \lim_{r\to\infty} C(r)x = Px$. As $(T(t))_{t\ge 0}$ is a contraction semigroup we have $\omega_0 \le 0$ and so $\lambda \in \rho(A)$ for every $\lambda > 0$ and

$$R(\lambda, A)x = \int_0^\infty e^{-\lambda s}T(s)x\,ds.$$

Integrating by parts yields

$$\lambda R(\lambda, A) = \lambda^2 \int_0^\infty s e^{-\lambda s}C(s)x\,ds.$$

We observe that $\lambda^2 \int_0^\infty s e^{-\lambda s}\,ds = 1$ so that for $p \in \mathcal{P}$ one obtains

$$\begin{aligned}
p(\lambda R(\lambda, A)x - Px) &= p\left(\lambda^2 \int_0^\infty s e^{-\lambda s}C(s)x\,ds - Px\right) \\
&= p\left(\lambda^2 \int_0^\infty s e^{-\lambda s}(C(s)x - Px)\,ds\right) \\
&= p\left(\int_0^\infty t e^{-t}(C(\tfrac{t}{\lambda})x - Px)\,dt\right) \\
&\le \int_0^\infty t e^{-t}p(C(\tfrac{t}{\lambda})x - Px)\,dt.
\end{aligned}$$

As observed above, we have that $C(r)x \to Px$ with respect to $\tau$ if $r \to \infty$. This yields that $p(C(\tfrac{t}{\lambda})x - Px) \to 0$ as $\lambda \to 0$ for every $t > 0$. The dominated convergence theorem allows us now to conclude that $p(\lambda R(\lambda, A)x - Px) \to 0$ as $\lambda \to 0$

For the converse implication (ii)$\Rightarrow$(i) let us assume that for $x \in X$ we have that $\lambda R(\lambda, A)x$ converges to some $a \in X$ for $\lambda \to 0$ with respect to $\tau$. By setting $\lambda = \tfrac{1}{r}$ we have

$$C(r)x = \lambda \int_0^{\frac{1}{\lambda}} T(s)x \, ds = \lambda \int_0^{\infty} \mathbf{1}_{[0,1]}(\lambda s) T(s)x \, ds. \tag{11.13}$$

As we need to show that $(T(t))_{t \geq 0}$ is $\tau$-mean ergodic it would suffice to show that

$$\tau \lim_{\lambda \to 0} \lambda \int_0^{\infty} \mathbf{1}_{[0,1]}(\lambda s) T(s)x \, ds = a.$$

For this purpose, we want to make use of Lemma 11.10. Therefore, we observe that for $\lambda > 0$ we have that

$$\int_0^{\infty} e^{-\lambda s} s^{-i\xi} \, ds = \lambda^{i\xi - 1} \int_0^{\infty} t^{-i\xi} e^{-t} \, dt = \lambda^{i\xi - 1} \Gamma(1 - i\xi) \neq 0, \quad \xi \in \mathbb{R},$$

because the $\Gamma$-function is known to vanish nowhere in $\mathbb{C} \setminus \mathbb{Z}_-$, see also [205, Chap. 2, Sects. 8 and 15]. Now, by assumption

$$\tau \lim_{\lambda \to 0} \lambda \int_0^{\infty} e^{-\lambda s} T(s)x \, ds = a = a \int_0^{\infty} e^{-s} \, ds.$$

Since the orbit $\{T(t)x : t \geq 0\}$ is $\| \cdot \|$-bounded and the map $t \mapsto T(t)x$ is $\tau$-continuous and the fact that $f(t) = e^{-t}$ is in $L^1(\mathbb{R}_{>0})$ we conclude by Lemma 11.10 the desired existence of the limit:

$$\tau \lim_{\lambda \to 0} \lambda \int_0^{\infty} \mathbf{1}_{[0,1]}(\lambda s) T(s)x \, ds = a \int_0^{\infty} \mathbf{1}_{[0,1]}(s) \, ds = a.$$

$$\square$$

The last result in this section is a bi-continuous version of [101, Chap. 5, Theorem 4.5] and describes $\tau$-mean ergodicity by means of a series of other properties.

**Theorem 11.11** *Let $(T(t))_{t \geq 0}$ be a bi-continuous contraction semigroup on $X$ with generator $(A, D(A))$. Consider the following statements:*

(a) *$(T(t))_{t \geq 0}$ is $\tau$-mean ergodic.*
(b) *The Cesáro means $(C(r))_{r > 0}$ converges in the $\tau$-weak operator topology as $r \to \infty$, i.e., $\varphi(C(r)x)$ converges as $r \to \infty$ for every $\phi \in (X, \tau)'$ and $x \in X$.*
(c) *For every $x \in X$ there exists an unbounded sequence $(r_n)_{n \in \mathbb{N}}$ in $\mathbb{R}_{\geq 0}$ such that $(C(r_n)x)_{n \in \mathbb{N}}$ has a $\tau$-weak accumulation point in $X$.*
(d) *For every $x \in X$ one has that $\overline{\mathrm{co}}^\tau \{T(t)x : t \geq 0\} \cap \mathrm{Fix}(T(t))_{t \geq 0} \neq \varnothing$.*
(e) *The fixed space $\mathrm{Fix}(T(t))_{t \geq 0} = \mathrm{Ker}(A)$ separates the dual fixed space $\mathrm{Fix}(T'(t))_{t \geq 0}$ in $(X, \tau)'$.*

*Then (a) $\Rightarrow$ (b) $\Rightarrow$ (c) and (a) $\Rightarrow$ (d) $\Rightarrow$ (e). Furthermore, if for every $\| \cdot \|$-bounded sequence $(x_n)_{n \in \mathbb{N}}$, which is $\tau$-weak convergent to $x$ it holds that $\lim_{n \to \infty} T(t)x_n = T(t)x$ with respect to the $\tau$-weak operator topology, then (c) $\Rightarrow$ (d). Finally, if the semigroup is* *equicontinuous, i.e., for every $p \in \mathcal{P}$ there exist $C \geq 1$ and $q \in \mathcal{P}$ such that $p(T(t)x) \leq Cq(x)$ for all $x \in X$ and $t \geq 0$, then (e) $\Rightarrow$ (a).*

**Proof**  The implication (a)$\Rightarrow$(b) is immediately obvious from the definition of $\tau$-mean ergodicity.

For (b)$\Rightarrow$(c) fix $x \in X$. By assumption $(C(r)x)_{r>0}$ converges to some $y \in X$ in the $\tau$-weak operator topology as $r \to \infty$. Now, let $(r_n)_{n\in\mathbb{N}}$ be an unbounded sequence, then $(C(r_n)x)_{n\in\mathbb{N}}$ also $\tau$-weak converges to $y \in X$ as $n \to \infty$. To see this, let $\varphi \in (X, \tau)'$ and $\varepsilon > 0$ be arbitrary. Then there exists $r_0 > 0$ such that

$$|\varphi(C(r)x) - \varphi(y)| < \varepsilon,$$

whenever $r \geq r_0$. By taking $N := \min\{n \in \mathbb{N} : r_m \geq r_0 \ \forall m \geq n\}$ we obtain

$$|\varphi(C(r_n)x) - \varphi(y)| < \varepsilon,$$

for all $n \geq N$. As we just showed that $(C(r_n)x)_{n\in\mathbb{N}}$ is $\tau$-weak convergent it also has a $\tau$-weak accumulation point.

To prove the implication (c)$\Rightarrow$(d) let us first assume the additional condition that for every $\|\cdot\|$-bounded sequence $(x_n)_{n\in\mathbb{N}}$, which is $\tau$-weakly convergent to $x$ it holds that $\lim_{n\to\infty} T(t)x_n = T(t)x$ with respect to the $\tau$-weak operator topology. Now, let $(r_n)_{n\in\mathbb{N}}$ be an unbounded sequence as predescribed in (c). Up to a subsequence, we can assume that $(C(r_n)x)_{n\in\mathbb{N}}$ is $\tau$-weakly convergent to some $y \in X$ for $n \to \infty$. By the additional assumption mentioned above, we have that $((I - T(t))C(r_n)x)_{n\in\mathbb{N}}$ converges to $y - T(t)y$ as $n \to \infty$ for all $t > 0$ with respect to the $\tau$-weak topology. Moreover, by Proposition 11.4(ii) we have that

$$\|(I - T(t))C(r_n)x\| \leq \frac{2t}{r_n}\|x\|,$$

for all $n \in \mathbb{N}$. Therefore, we have that $(I - T(t))C(r_n)x \to 0$ with respect to the norm as $n \to \infty$. Uniqueness of limits ensures that $y - T(t)y = 0$. As $t \geq 0$ is arbitrary, we conclude $y \in \mathrm{Fix}(T(t))_{t\geq0}$. On the other hand, by Proposition 11.4(i) we observe that $C(r_n)x \in \overline{\mathrm{co}}^{\tau}\{T(t)x : t \geq 0\}$ for all $n \in \mathbb{N}$. As the right-hand side is a $\tau$-weakly closed set in $X$ we conclude that $y \in \overline{\mathrm{co}}^{\tau}\{T(t)x : t \geq 0\}$. Hence, the implication (c)$\Rightarrow$(d) holds true.

(a)$\Rightarrow$(d): Fix $x \in X$. Since $(T(t))_{t\geq0}$ is $\tau$-mean ergodic we have to $C(r)x$ converges to some $y \in X$ with respect to $\tau$ whenever $r \to \infty$. For $(r_n)_{n\in\mathbb{N}}$ an unbounded sequence in $\mathbb{R}_{\geq0}$ one obtains that $(C(r_n)x)_{n\in\mathbb{N}}$ is a $\|\cdot\|$-bounded which is $\tau$-convergent to $y \in X$. As $(T(t))_{t\geq0}$ is bi-continuous, we also have that for any $t > 0$ one has that $((I - T(t))C(r_n)x)_{n\in\mathbb{N}}$ converges to $y - T(t)y$ with respect to $\tau$ as $n \to \infty$. By repeating the same argument as for the implication $(c) \Rightarrow (d)$ we obtain our desired implication.

Let us prove the implication (d)$\Rightarrow$(e). For this, fix $\varphi, \psi \in \mathrm{Fix}(T'(t))_{t\geq0}$ with $\varphi, \psi \in (X, \tau)'$ and $\varphi \neq \psi$. In particular, there exists $x_0 \in X$ such that $\varphi(x_0) \neq \psi(x_0)$. By assumption, there exists $x \in \overline{\mathrm{co}}^\tau \{T(t)x_0 : t \geq 0\} \cap \mathrm{Fix}(T(t))_{t\geq 0}$. If $y \in \mathrm{co}\{T(t)x_0 : t \geq 0\}$, then $y = \sum_{i=1}^k \lambda_i T(t_i)x_0$ with $\sum_{i=1}^k \lambda_i = 1$. For such a $y \in \mathrm{co}\{T(t)x_0 : t \geq 0\}$ we have

$$\varphi(y) = \varphi\left(\sum_{i=1}^k \lambda_i T(t_i)x_0\right) = \sum_{i=1}^k \lambda_i (T'(t_i)\varphi)(x_0) = \sum_{i=1}^k \lambda_i \varphi(x_0) = \varphi(x_0),$$

and similarly $\psi(y) = \psi(x_0)$. The $\tau$-continuity of $\varphi$ and $\psi$ yields that $\varphi(y) = \varphi(x_0)$ and $\psi(y) = \psi(x_0)$ for all $\overline{\mathrm{co}}^\tau \{T(t)x_0 : t \geq 0\}$. In particular, since $x \in \overline{\mathrm{co}}^\tau \{T(t)x_0 : t \geq 0\}$ we have

$$\varphi(x) = \varphi(x_0) \neq \psi(x_0) = \psi(x),$$

showing that indeed $\mathrm{Fix}(T(t))$ separates $\mathrm{Fix}(T'(t))_{t\geq 0}$.

For the final implication (e)$\Rightarrow$(a) assume that $(T(t))_{t\geq 0}$ is equicontinuous, i.e., for every $p \in \mathcal{P}$ there exists $K \geq 1$ and $q \in \mathcal{P}$ such that $p(T(t)x) \leq Kq(x)$ for all $x \in X$ and $t \geq 0$. From this we also obtain that $p(C(r)x) \leq Kq(x)$ for all $x \in X$ and $r > 0$. Let us consider the subspace of $X$ defined by

$$Y = \mathrm{Fix}(T(t))_{t\geq 0} \oplus \mathrm{Span}\{x - T(t)x : x \in X, t \geq 0\}.$$

That we are indeed dealing with a direct sum comes from the observations that $\mathrm{Fix}(T(t))_{t\geq 0} = \mathrm{Ran}(P)$, $\mathrm{Span}\{x - T(t)x : x \in X, t \geq 0\} \subseteq \mathrm{Ker}(P)$, and $\mathrm{Ker}(P) \cap \mathrm{Ran}(P) = \{0\}$. Now, we fix $\varphi \in (X, \tau)'$ that vanishes on $Y$. Then, for all $t \geq 0$ and $x \in X$ one has

$$\varphi(x - T(t)x) = 0 \iff (\varphi - T'(t)\varphi)(x) = 0.$$

In particular, $\varphi \in (X, \tau)' \cap \mathrm{Fix}(T'(t))_{t\geq 0}$. Since $\varphi(x) = 0$ for all $x \in \mathrm{Fix}(T(t))_{t\geq 0}$ as well, and since $\mathrm{Fix}(T(t))_{t\geq 0}$ separates the dual fixed space $\mathrm{Fix}(T'(t))_{t\geq 0}$ in $(X, \tau)'$, we conclude that $\varphi = 0$. Since $\varphi \in (X, \tau)'$ was arbitrary, we obtain that $X = \overline{Y}^\tau$. To prove now that $(T(t))_{t\geq 0}$ is $\tau$-mean ergodic it suffices to prove that for $x \in X$ there exists $y \in X$ such that

$$C(n)x = \frac{1}{n}\int_0^n T(t)x \, dt \to y, \tag{11.14}$$

as $n \to \infty$ with respect to $\tau$. Indeed, if (11.14) holds true, then for every $r > 0$ one has

$$C(r)x = \frac{1}{r}\int_0^n T(t)x \, dt + \frac{1}{r}\int_n^{n+\alpha} T(t)x \, dt,$$

where $n = \lfloor r \rfloor$ and $\alpha = r - n \in [0, 1)$. Therefore,

$$C(r)x = \frac{n}{r}C(n)x + \frac{1}{r}\int_0^\alpha T(t+n)x\,dt$$

$$= \left(1 - \frac{\alpha}{r}\right)C(n)x + \frac{\alpha}{r}T(n)C(\alpha)x \qquad (11.15)$$

$$= \left(1 - \frac{\alpha}{r}\right)\left(C(n)x + \frac{\alpha}{n}T(n)C(\alpha)x\right).$$

As $1 - \frac{\alpha}{r} \to 1$ for $r \to \infty$, $C(n) \to y$ with respect to $\tau$ if $n \to \infty$, and $\|\alpha C\alpha T(n)x\| \leq \|x\|$, we see that $C(r)x \to y$ with respect to $\tau$ as $r \to \infty$, and hence $n \to \infty$. So let us prove (11.14). Let $p \in \mathcal{P}$ and $\varepsilon > 0$ be arbitrary, and choose $y \in Y$ such that $q(x - y) < \frac{\varepsilon}{3K}$ (compare with the equicontinuity assumption). Since $y \in Y$, the $\|\cdot\|$-bounded sequence $(C(n)y)_{n\in\mathbb{N}}$ converges with respect to the norm, say to $y_0$. Hence, $(C(n)y)_{n\in\mathbb{N}}$ is also a Cauchy sequence with respect to the norm. In particular, there exists $N \in \mathbb{N}$ such that

$$\|C(n)y - C(m)y\| < \frac{\varepsilon}{3},$$

for all $n, m \geq N$. Combining this we obtain for $n, m \geq N$ that

$$p(C(n)x - C(m)x) \leq p(C(n)x - C(n)y) + p(C(n)y - C(m)y) + p(C(m)y - C(m)x)$$

$$\leq 2Cq(x - y) + \|C(n)y - C(m)y\|$$

$$\leq 2K\frac{\varepsilon}{3K} + \frac{\varepsilon}{3} = \varepsilon.$$

This shows that $(C(n)x)_{n\in\mathbb{N}}$ is a $\|\cdot\|$-bounded $\tau$-Cauchy sequence. By our general Assumption 2.1, we obtain convergence with respect to $\tau$ as desired. $\qquad\square$

## 11.2   Example: Autonomous Second-order Elliptic Operators

Now we discuss some application of the previous results to semigroups associated with second-order elliptic operators with possibly unbounded coefficients. In this section, we need more information on the differential operator defined in Chap. 1, i.e., the operator defined by (1.5.1). In particular, we want the reader to familiarize with some of the content covered in [169] and [170].

Let us consider the second-order elliptic partial differential operator (with possibly unbounded coefficients) defined as in Chap. 1.5. In particular, for $u \in C_b(\mathbb{R}^N)$ we consider the operator $(\mathcal{A}, D(\mathcal{A}))$ defined by

$$
\mathcal{A}u(x) = \sum_{i,j=1}^{N} q_{ij}(x)\frac{\partial^2}{\partial x_i \partial x_j}u(x) + \sum_{i=1}^{N} b_i(x)\frac{\partial}{\partial x_i}u(x) + c(x)u(x), \quad x \in \mathbb{R}^N,
$$

$$
D(\mathcal{A}) = \left\{ u \in C_b(\mathbb{R}^N) \cap \bigcap_{1 \leq p < \infty} W_{\mathrm{loc}}^{2,p}(\mathbb{R}^N) : \mathcal{A}u \in C_b(\mathbb{R}^N) \right\},
$$

subjected to the minimal conditions given by Assumption 1.23. The associated semigroup will be denoted by $(T(t))_{t\geq 0}$, cf. Theorems 1.24 and 1.25. It is noteworthy that as a consequence of Theorems 1.24 and 1.25, the associated semigroup $(T(t))_{t\geq 0}$ is bi-continuous on $C_b(\mathbb{R}^N)$ with respect to the compact-open topology $\tau_{\mathrm{co}}$. Our main result will characterize mean ergodicity of this bi-continuous semigroup.

**Theorem 11.12** *Suppose that $(T(t))_{t\geq 0}$ is conservative, i.e., $T(t)\mathbf{1} = \mathbf{1}$. Then the following statements are equivalent:*

(a) *The semigroup $(T(t))_{t\geq 0}$ is $\tau_{\mathrm{co}}$-mean ergodic.*
(b) *$(T(t))_{\geq 0}$ admits an invariant measure $\mu$, i.e., for all $t \geq 0$ and all $f \in C_b(\mathbb{R}^N)$ it holds that*

$$
\int_{\mathbb{R}^N} T(t)f \, d\mu = \int_{\mathbb{R}^N} f \, d\mu.
$$

*In this case*

$$
\tau_{\mathrm{co}}\lim_{r \to \infty} C(r)f = \overline{f} = \int_{\mathbb{R}^N} f \, d\mu,
$$

*for every $f \in C_b(\mathbb{R}^N)$.*

***Proof*** We start showing that the implication (b)$\Rightarrow$(a) is valid. Therefore, assume that $(T(t))_{t\geq 0}$ admits an invariant measure $\mu$. To show that $(T(t))_{t\geq 0}$ is $\tau_{\mathrm{co}}$-mean ergodic, we make use of Proposition 11.8. Let $f \in C_b(\mathbb{R}^N)$. Then we need to show that $u_\lambda := R(\lambda, \mathcal{A})f$ converges to some function $v \in C_b(\mathbb{R}^N)$ as $\lambda \to 0$ with respect to $\tau_{\mathrm{co}}$. We observe that for $u_\lambda \in D(\mathcal{A})$ the equation $\lambda u\lambda - \mathcal{A}u_\lambda = f$ is satisfied. Moreover, by using the Laplace transform representation of the resolvent, one obtains

$$
\|u_\lambda\|_\infty \leq \|f\|_\infty \quad \text{and} \quad \|\mathcal{A}u_\lambda\|_\infty \leq 2\lambda\|f\|_\infty. \tag{11.16}
$$

Since $u_\lambda \in D(\mathcal{A})$ one has that $u_\lambda \in W_{\mathrm{loc}}^{2,p}(\mathbb{R}^N)$ for all $1 < p < \infty$. In particular, the following estimate applies

$$
\|u_\lambda\|_{W_{\mathrm{loc}}^{2,p}(\mathbb{R}^N)} \leq C\left(\|\mathcal{A}u_\lambda\|_{L^p(B_{2R}(0))} + \|u_\lambda\|_{L^p(B_{2R}(0))}\right), \tag{11.17}
$$

for all $1 < p < \infty$, every $R > 0$ and some positive constant $C$ depending on $p, R$ and $A$, cf. [117, Theorem 9.11]. A combination of (11.16) and (11.17) leads to

$$\|u_\lambda\|_{W^{2,p}(B_R(0))} \le C_1 \|f\|_\infty, \tag{11.18}$$

for all $0 < \lambda \le \lambda_0$, showing that the family of functions $(u_\lambda)_{\lambda \in (0,\lambda_0]}$ is uniformly bounded in $W^{2,p}(B_R(0))$ for all $1 < p < \infty$ and each $R > 0$. Now, let $p > N$ and choose a sequence $(\lambda_n)_{n \in \mathbb{N}}$ in the interval $(0, \lambda_0]$ such that $\lambda_n \to 0$ for $n \to \infty$. By (11.18) and the Arzelà–Ascoli theorem, there exists a subsequence $(\lambda_{n_k})_{k \in \mathbb{N}}$ and $v \in C_b(\mathbb{R}^N)$ such that $u_{\lambda_{n_k}} \to v$ uniformly on $\overline{B_R(0)}$ for each $R > 0$ whenever $k \to \infty$. Hence, the sequence $(u_{\lambda_{n_k}})_{k \in \mathbb{N}}$ also converges to $v$ in $L^p(B_R(0))$. Since $C(r)f$ converges to $\overline{f}$ in $L^p(\mathbb{R}^N)$, it follows that $v = \overline{f}$ $\mu$-a.e. in $B_R(0)$ and hence almost everywhere in $B_R(0)$ with respect to the Lebesgue measure as $\mu$ and the Lebesgue measure are equivalent. We will now deduce from this that $u_\lambda \to f$ uniformly on compact sets for $\lambda \to 0$. To do so, assume for the sake of a contradiction that this assertion is not true. Then there exist a sequence $(\lambda_n)_{n \in \mathbb{N}}$ in $(0, \lambda_0]$ and two constants $R, M > 0$ such that $\|u_{\lambda_n} - \overline{f}\|_{C(\overline{B_R(0)})} > M$ for each $n \in \mathbb{N}$. Contradiction, as we previously showed that we can extract a subsequence $(u_{\lambda_{n_k}})_{k \in \mathbb{N}}$ which converges uniformly to $\overline{f}$ in $\overline{B_R(0)}$ as $k \to \infty$. Hence (b)$\Rightarrow$(a) has been proven.

For the converse implication (a)$\Rightarrow$(b) let us assume that $(T(t))_{t \ge 0}$ is $\tau_{co}$-mean ergodic. We have to show that it admits an invariant measure. To do so, fix $x \in \mathbb{R}^N$. The mean ergodic projection $P$ is a bounded operator on $C_b(\mathbb{R}^N)$ and $(T(t))_{t \ge 0}$ is positivity preserving semigroup (compare also with Theorem 1.25). Hence, the map $f \mapsto (Pf)(x)$ is a positive $\|\cdot\|_\infty$-continuous linear functional on $C_b(\mathbb{R}^N)$. By the Riesz representation theorem, there exists a finite positive Borel measure $\mu_x$ on $\mathbb{R}^N$ such that

$$(Pf)(x) = \int_{\mathbb{R}^N} f \, d\mu_x, \quad f \in C_c(\mathbb{R}^N).$$

We now want to show that indeed

$$\int_{\mathbb{R}^N} f \, d\mu_x = \int_{\mathbb{R}^N} T(t)f \, d\mu_x, \tag{11.19}$$

for all $f \in C_b(\mathbb{R}^N)$. Since $(T(t))_{t \ge 0}$ is positivity preserving, also $P$ enjoys this property, i.e., $Pf \ge 0$ if $f \ge 0$. By using (11.3) we observe that for positive $f \in C_c(\mathbb{R}^N)$ one has

$$\int_{\mathbb{R}^N} f \, d\mu_x = (Pf)(x) = (PT(t)f)(x) \ge P(\theta_n T(t)f)(x) = \int_{\mathbb{R}^N} \theta_n T(t)f \, d\mu_x,$$

for every $n \in \mathbb{N}$ and $\theta_n \in C_c(\mathbb{R}^N)$ that satisfies $0 \le \theta_n \le 1$ and $\theta_n \equiv 1$ on $B_n(0)$ for every $n \in \mathbb{N}$. By taking the limit $n \to \infty$ we obtain

$$\int_{\mathbb{R}^N} f \, d\mu_x \ge \int_{\mathbb{R}^N} T(t)f \, d\mu_x, \quad f \in C_c(\mathbb{R}^N), \ t > 0. \tag{11.20}$$

Let $f \in C_b(\mathbb{R}^N)$ be positive. Since the sequence $(\theta_n f)_{n \in \mathbb{N}}$ is $\| \cdot \|_\infty$-bounded and $\tau_{co}$-convergent to $f$, the sequence $(T(t)(\theta_n f))_{n \in \mathbb{N}}$ is also $\| \cdot \|_\infty$-bounded as well as $\tau_{co}$-convergent to $T(t)f$ for $n \to \infty$. By choosing $\theta_n f$ in (11.20) instead of $f$ and taking the limit $n \to \infty$ we deduce that

$$\int_{\mathbb{R}^N} f \, d\mu_x \geq \int_{\mathbb{R}^N} T(t)f \, d\mu_x, \quad f \in C_b(\mathbb{R}^N), \ f \geq 0, \ t > 0. \tag{11.21}$$

Now let $f \in C_b(\mathbb{R}^N)$ be arbitrary. Then we can replace $f$ in (11.21) by $f + \|f\|_\infty$. Since $T(t)\|f\|_\infty = \|f\|_\infty$ we deduce that

$$\int_{\mathbb{R}^N} f \, d\mu_x \geq \int_{\mathbb{R}^N} T(t)f \, d\mu_x, \quad f \in C_b(\mathbb{R}^N), \ t > 0. \tag{11.22}$$

By replacing $f$ by $-f$ in (11.22) we obtain (11.19). We also observe that

$$\mu_x(\mathbb{R}^N) = \int_{\mathbb{R}^N} \mathbf{1} \, d\mu_x = (P\mathbf{1})(x) = \lim_{r \to \infty} \frac{1}{r} \int_0^r (T(s)\mathbf{1})(x) \, ds = \lim_{r \to \infty} \frac{1}{r} \int_0^r 1 \, ds = 1.$$

Hence, $\mu_x$ is a probability measure and (11.19) shows that $\mu_x$ is an invariant measure for $(T(t))_{t \geq 0}$. By the uniqueness of the invariant measure it follows that $\mu_x = \mu_y =: \mu$ for all $x, y \in \mathbb{R}^N$.

**Example 11.13** As a final conclusion, we want to show that there are bi-continuous semigroups that are not $\tau$-mean ergodic. In particular, mean ergodicity goes wrong already for simple examples like the left translation semigroup, see Sect. 2.2.1. We know that the left translation semigroup is bi-continuous on $C_b(\mathbb{R})$ with respect to $\tau_{co}$. Let us show that this semigroup is not $\tau_{co}$-mean ergodic. To do so, we can adapt some ideas from the second part of the proof of Theorem 11.12.

For the sake of a contradiction, assume that the left translation semigroup $(T(t))_{t \geq 0}$ is $\tau_{co}$-mean ergodic. Then, the mean ergodic projection $P$ can be identified with a positive functional on $C_b(\mathbb{R})$ since the kernel of the operator $Au = u'$ consists only of constant functions. Therefore, there exists a finite positive Borel measure $\mu$ such that

$$\int_{\mathbb{R}} f(t + \cdot) \, d\mu = \int_{\mathbb{R}} f \, d\mu, \quad f \in C_b(\mathbb{R}), \ t > 0. \tag{11.23}$$

By positive and monotone convergence, formula (11.23) extends to any bounded Borel measurable function $f$. Hence, writing (11.23) with $t = n \in \mathbb{N}$ and $f = \mathbf{1}_{[0,1)}$, we obtain $\mu([n, n+1)) = \mu([0, 1))$. It follows that

$$\infty > \mu([0, \infty)) = \sum_{n=0}^{\infty} \mu([n, n+1)) = \sum_{n=0}^{\infty} \mu([0, 1)),$$

which implies that $\mu([0, \infty)) = 0$. Applying the same argument to the function $f_{-k} = \mathbf{1}_{[-k,-k+1)}$, we obtain $\mu([-k, -k + 1)) = 0$ for all $k \in \mathbb{N}$ so that $\mu((-\infty, 0)) = 0$. This yields that $0 = \mu(\mathbb{R}) = P\mathbf{1}$. But this is a contradiction since it implies that $Pf = 0$ for all $f \in C_b(\mathbb{R})$ so that $\mathrm{Ker}(A) = \{0\}$, which is not the case.

## Notes on This Chapter

The presented work is originally due to Albanese, Lorenzi, and Manco [7]. These authors also present an application to non-autonomous evolution equations. However, we decided to stay within the time-independent and hence autonomous setting. For more information, the interested reader is kindly requested to read [7]. Mean ergodic theorems for $C_0$-semigroups can be found in [101, Chap. V, Sect. 4] or [80]. More research on mean ergodicity, for example, on locally convex spaces has been done by [3–5]. Further research on mean ergodic semigroups has also been done in different settings. For multi-Banach spaces we refer to [144], whereas ergodic results on Riesz spaces have been formulated here [39, 133, 134, 165]. Perturbations of mean ergodic semigroups have also been discussed [51].

# Bi-continuous Semigroups for Flows on Infinite Networks  **12**

Consider a huge network of possibly unknown size but known structure. To model this situation, we may take an infinite graph and equip it with the appropriate combinatorial assumptions. Along the edges of the network some transport processes take place that are coupled in the vertices in which the edges meet. This means that we consider each edge as an interval and describe functions on it, that is, we consider a *metric graph*. Such systems of partial differential equations on a metric graph are also known as *quantum graphs*. The transport processes (or *flows*) on the edges are given by partial differential equations of the form $\frac{\partial}{\partial t} u_j(t, x) = c_j \frac{\partial}{\partial x} u_j(t, x)$ and are interlinked in the common nodes via some prescribed transmission conditions.

When studying problems in infinite graphs, the flow problem in the $L^\infty$-setting should be interesting for applications as well. Von Below and Lubary [233, 234], for example, study eigenvalues of the Laplacian on infinite networks in an $L^\infty$-setting. Also the celebrated existence, uniqueness, and stability results for linear transport equations by DiPerna and Lions [86] are proven in the $L^p$-setting for all $1 \leq p \leq \infty$. By Lotz' theorem [171, Theorem 3], every strongly continuous semigroup on an $L^\infty$-space is automatically uniformly continuous which means that semigroups on such spaces, which are not strongly continuous, are of special interest and a different approach is needed. In this chapter, we consider the transport problem on the state space $L^\infty\left([0, 1], \ell^1\right)$ where the obtained operator semigroup is not strongly continuous but bi-continuous.

  237

C. Budde, *Bi-Continuous Operator Semigroups*, Frontiers in Mathematics,
https://doi.org/10.1007/978-3-032-12948-2_12

## 12.1    Preliminaries

### 12.1.1  Bochner $L^p$-spaces, Duality, and Translation Semigroups

We consider Bochner spaces of the form $L^p([0, 1], X)$ where $X$ is a Banach space. For more information, we refer to Appendix A. From [85, Chap. IV, Sect. 1] or Theorem D.4 we obtain that for $1 < p < \infty$ and $\frac{1}{p} + \frac{1}{q} = 1$ one has

$$L^p([0, 1], X)' \cong L^q([0, 1], X') \quad \text{and} \quad L^1([0, 1], X)' \cong L^\infty([0, 1], X') \qquad (12.1)$$

whenever $X'$ has the Radon–Nikodym property. If this is not the case, one only has an isometric inclusion $L^q([0, 1], X') \hookrightarrow L^p([0, 1], X)'$ for $1 \le p < \infty$. Recall that the pairing is defined by

$$\langle f, g \rangle := \int_0^1 \langle f(s), g(s) \rangle_X \, ds, \quad f \in L^q([0, 1], X'), \ g \in L^p([0, 1], X), \qquad (12.2)$$

and $\langle \cdot, \cdot \rangle_X$ denotes the dual pairing between $X$ and $X'$.

**Example 12.1**  It is known that the space $\ell^1$ has the Radon–Nikodym property while the spaces $c_0$, $c$, and $\ell^\infty$ do not. Recall that $(\ell^1)' = \ell^\infty$ while $(c_0)' \cong \ell^1$ as well as $(c)' \cong \ell^1$ (that is, the space $\ell^1$ does not have a unique predual space). By (12.1) we obtain that

$$L^\infty([0, 1], \ell^1) \cong L^1([0, 1], c_0)' \cong L^1([0, 1], c)'.$$

while $L^\infty([0, 1], \ell^\infty)$ is only isomorphic to a subspace of $L^1([0, 1], \ell^1)'$.

We would like to study the left-translation semigroup $(T_l(t))_{t \ge 0}$ on the space $X = L^\infty([0, 1], \ell^1)$, which is not strongly continuous. By using duality arguments we will show that it is, however, a bi-continuous semigroup on $X$. We start by showing that the right-translation semigroup on $L^1([0, 1], c_0)$ is strongly continuous.

**Lemma 12.2**  *The right-translation semigroup* $(T_r(t))_{t \ge 0}$, *defined by*

$$T_r(t)f(s) := \begin{cases} f(s - t), & s - t \ge 0, \\ 0, & s - t < 0, \end{cases}$$

*for* $t \ge 0$ *and* $s \in [0, 1]$, *is strongly continuous on* $L^1([0, 1], c_0)$.

***Proof***  Let $\mathbf{x} = (x_n)_{n \in \mathbb{N}} \in c_0$ and let $f := \mathbf{x} \cdot \chi_\Omega$ for a measurable subset $\Omega \subseteq [0, 1]$. We first show that $T_r(t)f \to f$ with respect to the norm on $L^1([0, 1], c_0)$ as $t \to 0$:

$$\int_0^1 \|T_r(t)f(s) - f(s)\|_{c_0} \, ds = \int_0^1 \|f(s-t)\chi_{[s-t\geq 0]} - f(s)\|_{c_0} \, ds$$

$$= \|\mathbf{x}\|_{c_0} \cdot \int_0^1 |\chi_\Omega(s-t) - \chi_\Omega(s)| \, ds$$

$$= \|\mathbf{x}\|_{c_0} \cdot \int_0^1 |\chi_{(\Omega+t)\Delta\Omega}(s)| \, ds$$

$$= \|\mathbf{x}\|_{c_0} \cdot \lambda^1 \left((\Omega + t)\Delta\Omega\right) \to 0 \text{ as } t \to 0,$$

where $\lambda^1$ is the one-dimensional Lebesgue measure on $[0, 1]$ and $\Delta$ the symmetric difference of sets defined by $A\Delta B := (A \cup B) \setminus (A \cap B)$. Since every function $f \in L^1([0, 1], c_0)$ is an increasing limit of linear combinations of functions of the form $\mathbf{x} \cdot \chi_\Omega$ for some $\mathbf{x} \in c_0$ and measurable set $\Omega \subseteq [0, 1]$, the vector-valued version of Beppo Levi's monotone convergence theorem [229, Proposition 2.6] yields the result. $\qquad\square$

Now note that the left-translation semigroup on $L^\infty([0, 1], \ell^1) = L^1([0, 1], c_0)'$ is the adjoint semigroup of the right-translation semigroup on $L^1([0, 1], c_0)$, see Sect. 12.1.1 and [101, II.5.14].

**Lemma 12.3** *The left-translation semigroup $(T_l(t))_{t\geq 0}$, defined by*

$$T_l(t)f(s) := \begin{cases} f(s+t), & s+t \leq 1, \\ 0, & s+t > 1, \end{cases}$$

*for $t \geq 0$ and $s \in [0, 1]$, is bi-continuous on $L^\infty([0, 1], \ell^1)$ with respect to the weak*-topology.*

**Proof** We saw in Sect. 2.1.2 that the dual space $Y'$ of any Banach space $Y$ satisfies Assumption 2.1 for the weak*-topology. Moreover, in Sect. 2.2.3 we saw that the dual semigroup $(T'(t))_{t\geq 0}$ on $Y'$ is bi-continuous with respect to the weak*-topology whenever $(T(t))_{t\geq 0}$ is a strongly continuous semigroup on $Y$. Thus, Example 12.1 and Lemma 12.2 imply the result. $\qquad\square$

**Remark 12.4** Let us stress here that, given a strongly continuous semigroup, its adjoint semigroup on the dual space is not necessarily strongly continuous. Indeed, the translation semigroup is not strongly continuous in the sup-norm of the $L^\infty$-space and the largest subspace where it is strongly continuous is $C_{ub}$, the space of all bounded, uniformly continuous functions (the so-called *sun dual*, see [101, Sect. II.2.5]). Since we are interested in the solutions taking place in $X = L^\infty([0, 1], \ell^1)$, the classical theory on adjoint semigroups does not apply. The above result, however, shows that the theory of bi-continuous semigroups is appropriate here.

### 12.1.2 Infinite Networks, Metric Graphs

We use the notation introduced in [151] for finite networks and expanded in [87] to infinite networks. A network is modeled with an *infinite directed graph $G = (V, E)$* with a set of *vertices $V = \{v_i : i \in I\}$* and a set of *directed edges $E = \{e_j : j \in J\} \subseteq V \times V$* for some countable sets $I, J \subseteq \mathbb{N}$. For a directed edge $e = (v_i, v_k)$ we call $v_i$ the *tail* and $v_k$ the *head* of e. Further, the edge e is an *outgoing edge* of the vertex $v_i$ and an *incoming edge* for the vertex $v_k$. We assume that the graph $G$ is *simple*, i.e., there are no loops or multiple edges, and *locally finite*, i.e., each vertex only has finitely many incident edges.

The graph $G$ is also assumed to be *weighted* that is equipped with some weights $0 \leq w_{ij} \leq 1$ such that

$$\sum_{i \in J} w_{ij} = 1 \text{ for all } j \in J. \tag{12.3}$$

The structure of a graph can be described by its incidence and/or adjacency matrices. We shall only use the so-called *weighted (transposed) adjacency matrix of the line graph* $\mathbb{B} = (\mathbb{B}_{ij})_{i,j \in J}$ defined as

$$\mathbb{B}_{ij} := \begin{cases} w_{ij} & \text{if } \xrightarrow{e_j} v \xrightarrow{e_i}, \\ 0 & \text{otherwise.} \end{cases} \tag{12.4}$$

By (12.3), the matrix $\mathbb{B}$ defines a stochastic operator on $\ell^1$. From that it follows that $r(\mathbb{B}) = \|\mathbb{B}\| = 1$. It reflects many properties of the graph $G$. For example, $\mathbb{B}$ is irreducible iff the graph $G$ is strongly connected, cf. [87, Proposition 4.9].

We identify every edge of our graph with the unit interval, $e_j \equiv [0, 1]$ for each $j \in J$, and parametrize it contrary to its direction, so that it is assumed to have its tail at the endpoint 1 and its head at the endpoint 0. For simplicity we use the notations $e_j(1)$ and $e_j(0)$ for the tail and the head, respectively. In this way, we obtain a *metric graph.*

For the unexplained terminology we refer, for example, to [33, Sect. 18] and [87].

## 12.2  Transport Problems in (in)finite Metric Graphs

We now consider a transport process (or a flow) along the edges of an infinite network, modeled by a metric graph $G$. The distribution of material along the edge $e_j$ at time $t \geq 0$ is described by a function $u_j(x, t)$ for $x \in [0, 1]$. The material is transported along the edge $e_j$ with constant velocity $c_j > 0$ and is absorbed according to the absorption rate $q_j(x)$. We assume that $q \in L^\infty([0, 1], \ell^\infty)$ and

$$0 < c_{\min} \leq c_j \leq c_{\max} < \infty \tag{12.5}$$

for all $j \in J$. Let $C := \mathrm{diag}(c_j)_{j \in J}$ be a diagonal velocity matrix and define another weighted adjacency matrix of the line graph by

$$\mathbb{B}^C := C^{-1} \mathbb{B} C.$$

In the vertices, the material gets redistributed according to some prescribed rules. This is modeled in the boundary conditions by using the adjacency matrix $\mathbb{B}^C$. The flow process on $G$ is thus given by the following infinite system of equations:

$$\begin{cases} \frac{\partial}{\partial t} u_j(x,t) = c_j \frac{\partial}{\partial x} u_j(x,t) + q_j(x) u_j(x,t), & x \in (0,1),\ t \geq 0, \\ u_j(1,t) = \sum_{k \in J} \mathbb{B}^C_{jk} u_k(0,t), & t \geq 0, \\ u_j(x,0) = f_j(x), & x \in (0,1), \end{cases} \tag{12.6}$$

for every $j \in J$, where $f_j(x)$ are the initial distributions along the edges.

One can give different interpretations to the weights $w_{ij}$, i.e, entries of the matrix $\mathbb{B}$, resulting in different transport problems. The two most obvious are the following:

(a)  $w_{ij}$ is the proportion of the material arriving from edge $\mathrm{e}_j$ leaving on edge $\mathrm{e}_i$.
(b)  $w_{ij}$ is the proportion of the material arriving from vertex $\mathrm{e}_j(0) = \mathrm{e}_i(1)$ leaving on edge $\mathrm{e}_i$.

Note that in both situations (12.3) represents a conservation of mass and the assumption on local finiteness of the graph guarantees that all the sums are finite. While the latter situation is the most common one (see, e.g., [33, 87, 151]) the first one was considered for finite networks in [38, Sect. 5]. Here, we will not give any particular interpretation and will treat all the cases simultaneously.

**Remark 12.5**  By replacing in (12.6) the graph matrix $\mathbb{B}^C$ with some other matrix, one obtains a more general initial-value problem that does not necessarily consider a process in a physical network. Such a problem from population dynamics was, for example, studied in [25]. Furthermore, a question when can such a general problem be identified with a corresponding problem on a metric graph was raised in [24].

**Remark 12.6**  Note that we could also consider space-dependent velocities $c_j(x)$ and then either renorm the space as in [179]:

$$\|f\|_C := \sum_{j \in J} \int_0^1 \frac{|f_j(s)|}{c_j(s)}\, ds, \quad f \in \mathrm{L}^\infty\left([0,1], \ell^1\right)$$

or change the variables appropriately to obtain an equivalent problem with constant velocities.

## 12.2.1 The Simple Case

First we assume that all the velocities are the same and there is no absorption: $c_j = 1$ and $q_j(\cdot) = 0$ for each $j \in J$. As the state space we choose $X := L^\infty\left([0, 1], \ell^1\right)$ equipped with the norm:

$$\|f\|_X := \operatorname*{ess\ sup}_{s \in [0,1]} \|f(s)\|_{\ell^1}.$$

On the Banach space $X$ we define the operator $(A, D(A))$ by

$$A := \operatorname{diag}\left(\frac{\mathrm{d}}{\mathrm{d}x}\right)$$

$$D(A) := \left\{v \in W^{1,\infty}\left([0, 1], \ell^1\right) : v(1) = \mathbb{B}v(0)\right\}.$$

(12.7)

Observe that the corresponding abstract Cauchy problem

$$\begin{cases} \dot{u}(t) = Au(t), & t \geq 0, \\ u(0) = (f_j)_{j \in J} \end{cases}$$

(12.8)

on $X$ is equivalent to the flow problem (12.6) in case when all the velocities equal 1. This problem was considered by Dorn [87] on the state space $L^1\left([0, 1], \ell^1\right)$ where an explicit formula for the solution semigroup in terms of a shift and the matrix $\mathbb{B}$ was derived. We will use this formula and show that it yields a bi-continuous semigroup on $L^\infty\left([0, 1], \ell^1\right)$. For that we have to check all the assertions from Definition 2.4 which we do in several steps.

**Lemma 12.7** *The semigroup* $(T(t))_{t \geq 0}$ *on* $X = L^\infty\left([0, 1], \ell^1\right)$, *defined by*

$$T(t)f(s) = \mathbb{B}^n f(t + s - n), \quad n \leq t + s < n + 1, \ f \in X, \ n \in \mathbb{N}_0,$$

(12.9)

*is strongly continuous with respect to the weak*-topology.*

**Proof** The semigroup property is easy to verify. Observe that for any $f \in X$, $g \in L^1$ $([0, 1], c_0)$, and $t \in (0, 1]$ we have

$$|\langle T(t)f - f, g\rangle| =$$

$$= \left|\int_0^1 \langle T(t)f(s) - f(s), g(s)\rangle \,\mathrm{d}s\right|$$

$$\leq \left|\int_0^{1-t} \langle f(s + t) - f(s), g(s)\rangle \,\mathrm{d}s\right| + \int_{1-t}^1 |\langle \mathbb{B}f(s + t - 1) - f(s), g(s)\rangle| \,\mathrm{d}s$$

$$\leq \left|\int_0^1 \langle T_l(t)f(s) - f(s), g(s)\rangle \,\mathrm{d}s\right| + \int_{1-t}^1 |\langle \mathbb{B}f(s + t - 1) - f(s), g(s)\rangle| \,\mathrm{d}s.$$

Now, notice that the second summand vanishes since $\lambda^1 ([1 - t, 1]) \to 0$ as $t \to 0$. Here, $\lambda^1$ is the one-dimensional Lebesgue measure on the unit interval $[0, 1]$. By Lemma 12.3, the left-translation semigroup is bi-continuous on $X$, which means, in particular, that it is strongly continuous with respect to the weak*-topology and hence the first summand also vanishes as $t \to 0$. $\qquad\square$

**Lemma 12.8** *The semigroup* $(T(t))_{t\geq 0}$, *defined by* (12.9), *is a contraction semigroup on* $X$.

***Proof*** Let $f \in X$ and $t \geq 0$. Then there exists $n \in \mathbb{N}_0$ such that $n \leq t < n + 1$. This means that for $s \in [0, 1]$ one has $n \leq s + t < n + 2$. By (12.9), we can make the following estimate:

$$\|T(t)f\|_X =$$
$$= \operatorname*{ess\,sup}_{s\in[0,1]} \|T(t)f(s)\|_{\ell^1}$$
$$\leq \max \left\{ \operatorname*{ess\,sup}_{s\in[0,n+1-t)} \left\|\mathbb{B}^n f(s + t - n)\right\|_{\ell^1}, \quad \operatorname*{ess\,sup}_{s\in[n+1-t,1]} \left\|\mathbb{B}^{n+1} f(s + t - n - 1)\right\|_{\ell^1} \right\}.$$

Since $\|\mathbb{B}^n\| = \|\mathbb{B}\|^n = 1$, we have

$$\left\|\mathbb{B}^n f(s + t - n)\right\|_{\ell^1} \leq \left\|\mathbb{B}^n\right\| \cdot \|f\|_X = \|f\|_X$$

and hence $\|T(t)f\|_X \leq \|f\|_X$. $\qquad\square$

**Lemma 12.9** *The semigroup* $(T(t))_{t\geq 0}$, *defined by* (12.9), *is locally bi-equicontinuous with respect to the weak*-topology on* $X = L^\infty \left([0, 1], \ell^1\right)$.

***Proof*** Let $(f_n)_{n\in\mathbb{N}}$ be a sequence of functions in $X$ that is $\|\cdot\|_X$-bounded and converges to $0$ with respect to the weak*-topology. By Definition 2.4, we need to show that $(T(t)f_n)_{n\in\mathbb{N}}$ converges to $0$ with respect to the weak*-topology uniformly on compact intervals $[0, t_0]$. To this end, fix $t_0 > 0$. For every $t \in [0, t_0]$ one can find $k \in \mathbb{N}_0$, $0 \leq k \leq \lceil t_0 \rceil$, such that $k \leq s + t < k + 1$ for all $s \in [0, 1]$ and by (12.9) we have

$$T(t)f_n(s) = \mathbb{B}^k f_n(s + t - k),$$

hence

$$\langle T(t) f_n, g \rangle = \int_0^1 \langle T(t) f_n(s), g(s) \rangle \, ds$$

$$= \int_0^1 \left\langle \mathbb{B}^k f_n(s + t - k), g(s) \right\rangle ds$$

$$= \int_0^1 \left\langle T_l(t - k) f_n(s), \left( \mathbb{B}^k \right)' g(s) \right\rangle ds$$

for any $g \in L^1 \left( [0, 1], c_0 \right)$. Observe that our assumption on local finiteness of the graph implies that both $\mathbb{B}$ and $\mathbb{B}'$ only have finitely many elements in each row or column. Therefore, $\mathbb{B}'$ leaves the space $c_0$ invariant. Moreover, there are only finitely many choices of the integer $k \in [0, \lceil t_0 \rceil]$ and we have

$$\{ T_l(t - k) : \ 0 \leq t \leq t_0 \} = \{ T_l(\tau) : \ 0 \leq \tau \leq 1 \}.$$

Since, by Lemma 12.3, the left-translation semigroup $(T_l(t))_{t \geq 0}$ is bi-continuous, hence locally bi-equicontinuous with respect to the weak*-topology on $X$, also $\langle T(t) f_n, g \rangle$ tends to 0 uniformly on $[0, t_0]$. This finishes the proof.                                $\square$

Let us here recall the explicit expression of the resolvent of operator $(A, D(A))$ defined by (12.7) which was obtained in [87, Theorem 18]. This result does not rely on the Banach space and remains the same by taking $X = L^\infty \left( [0, 1], \ell^1 \right)$.

**Proposition 12.10**  *For* $\mathrm{Re}(\lambda) > 0$ *the resolvent* $R(\lambda, A)$ *of the operator* $(A, D(A))$ *defined by (12.7) is given by*

$$(R(\lambda, A) f)(s) = \sum_{k=0}^{\infty} e^{-\lambda k} \int_0^1 e^{-\lambda(t+1-s)} \mathbb{B}^{k+1} f(t) \, dt + \int_s^1 e^{\lambda(s-t)} f(t) \, dt,$$

$f \in X, s \in [0, 1]$.

We are now in the state to prove the first-generation theorem.

**Theorem 12.11**  *The operator* $(A, D(A))$, *defined in (12.7), generates a bi-continuous contraction semigroup* $(T(t))_{t \geq 0}$ *on* $X$ *with respect to the weak*-topology. This semigroup is given by (12.9).*

***Proof*** By Lemmas 12.7, 12.8 as well as Lemma 12.9, the semigroup $(T(t))_{t \geq 0}$ defined by (12.9) is a bi-continuous (contraction) semigroup with respect to the weak*-topology. It remains to show that $(A, D(A))$, given in (12.7), is the generator of this semigroup. Let $(C, D(C))$ be the generator of $(T(t))_{t \geq 0}$. For $f \in D(A)$ and $s \in [0, 1]$ we have $T(t) f \in D(A)$. By (3.3.1), the resolvent of $C$ is the Laplace transform of the semigroup $(T(t))_{t \geq 0}$, that is, for $\lambda > \omega_0(T)$ we have

$$R(\lambda, C)f(s) = \int_0^\infty e^{-\lambda t} T(t) f(s)\, dt$$

$$= \int_0^{1-s} e^{-\lambda t} f(t+s)\, dt + \sum_{n=1}^\infty \int_{n-s}^{n-s+1} e^{-\lambda t} \mathbb{B}^n f(t+s-n)\, dt$$

$$= \int_s^1 e^{-\lambda(\xi-s)} f(\xi)\, d\xi + \sum_{n=1}^\infty \int_0^1 e^{-\lambda(\xi-s-n)} \mathbb{B}^n f(\xi)\, d\xi.$$

By Proposition 12.10, the resolvent operators $R(\lambda, A)$ and $R(\lambda, C)$ coincide on the bi-dense set $D(A)$, so we may conclude that $C = A$. $\qquad\square$

**Corollary 12.12** *If all $c_j = 1$ and $q_j(\cdot) = 0$, $j \in J$, the flow problem* (12.6) *is well-posed on* $X = \mathrm{L}^\infty\left([0, 1], \ell^1\right)$.

**Remark 12.13** All the obtained results also hold for finite networks. If $G = (V, E)$ is a finite network with $|E| = m < \infty$, we have $\ell^1\left(\{1, \ldots, m\}\right) \cong \mathbb{C}^m$, hence we consider our semigroups on the space $X = \mathrm{L}^\infty\left([0, 1], \mathbb{C}^m\right)$.

### 12.2.2 The Rationally Dependent Case

We now consider the case when the velocities $c_j$ appearing in (12.6) are not all equal to 1 and define on $X := \mathrm{L}^\infty\left([0, 1], \ell^1\right)$ the operator

$$A_C := \mathrm{diag}\left(c_j \cdot \frac{\mathrm{d}}{\mathrm{d}x}\right),$$

$$D(A_C) := \left\{ f \in \mathrm{W}^{1,\infty}\left([0, 1], \ell^1\right) : \ f(1) = \mathbb{B}^C f(0) \right\}. \tag{12.10}$$

We assume, however, that the velocities are linearly dependent over $\mathbb{Q}$: $\frac{c_i}{c_j} \in \mathbb{Q}$ for all $i, j \in J$, with a finite common multiplier, that is:

$$\text{there exists } 0 < c \in \mathbb{R} \text{ such that } \ell_j := \frac{c}{c_j} \in \mathbb{N} \text{ for all } j \in J. \tag{12.11}$$

This enables us to use the procedure that was introduced in the proof of [151, Theorem 4.5] and carried out in detail in [26, Sect. 3]. We construct a new directed graph $\widetilde{G}$ by adding $\ell_j - 1$ vertices on edge $e_j$ for all $j \in J$. The newly obtained edges inherit the direction of the original edge and are parametrized as unit intervals $[0, 1]$. We can thus consider a new problem on $\widetilde{G}$ with corresponding functions $\widetilde{u}_j$ and velocities $\widetilde{c}_j := c$ for each $j \in \widetilde{J}$. After appropriately correcting the initial and boundary conditions the new problem is equivalent to the original one. Since all the velocities on the edges of the new graph are equal, we can treat this case by rescaling to 1 and use the results from Sect. 12.2.1. Moreover, since

(12.5) and (12.11) hold, the procedure described in [26, Sect. 3] for the finite case can be as performed in the infinite case as well. Hence, we even obtain an isomorphism between the corresponding semigroups.

**Theorem 12.14** *Let Assumptions* (12.5) *and* (12.11) *on the velocities* $c_j$ *hold. Then the operator* $(A_C, D(A_C))$, *defined in* (12.10), *generates a bi-continuous contraction semigroup* $(T_C(t))_{t \geq 0}$ *on* $X$ *with respect to the weak*-topology. Moreover, there exists a lattice isomorphism* $S \colon X \to X$ *such that*

$$T_C(ct)f = ST(t)S^{-1}f, \tag{12.12}$$

*where the semigroup* $(T(t))_{t \geq 0}$ *is given by* (12.9).

Finally, we also allow non-zero absorption functions $q_j$. Notice that (12.6) is equivalent to the abstract Cauchy problem associated to the following operator:

$$\begin{aligned}
A &:= \operatorname{diag}\left(c_j \cdot \frac{\mathrm{d}}{\mathrm{d}x} + M_{q_j}\right), \\
D(A) &:= \left\{ f \in \mathrm{W}^{1,\infty}\left([0,1], \ell^1\right) : f(1) = \mathbb{B}^C f(0) \right\},
\end{aligned} \tag{12.13}$$

where $M_{q_j}$ denotes the multiplication operator with the function $q_j$. We can split the operator $A$ into a sum of operators

$$A = \operatorname{diag}\left(c_j \cdot \frac{\mathrm{d}}{\mathrm{d}x}\right) + \operatorname{diag}\left(M_{q_j}\right) =: A_C + M_q.$$

We already know that the operator $(A_C, D(A))$ generates a bi-continuous semigroup on $X$. In order to show that $(A, D(A))$ is a generator as well we shall apply the perturbation theory for bi-continuous semigroups.

**Theorem 12.15** *Let Assumptions* (12.5) *and* (12.11) *hold and let* $q \in \mathrm{L}^\infty\left([0,1], \ell^\infty\right)$. *Then the operator* $(A, D(A))$, *defined by* (12.13), *generates a bi-continuous semigroup on* $X$ *with respect to the weak*-topology.*

***Proof*** By Theorem 12.14, $(A_C, D(A))$ is the generator of a bi-continuous semigroup. Since $q \in \mathrm{L}^\infty\left([0,1], \ell^\infty\right)$, the multiplication operator $M_q$ is a bounded operator on $\mathrm{L}^1\left([0,1], c_0\right)$ and its dual $M_q' = M_q$ is again a multiplication operator. Therefore, we can apply Proposition 7.10, yielding that $(A, D(A)) = (A_C + M_q, D(A))$ generates a bi-continuous semigroup on $X = \mathrm{L}^1\left([0,1], c_0\right)'$. $\qquad\square$

**Remark 12.16** Note that the bi-continuous semigroup obtained in Theorem 12.15 is given by the Dyson–Phillips series, see Theorem 7.9. The semigroup can also be represented by means of the Lie–Trotter product formula for bi-continuous semigroups, cf. Corollary 6.10. Let $(S(t))_{t\geq 0}$ denote the semigroup generated by $(A, D(A))$. Then

$$S(t)x = \tau\lim_{n\to\infty}\left[T_C\left(\tfrac{t}{n}\right)e^{\frac{tM_q}{n}}\right]^n x,$$

for all $x \in X$ and uniformly on compact intervals, where we denote by $(T_C(t))_{t\geq 0}$ the bi-continuous semigroup generated by $(A_C, D(A))$ given in (12.12).

**Corollary 12.17**  *If Assumptions (12.5) and (12.11) on the velocities $c_j$ hold and $q \in L^\infty$ $([0, 1], \ell^\infty)$, then the flow problem (12.6) is well-posed on $X = L^\infty\left([0, 1], \ell^1\right)$.*

### 12.2.3 The General Case for Finite Networks

We finally consider the case of general $c_j \in \mathbb{R}$ but restrict ourselves to *finite* graphs, i.e., we work on the Banach space $X = L^\infty\left([0, 1], \mathbb{C}^m\right)$, where $m$ denotes the number of edges in the graph. In [33, Corollary 18.15], the Lumer–Phillips generation theorem for positive strongly continuous semigroups is applied to show that the transport problem with general $c_j \in \mathbb{R}$ is well-posed on $X = L^1\left([0, 1], \mathbb{C}^m\right)$. We will use the Lumer–Phillips result for bi-continuous semigroups, see Sect. 6.4, for the general case for infinite networks. For the finite network case, we will make use of the second Trotter–Kato approximation theorem, cf. Theorem 6.6.

Let
$$E_\lambda(s) := \mathrm{diag}\left(e^{(\lambda/c_j)s}\right), \quad s \in [0, 1], \quad \text{and} \quad \mathbb{B}_\lambda^C := E_\lambda(-1)\mathbb{B}^C.$$

By using this notation one can write an explicit expression for the resolvent of the operator $A_C$ defined in (12.10), see also [33, Proposition 18.12].

**Lemma 12.18**  *For $\mathrm{Re}(\lambda) > 0$ the resolvent $R(\lambda, A_C)$ of the operator $A_C$, given in (12.10), equals*

$$R(\lambda, A_C) = \left(I_X + E_\lambda(\cdot)\left(1 - \mathbb{B}_\lambda^C\right)^{-1}\mathbb{B}_\lambda^C \otimes \delta_0\right)R_\lambda,$$

*where $\delta_0$ denotes the point evaluation at $0$ and*

$$(R_\lambda f)(s) = \int_s^1 E_\lambda(s - t)C^{-1}f(t)\,\mathrm{d}t, \quad s \in [0, 1], \ f \in L^\infty\left([0, 1], \mathbb{C}^m\right).$$

**Theorem 12.19**  *The operator $(A_C, D(A_C))$, defined in (12.10), generates a bi-continuous semigroup $(T_C(t))_{t\geq 0}$ on $X = L^\infty\left([0, 1], \mathbb{C}^m\right)$.*

***Proof*** We first show that the operator $A_C$ is bi-densely defined. Take any $f \in L^\infty$ $([0, 1], \mathbb{C}^m)$. For $n \in \mathbb{N}$ let $\Omega_n := \left[\frac{1}{n}, 1 - \frac{1}{n}\right] \subseteq [0, 1]$ and define $f_n : [0, 1] \to \mathbb{C}^m$ by a linear truncation of $f$ outside $\Omega_n$, i.e.:

$$f_n(x) := \begin{cases} nf\left(\frac{1}{n}\right) x, & x \in \left[0, \frac{1}{n}\right], \\ f(x), & x \in \left[\frac{1}{n}, 1 - \frac{1}{n}\right], \\ nf\left(1 - \frac{1}{n}\right)(1 - x), & x \in \left[1 - \frac{1}{n}, 1\right]. \end{cases}$$

Observe that $f_n$ is Lipschitz for each $n \in \mathbb{N}$ and hence $f_n \in W^{1,\infty}([0, 1], \mathbb{C}^m)$. Moreover $f_n(1) = f_n(0) = 0$ for each $n \in \mathbb{N}$ implying that $f_n(1) = \mathbb{B}^C f_n(0)$, hence $f_n \in D(A_C)$. Furthermore, one has that $\sup_{n \in \mathbb{N}} \|f_n\|_\infty \leq \|f\|_\infty < \infty$ and $f_n \to f$ as $n \to \infty$ with respect to the weak*-topology since

$$\left| \int_0^1 \langle (f_n(x) - f(x)), g(x) \rangle \, dx \right| \leq 2 \|f\|_\infty \lambda^1 \left( \left[0, \frac{1}{n}\right] \cup \left[1 - \frac{1}{n}, 1\right] \right) \|g\|_1$$

$$= \frac{4}{n} \|f\|_\infty \|g\|_1$$

for each $g \in L^1([0, 1], \mathbb{C}^m)$.

We now define a sequence of operators $A_n$ approximating $A_C$ in the following way. For each $c_j \in \mathbb{R}$ there exists a sequence $\left(c_j^{(n)}\right)_{n \in \mathbb{N}}$ in $\mathbb{Q}$ such that $\lim_{n \to \infty} c_j^{(n)} = c_j$. Since the network is finite, for each $n \in \mathbb{N}$ the velocities $c_j^{(n)}$, $j \in J$, satisfy condition (12.11) and by Proposition 12.14 we obtain a bi-continuous contraction semigroup $(T_n(t))_{t \geq 0}$ generated by

$$A_n := \mathrm{diag}\left(c_j^{(n)} \cdot \frac{\mathrm{d}}{\mathrm{d}x}\right),$$

$$D(A_n) := \left\{ f \in W^{1,\infty}([0, 1], \mathbb{C}^m) : f(1) = \mathbb{B}^{C_n} f(0) \right\}, \tag{12.14}$$

where $C_n := \mathrm{diag}\left(c_j^{(n)}\right)$. Moreover, all semigroups $(T_n(t))_{t \geq 0}$, $n \in \mathbb{N}$, are similar and thus uniformly bi-continuous of type 0.

Observe that the general assumptions of Theorem 6.6 are satisfied. Let us now check the assumptions of assertion (b). Let $R := R(\lambda, A_C)$ and observe that $R : L^\infty([0, 1], \mathbb{C}^m) \to D(A_C)$ is a bijection. By above, $\mathrm{Ran}(R)$ is bi-dense in $L^\infty([0, 1], \mathbb{C}^m)$. For every $n \in \mathbb{N}$, replacing $c_j$ by $c_j^{(n)}$ for all $j \in J$, Lemma (12.18) yields an explicit expression for $R(\lambda, A_n)$.

It is easy to see that $R(\lambda, A_n)f \overset{\|\cdot\|_\infty}{\to} Rf$ for $f \in D(A_C)$ as $n \to \infty$. Applying Theorem 6.6 gives us a bi-continuous semigroup $(T_C(t))_{t \geq 0}$ with generator $(B, D(B))$. Note that, since in our case $R = R(\lambda, A_C)$ is a resolvent, by Remark 6.7 we have $R = R(\lambda, A_C) = R(\lambda, B)$ for $\lambda \in \rho(A_C)$ and by the uniqueness of the Laplace transform we conclude that $(B, D(B)) = (A_C, D(A_C))$. $\qquad\square$

**Corollary 12.20**  *The flow problem (12.6) is well-posed on* $X = \mathrm{L}^\infty\left([0, 1], \mathbb{C}^m\right)$.

**Remark 12.21**  In the same manner, by using the original strongly continuous version of the Trotter–Kato theorem (see [101, Sect. III.4b]), one can deduce the well-posedness of the problem on $X = \mathrm{L}^1\left([0, 1], \mathbb{C}^m\right)$.

### 12.2.4 The General Case for Infinite Networks

For the case of an infinite network, we will use the Lumer–Phillips generation theorem for bi-continuous semigroups, see Sect. 6.4. To do so, we first want to characterize $\Gamma_\tau$-dissipative operators, cf. Definition 6.24 in an alternative way. In the case of $C_0$-semigroups, this is done by the so-called duality set, cf. [101, Chap. II, Proposition 3.23]. For bi-continuous semigroups, we will make use of [6, Proposition 4.2] and characterize bi-dissipative operators by means of the so-called subdifferential which can be seen as the locally convex equivalent to the duality set. For $p \in \mathcal{P}$ and $x \in X$ one defines the set

$$\mathcal{J}(x, p) := \left\{\varphi \in X' : \ \mathrm{Re}\,\langle y, \varphi\rangle \leq p(y)\ \forall y \in X,\ \langle x, \varphi\rangle = p(x)\right\}.$$

By the Hahn–Banach theorem $\mathcal{J}(x, p)$ is non-empty, see also [6, Sect. 4]. The following result characterizes $\Gamma_\tau$-dissipative operators by means of $\mathcal{J}(x, p)$, cf. [6, Proposition 4.2].

**Proposition 12.22**  *Let* $(A, D(A))$ *be an operator on a bi-admissible space* $(X, \|\cdot\|, \tau)$. *Then the following assertions are equivalent:*

(a)  $(A, D(A))$ *is* $\Gamma_\tau$-*dissipative.*
(b)  *There exists a norming fundamental system of seminorms* $\Gamma$ *and such that for all* $p \in \Gamma$ *and* $x \in D(A)$ *there exists* $\varphi \in \mathcal{J}(x, p)$ *such that* $\mathrm{Re}\,\langle Ax, \varphi\rangle \leq 0.$

In Proposition 12.22, we added the bi-admissibility for the sake of completeness and to stay self-contained. However, only the locally convex topology plays an actual role in that statement. We will use Proposition 12.22 in order to improve the results from Sect. 12.2.3 concerning bi-continuous semigroups for flows in infinite networks. In fact, we show that the generation result mentioned there also holds for infinite networks and that the restriction to finite networks is not longer necessary.

**Lemma 12.23**  *The operator* $(A_C, D(A_C))$ *defined by (12.10) is bi-dissipative.*

***Proof***  We use Proposition 12.22. From [180, Lemma 23.2] we obtain the equality $(X, \tau_{\mathrm{w}*})' = (Y', \sigma(Y', Y))' = Y = \mathrm{L}^1\left([0, 1], c_0\right).$

Now let $\Gamma$ be a fundamental system for $\tau_{w*}$ and $p \in \Gamma$ and $f \in D(A_C)$. We define $\chi := \left(\chi_{\{f \neq 0\}}\right) = \left(\chi_{\{f_n \neq 0\}}\right)_{n \in \mathbb{N}}$, where $\chi_{\{f \neq 0\}}$ denotes the characteristic function of the set $\{f \neq 0\}$, and observe that actually $\chi \in J(f, p)$, so we only have to show that $\mathrm{Re}\,\langle A_C f, \chi \rangle \leq 0$. Making use of (12.10), we obtain

$$
\begin{aligned}
\mathrm{Re}\,\langle A_C f, \chi \rangle &\leq \left| \int_0^1 \langle (Af)(x), \chi_{\{f \neq 0\}}(x) \rangle \, \mathrm{d}x \right| \leq \left| \int_0^1 \sum_{n \in \mathbb{N}} c_n f_n'(x) \chi_{\{f_n \neq 0\}}(x) \, \mathrm{d}x \right| \\
&= \left| \sum_{n \in \mathbb{N}} \int_0^1 c_n f_n'(x) \chi_{\{f_n \neq 0\}}(x) \, \mathrm{d}x \right| = \langle Cf(1) - Cf(0), \mathbf{1} \rangle \\
&\leq \langle \mathbb{B}_C f(0) - f(0), C\mathbf{1} \rangle \\
&= \langle f(0), (\mathbb{B}_C^* - I)C\mathbf{1} \rangle = 0,
\end{aligned}
$$

which concludes the proof. $\qquad\square$

## Notes on This Chapter

Transport problems on graphs were considered by Dorn et al. [87–89] on the state space $L^1\left([0, 1], \ell^1\right)$ applying the theory of strongly continuous operator semigroups. A semigroup approach to flows in finite metric graphs was first presented by Kramar and Sikolya [151] and further used in [24, 33, 38, 88, 100, 179] while transport processes in infinite networks were also studied in [25, 27]. All these results were obtained in the $L^1$-setting. Mátrai and Sikolya [179] went further and considered the space of bounded Borel measures. This chapter is based on [61].

# Bi-continuous Semigroups in Control Theory  13

In this chapter, we want to discuss *final state observability estimates*. Let $X$ be a Banach space and $(S(t))_{t\geq 0}$ an operator semigroup on $X$. Moreover, let $Y$ be another Banach space and $C \in \mathscr{L}(X, Y)$. We call $C$ the observation operator. We say that such a semigroup satisfies a *final state observability estimates* with respect to some Banach space $Z$ of functions on $[0, T]$ with values in Y, if there exists $C_{\mathrm{obs}} \geq 0$ such that

$$\|S(T)x\|_X \leq C_{\mathrm{obs}} \|CS(\cdot)x\|_Z \qquad (x \in X).$$

Typical applications arise from evolution equations on some function space over (a subset of) $\mathbb{R}^d$, where $C$ is a restriction operator to a suitable subset $\Omega$ of $\mathbb{R}^d$ such that one wants to control the final state on all of $\mathbb{R}^d$ by just measuring the evolution on the subset $\Omega$.

Classically, the space $\mathcal{Z}$ is chosen to be some $L^p$-space with $p \in [1, \infty]$. In this case, the final state observability estimate is of the form:

$$\|S(T)x\|_X \leq \begin{cases} C_{\mathrm{obs}} \left( \int_0^T \|CS(t)x\|_Y^p \, dt \right)^{1/p} & \text{if } p \in [1, \infty), \\ C_{\mathrm{obs}} \operatorname{ess\,sup}_{t \in [0,T]} \|CS(t)x\|_Y & \text{if } p = \infty, \end{cases} \qquad (x \in X).$$

Clearly, in order to formulate this final state observability estimate one needs some regularity of the semigroup. Indeed, we require measurability of $t \mapsto \|CS(t)x\|_Y$ for all $x \in X$. Of course, strong continuity of $(S(t))_{t\geq 0}$ yields continuity of these maps and is therefore sufficient, but also weaker regularities are suitable. In [46], dual semigroups of strongly continuous semigroups were considered which yield sufficient regularity.

As we saw already in Chap. 2, the dual semigroup is a particular example of a bi-continuous semigroup. In this chapter, we consider final state observability estimates for bi-continuous semigroups. Especially, we want to apply the theory to the diffusion semigroup (Sect. 2.2.2) as well as to the Ornstein–Uhlenbeck semigroup on $C_b(\mathbb{R}^d)$, see also Sect. 2.2.5.

© The Author(s), under exclusive license to Springer Nature Switzerland AG 2026     251
C. Budde, *Bi-Continuous Operator Semigroups*, Frontiers in Mathematics,
https://doi.org/10.1007/978-3-032-12948-2_13

## 13.1   Final State Observability Estimates for Bi-continuous Semigroups

The final state observability estimate rests on the following abstract theorem. It provides a sufficient criterion stating that an abstract uncertainty principle (also called spectral inequality), see (UP), together with a dissipation estimate, see (DISS), yields a final state observability estimate, and has its roots in [167], see also [46, 115, 183, 192].

**Theorem 13.1** ([46, Theorem A.1]) *Let $X$ and $Y$ be Banach spaces, $C \in \mathcal{L}(X, Y)$, $(S(t))_{t \geq 0}$ a semigroup on $X$, $M \geq 1$ and $\omega \in \mathbb{R}$ such that $\|S(t)\|_{\mathcal{L}(X)} \leq M e^{\omega t}$ for all $t \geq 0$, and assume that for all $x \in X$ the map $t \mapsto \|CS(t)x\|_Y$ is measurable. Further, let $\lambda^* \geq 0$, $(P_\lambda)_{\lambda > \lambda^*}$ in $\mathcal{L}(X)$, $p \in [1, \infty]$, $d_0, d_1, d_3, \gamma_1, \gamma_2, \gamma_3, T > 0$ with $\gamma_1 < \gamma_2$, and $d_2 \geq 1$, and assume that*

$$\forall x \in X \; \forall \lambda > \lambda^*: \quad \|P_\lambda x\|_X \leq d_0 e^{d_1 \lambda^{\gamma_1}} \|C P_\lambda x\|_Y \tag{UP}$$

*and*

$$\forall x \in X \; \forall \lambda > \lambda^* \; \forall t \in (0, T/2]: \quad \|(I - P_\lambda)S(t)x\|_X \leq d_2 e^{-d_3 \lambda^{\gamma_2} t^{\gamma_3}} \|x\|_X. \tag{DISS}$$

*Then there exists $C_{\mathrm{obs}} \geq 0$ such that for all $x \in X$ we have*

$$\|S(T)x\|_X \leq \begin{cases} C_{\mathrm{obs}} \left( \int_0^T \|CS(t)x\|_Y^p \, dt \right)^{1/p} & \text{if } p \in [1, \infty), \\ C_{\mathrm{obs}} \, \mathrm{ess} \, \sup_{t \in [0, T]} \|CS(t)x\|_Y & \text{if } p = \infty. \end{cases}$$

**Remark 13.2**  The constant $C_{\mathrm{obs}}$ is explicit in all parameters and of the form:

$$C_{\mathrm{obs}} = \frac{C_1}{T^{1/r}} \exp \left( \frac{C_2}{T^{\frac{\gamma_1 \gamma_3}{\gamma_2 - \gamma_1}}} + C_3 T \right),$$

with $T^{1/r} = 1$ if $r = \infty$, and suitable constants $C_1, C_2, C_3 \geq 0$ depending on the parameters; see [46, Theorem A.1] as well as [114, Theorem 2.1].

In order to obtain a version Theorem 13.1 for bi-continuous semigroups, we have to discuss the measurability of the map $[0, \infty) \ni t \mapsto \|CS(t)x\|_Y$ for all $x \in X$.

**Lemma 13.3**  *Let $(X, \|\cdot\|_X, \tau_X)$ and $(Y, \|\cdot\|_Y, \tau_Y)$ be two bi-admissible spaces, $(S(t))_{t \geq 0}$ a bi-continuous semigroup on $X$ and $C \in \mathcal{L}(X, Y)$ sequentially continuous on $\|\cdot\|_X$-bounded sets. Then the map $[0, \infty) \ni t \mapsto \|CS(t)x\|_Y$ is measurable for all $x \in X$.*

**Proof**  Since $[0, \infty) \ni t \mapsto |\psi(CS(t)x)|$ is continuous for all $\psi \in (Y, \tau_Y)'$ by the assumptions on $C$ and the exponential boundedness of $(S(t))_{t \geq 0}$ and the fact that

$$\|CS(t)x\|_Y = \sup_{\substack{\psi \in (Y, \tau_Y)' \\ \|\psi\| \le 1}} |\psi(CS(t)x)|$$

for all $t \ge 0$ one obtains that $[0, \infty) \ni t \mapsto \|CS(t)x\|_Y$ is lower semicontinuous and hence measurable. $\qquad\square$

In view of Lemma 13.3, we can apply Theorem 13.1 to obtain the following.

**Theorem 13.4** *Let $(X, \|\cdot\|_X, \tau_X)$ and $(Y, \|\cdot\|_Y, \tau_Y)$ be two bi-admissible spaces, $(S(t))_{t\ge 0}$ a bi-continuous semigroup on $X$ and $C \in \mathscr{L}(X, Y)$ sequentially continuous on $\|\cdot\|_X$-bounded sets. Further, let $\lambda^* \ge 0$, $(P_\lambda)_{\lambda > \lambda^*}$ a family in $\mathscr{L}(X)$, $p \in [1, \infty]$, $d_0, d_1, d_3, \gamma_1, \gamma_2, \gamma_3, T > 0$ with $\gamma_1 < \gamma_2$, and $d_2 \ge 1$, and assume that*

$$\forall x \in X \ \forall \lambda > \lambda^*: \quad \|P_\lambda x\|_X \le d_0 e^{d_1 \lambda^{\gamma_1}} \|C P_\lambda x\|_Y \tag{UP'}$$

*and*

$$\forall x \in X \ \forall \lambda > \lambda^* \ \forall t \in (0, T/2]: \quad \|(I - P_\lambda)S(t)x\|_X \le d_2 e^{-d_3 \lambda^{\gamma_2} t^{\gamma_3}} \|x\|_X. \tag{DISS'}$$

*Then there exists $C_{\mathrm{obs}} \ge 0$ such that for all $x \in X$ we have*

$$\|S(T)x\|_X \le \begin{cases} C_{\mathrm{obs}} \left( \int_0^T \|CS(t)x\|_Y^p \, \mathrm{d}t \right)^{1/p} & \text{if } p \in [1, \infty), \\ C_{\mathrm{obs}} \ \mathrm{ess\,sup}_{t \in [0, T]} \|CS(t)x\|_Y & \text{if } p = \infty. \end{cases}$$

**Remark 13.5** The statements in Theorems 13.1 and 13.4 can be generalized in the sense that one can obtain an estimate with an $L^p$-norm of $t \mapsto \|CS(t)x\|_Y$ on a measurable subset $E \subseteq [0, T]$ with positive Lebesgue measure, see also [45]. However, in this case, the constant $C_{\mathrm{obs}}$ is not explicit anymore.

## 13.2  Two Examples of Bi-continuous Semigroups

In this section, we consider final state observability for two important examples: the Gauss–Weierstrass semigroup on $C_b(\mathbb{R}^d)$ and the Ornstein–Uhlenbeck semigroup on $C_b(\mathbb{R}^d)$. We begin with the study of restriction operators on $C_b(\mathbb{R}^d)$, restricting functions to suitable subsets, and relate this to an abstract uncertainty principle.

### 13.2.1  Restriction Operators on $C_b(\mathbb{R}^d)$ and the Uncertainty Principle

Let $\Omega \subseteq \mathbb{R}^d$ be non-empty, $C: C_b(\mathbb{R}^d) \to C_b(\Omega)$ the restriction operator defined by $Cf := f|_\Omega$ for $f \in C_b(\mathbb{R}^d)$. Then obviously $C \in \mathscr{L}(C_b(\mathbb{R}^d), C_b(\Omega))$. We know from Sect. 2.1.1

that $(C_b(\Omega), \|\cdot\|_\infty, \tau_{co})$ is a bi-admissible space if $\Omega$ is locally compact. Let us show that the operator $C$ fits in our framework. The following result is left to the reader and uses the properties of subspace topologies.

**Lemma 13.6** $C\colon (C_b(\mathbb{R}^d), \tau_{co}) \to (C_b(\Omega), \tau_{co})$ *is continuous.*

We now use the operator $C$ to provide an uncertainty principle based on the well-known Logvinenko–Sereda theorem. Let $\eta \in C_c([0, \infty))$ such that $\mathbf{1}_{[0,1/2]} \le \eta \le \mathbf{1}_{[0,1]}$. Moreover, for $\lambda > 0$ let $\chi_\lambda \colon \mathbb{R}^d \to \mathbb{R}$ be defined by $\chi_\lambda := \eta(|\cdot|/\lambda)$. This allows us to also define $P_\lambda \in \mathscr{L}(C_b(\mathbb{R}^d))$ by $P_\lambda f := (\mathcal{F}^{-1}\chi_\lambda) * f$, where $\mathcal{F}$ denotes the Fourier transformation. By Young's inequality and scaling properties of the Fourier transformation, we have

$$\|P_\lambda\| \le \left\|\mathcal{F}^{-1}\chi_\lambda\right\|_{L^1(\mathbb{R}^d)} = \left\|\mathcal{F}^{-1}\chi_1\right\|_{L^1(\mathbb{R}^d)} \quad (\lambda > 0).$$

Note that for all $f \in C_b(\mathbb{R}^d)$ and $\lambda > 0$ we have $\mathcal{F}P_\lambda f = \chi_\lambda \mathcal{F} f$ and therefore $\operatorname{supp}(\mathcal{F}P_\lambda f) \subseteq \overline{B}_\lambda(0) \subseteq [-\lambda, \lambda]^d$.

**Definition 13.7** Let $\Omega \subseteq \mathbb{R}^d$. Then $\Omega$ is called *thick* if $\Omega$ is measurable and there exist $L \in (0, \infty)^d$ and $\rho \in (0, 1]$ such that

$$\lambda^d(\Omega \cap (x + (0, L))) \ge \rho\lambda^d((0, L)) \quad (x \in \mathbb{R}^d),$$

where $\lambda^d$ denotes the $d$-dimensional Lebesgue measure, and $(0, L) := \prod_{j=1}^{d}(0, L_j)$ is the hypercube with sidelengths contained in $L$.

Thus, a measurable set $\Omega \subseteq \mathbb{R}^d$ is thick (with parameters $L$ and $\rho$) provided the portion of $\Omega$ in every hypercube with sidelengths contained in $L$ is at least $\rho$.

By the Logvinenko–Sereda theorem, see, for example, [149, 168], if $\Omega \subseteq \mathbb{R}^d$ is a thick set, then there exist $d_0, d_1 > 0$ such that

$$\|P_\lambda f\|_{C_b(\mathbb{R}^d)} \le d_0 e^{d_1\lambda} \|C P_\lambda f\|_{C_b(\Omega)} \quad (\lambda > 0,\, f \in C_b(\mathbb{R}^d)). \tag{LS}$$

Thus, (LS) yields an estimate of the form (UP').

## 13.2.2 The Gauss–Weierstrass Semigroup on $C_b(\mathbb{R}^d)$

In this section, we consider the Gauss–Weierstrass semigroup that we already encountered in Sect. 2.2.2 for the one-dimensional problem and in Sect. 5.2.3 for the higher-dimensional setting. In fact, we saw already that this semigroup is a bi-continuous contraction semigroup on the bi-admissible space $(C_b(\mathbb{R}^d), \|\cdot\|_\infty, \tau_{co})$.

For $\lambda > 0$ let $P_\lambda \in \mathscr{L}(C_b(\mathbb{R}^d))$ be defined as in Sect. 13.2.1. It is know from the work of Bombach, Gallaun, Seifert, and Tautenhahn, cf. [46, Proposition 3.2], that there exist $d_2 \geq 1$ and $d_3 > 0$ such that

$$\|(I - P_\lambda)S_t f\|_{C_b(\mathbb{R}^d)} \leq d_2 e^{-d_3\lambda^2 t} \|f\|_{C_b(\mathbb{R}^d)} \quad (\lambda > 0, t \geq 0, f \in C_b(\mathbb{R}^d)), \qquad \text{(DISS(GW))}$$

i.e., a dissipation estimate (DISS') is fulfilled.

Thus, if $\Omega \subseteq \mathbb{R}^d$ is thick, then (LS) and (DISS(GW)) provide the estimates (UP') and (DISS') and so Theorem 13.4 yields a final state observability estimate for the Gauss–Weierstrass semigroup on $C_b(\mathbb{R}^d)$.

### 13.2.3 The Ornstein–Uhlenbeck Semigroup on $C_b(\mathbb{R}^d)$

In Sect. 2.2.5, we already discussed the Ornstein–Uhlenbeck semigroup on infinite-dimensional, separable Hilbert spaces $\mathcal{H}$. We will restrict ourself here to the finite-dimensional case $\mathcal{H} = \mathbb{R}^d$. For the sake of completeness, we will recall here the explicit form of the Ornstein–Uhlenbeck semigroup in this case.

Let $M \colon (0, \infty) \times \mathbb{R}^d \times \mathbb{R}^d \to \mathbb{R}$ be given by

$$M_t(x, y) := M(t, x, y) := \frac{1}{\pi^{d/2}(1 - e^{-2t})^{d/2}} e^{-|y - e^{-t}x|^2/(1 - e^{-2t})} \quad (t > 0, x, y \in \mathbb{R}^d),$$

the so-called *Mehler kernel*. For $t \geq 0$ we define the family $(S(t))_{\geq 0}$ of bounded linear operators on $C_b(\mathbb{R}^d)$ by

$$S(t)f := \begin{cases} f & \text{if } t = 0, \\ \displaystyle\int_{\mathbb{R}^d} M_t(\cdot, y) f(y)\, dy & \text{if } t > 0. \end{cases}$$

Since $\int_{\mathbb{R}^d} M_t(\cdot, y)\, dy = 1$ for all $t > 0$, we have $\|S(t)f\|_\infty \leq \|f\|_\infty$ for all $f \in C_b(\mathbb{R}^d)$ and $t \geq 0$. We know from Sect. 2.2.5 that $(S(t))_{t \geq 0}$ yields a bi-continuous semigroup on the bi-admissible space $(C_b(\mathbb{R}^d), \|\cdot\|_\infty, \mathcal{T}_{co})$.

Define $k \colon (0, 1) \times \mathbb{R}^d \to \mathbb{R}$ by

$$k_s(x) := \frac{1}{\pi^{d/2}(1 - s^2)^{d/2}} e^{-|x|^2/(1 - s^2)}.$$

Let $s \in (0, 1)$ and observe that

$$M_{\ln\frac{1}{s}}(\tfrac{1}{s}x, y) = \frac{1}{\pi^{d/2}(1 - s^2)^{d/2}} e^{-|y - x|^2/(1 - s^2)} = k_s(x - y) \quad x, y \in \mathbb{R}^d.$$

Hence,

$$\left(S(\ln \tfrac{1}{s})f\right)(\tfrac{1}{s}\cdot) = k_s * f,$$

for all $f \in C_b(\mathbb{R}^d)$. For $\lambda > 0$ let $P_\lambda \in \mathscr{L}(C_b(\mathbb{R}^d))$ as in Sect. 13.2.2. Since

$$\mathcal{F}k_s(\xi) = e^{-(1-s^2)|\xi|^2/4} =: h_s(\xi) \quad (\xi \in \mathbb{R}^d),$$

for $\lambda > 0$ and $s \in (0, 1)$ we conclude that

$$\left((I - P_\lambda)S(\ln \tfrac{1}{s})f\right)(\tfrac{1}{s}\cdot) = (I - P_{\lambda/s})\left(S(\ln \tfrac{1}{s})f(\tfrac{1}{s}\cdot)\right) = \mathcal{F}^{-1}\left((1 - \chi_{\lambda/s})h_s\right) * f \quad (f \in C_b(\mathbb{R}^d)).$$

The following result gives us a necessary estimate to obtain a desired final state observability estimate.

**Lemma 13.8** *There exist $d_2 \geq 1$ and $d_3 > 0$ such that for $\lambda > 0$ and $s \in (0, 1)$ we have*

$$\left\|\mathcal{F}^{-1}\left((1 - \chi_\lambda)h_s\right)\right\|_{L^1(\mathbb{R}^d)} \leq d_2 e^{-d_3\lambda^2(1-s^2)}.$$

***Proof*** Let $\lambda > 0$, $s \in (0, 1)$, and define

$$\sigma_{s,\lambda} := (1 - \chi_{\sqrt{1-s^2}\lambda})h_s\left(\tfrac{1}{\sqrt{1-s^2}}\cdot\right) = (1 - \chi_{\sqrt{1-s^2}\lambda})e^{-|\cdot|^2/4}.$$

Then by a linear substitution we obtain

$$\left\|\mathcal{F}^{-1}\left((1 - \chi_\lambda)h_s\right)\right\|_{L^1(\mathbb{R}^d)} = \left\|\mathcal{F}^{-1}\sigma_{s,\lambda}\right\|_{L^1(\mathbb{R}^d)}.$$

Let $\alpha \in \mathbb{N}_0^d$, $|\alpha| \leq d + 1$. Then

$$\left|\partial^\alpha \sigma_{s,\lambda}\right| \leq \mathbf{1}_{\{|\cdot|\geq\sqrt{1-s^2}\lambda/2\}}\left|\partial^\alpha e^{-|\cdot|^2/4}\right| + \sum_{\beta\in\mathbb{N}_0^d,\beta<\alpha} \binom{\alpha}{\beta}\left|\partial^{\alpha-\beta}(1 - \chi_{\sqrt{1-s^2}\lambda})\right|\left|\partial^\beta e^{-|\cdot|^2/4}\right|.$$

There exists $K \geq 0$ such that for all $\beta \in \mathbb{N}_0^d$ with $\beta \leq \alpha$ and all $\xi \in \mathbb{R}^d$ we have

$$\left|\partial^\beta e^{-|\cdot|^2/4}(\xi)\right| \leq K(1 + |\xi|)^{|\beta|}e^{-|\xi|^2/4}.$$

Let

$$C_1 := \sup_{\substack{\beta\in\mathbb{N}_0^d \\ \beta\leq\alpha \\ \xi\in\mathbb{R}^d}} K(1 + |\xi|)^{|\beta|}e^{-|\xi|^2/16}.$$

Then, for $\beta \leq \alpha$ and $|\xi| \geq \sqrt{1 - s^2}\lambda/2$ we have

$$\left|\partial^\beta e^{-|\cdot|^2/4}(\xi)\right| \leq C_1 e^{-|\xi|^2/16}e^{-(1-s^2)\lambda^2/32}.$$

Further, for $\beta < \alpha$ and $\xi \in \mathbb{R}^d$ we have

$$\left| \partial^{\alpha-\beta} \left( 1 - \chi_{\sqrt{1-s^2}\lambda} \right)(\xi) \right|$$

$$\leq (\sqrt{1-s^2}\lambda)^{-|\alpha-\beta|} \left| \partial^{\alpha-\beta} \chi_1 \left( \frac{\xi}{\sqrt{1-s^2}\lambda} \right) \right| \mathbf{1}_{\left\{ \frac{\sqrt{1-s^2}\lambda}{2} \leq |\cdot| \leq \sqrt{1-s^2}\lambda \right\}}(\xi)$$

$$\leq C_2 (\sqrt{1-s^2}\lambda)^{-|\alpha-\beta|} \mathbf{1}_{\left\{ \frac{\sqrt{1-s^2}\lambda}{2} \leq |\cdot| \leq \sqrt{1-s^2}\lambda \right\}}(\xi)$$

where $C_2 := \max_{\beta < \alpha} \left\| \partial^{\alpha-\beta} \chi_1 \right\|_{C_b(\mathbb{R}^d)}$. Hence, there exists $C \geq 0$ (which is independent of $s$ and $\lambda$) such that if $\sqrt{1-s^2}\lambda \geq 1$, then for all $\xi \in \mathbb{R}^d$ we have

$$\left| \partial^\alpha \sigma_{s,\lambda} \right| \leq C e^{-|\xi|^2/16} e^{-(1-s^2)\lambda^2/32}.$$

Therefore, increasing $C$, for all $x \in \mathbb{R}^d$ we obtain

$$\left| x^\alpha \mathcal{F}^{-1} \sigma_{s,\lambda}(x) \right| = \left| \mathcal{F}^{-1}(\partial^\alpha \sigma_{s,\lambda})(x) \right| \leq C e^{-(1-s^2)\lambda^2/32}. \tag{13.2.1}$$

By choosing $j \in \{1, \ldots, d\}$ and $\alpha := (d+1)e_j$ for the $j$-th canonical unit vector $e_j$, we observe $\|x\|_\infty^{d+1} \left| \mathcal{F}^{-1} \sigma_{s,\lambda}(x) \right| \leq C e^{-(1-s^2)\lambda^2/32}$ and hence

$$\left| \mathcal{F}^{-1} \sigma_{s,\lambda}(x) \right| \leq C e^{-(1-s^2)\lambda^2/32} |x|^{-d-1} \tag{13.2.2}$$

for all $x \in \mathbb{R}^d \setminus \{0\}$, where we increased $C$.

Therefore, if $\sqrt{1-s^2}\lambda \geq 1$, we can conclude by (13.2.1) for $\alpha = 0$ and (13.2.2) that

$$\left\| \mathcal{F}^{-1}\big( (1-\chi_\lambda) h_s \big) \right\|_{L^1(\mathbb{R}^d)} = \left\| \mathcal{F}^{-1} \sigma_{s,\lambda} \right\|_{L^1(\mathbb{R}^d)}$$

$$\leq C e^{-(1-s^2)\lambda^2/32} \left( \int_{B[0,1]} 1 \, dx + \int_{\mathbb{R}^d \setminus B[0,1]} |x|^{-d-1} \, dx \right)$$

$$\leq C e^{-(1-s^2)\lambda^2/32},$$

where we increased $C$ again.

It remains to prove the estimate for the case $\sqrt{1-s^2}\lambda < 1$. Note that

$$\left\| \mathcal{F}^{-1}(\chi_\lambda h_s) \right\|_{L^1(\mathbb{R}^d)} = \left\| \mathcal{F}^{-1}\chi_\lambda * \mathcal{F}^{-1} h_s \right\|_{L^1(\mathbb{R}^d)}$$

$$\leq \left\| \mathcal{F}^{-1}\chi_\lambda \right\|_{L^1(\mathbb{R}^d)} \|k_s\|_{L^1(\mathbb{R}^d)} = \left\| \mathcal{F}^{-1}\chi_1 \right\|_{L^1(\mathbb{R}^d)},$$

where the last equality follows form scaling properties of the Fourier transformation and the fact that $k_s$ is normalized in $L^1(\mathbb{R}^d)$.

Thus, for $\sqrt{1-s^2}\lambda < 1$ we obtain

$$\left\| \mathcal{F}^{-1}(\chi_\lambda h_s) \right\|_{L^1(\mathbb{R}^d)} \leq \left\| \mathcal{F}^{-1}\chi_1 \right\|_{L^1(\mathbb{R}^d)} e^{1/32} e^{-(1-s^2)\lambda^2/32},$$

which ends the proof.                                                           $\square$

In view of Lemma 13.8, we obtain the dissipation estimate (DISS') as follows. Note that for $t \geq 0$ we have $e^{2t} - 1 \geq 2t$. Let $t > 0$ and $\lambda > 0$, and set $s := e^{-t} \in (0, 1)$. Then, for $f \in C_b(\mathbb{R}^d)$, Young's inequality and Lemma 13.8 yield

$$
\begin{aligned}
\|(I - P_\lambda)S(t)f\|_{C_b(\mathbb{R}^d)} &= \left\|\left((I - P_\lambda)S(\ln \tfrac{1}{s})f\right)(\tfrac{1}{s} \cdot)\right\|_{C_b(\mathbb{R}^d)} \\
&\leq \left\|\mathcal{F}^{-1}\left((1 - \chi_{\lambda/s})h_s\right)\right\|_{L^1(\mathbb{R}^d)} \|f\|_{C_b(\mathbb{R}^d)} \\
&\leq d_2 e^{-d_3 \lambda^2 s^{-2}(1-s^2)} \|f\|_{C_b(\mathbb{R}^d)} = d_2 e^{-d_3 \lambda^2 (e^{2t}-1)} \|f\|_{C_b(\mathbb{R}^d)} \\
&\leq d_2 e^{-2d_3 \lambda^2 t} \|f\|_{C_b(\mathbb{R}^d)} . \qquad\qquad\qquad \text{(DISS(OU))}
\end{aligned}
$$

Thus, if $\Omega \subseteq \mathbb{R}^d$ is thick and $C \colon C_b(\mathbb{R}^d) \to C_b(\Omega)$ is the restriction map as in Sect. 13.2.2, then (LS) and (DISS(OU)) provide the estimates (UP') and (DISS') and so Theorem 13.4 yields a final state observability estimate for the Ornstein–Uhlenbeck semigroup on $C_b(\mathbb{R}^d)$.

## Notes on This Chapter

The work presented here was developed by Kruse and Seifert [157]. Final state observability estimates have been studied in various contexts due to its relation to null-controllability, see, e.g., [45, 46, 65, 114, 115, 167, 183, 192, 230, 242] and references therein. In [157], also cost-uniform approximate null-controllability of a control system sharing only weak continuity properties such as bi-continuity is discussed, and a relation to a final state observability estimate for the dual system is established, thus generalizing the well-known duality in Hilbert and Banach spaces [65, 92, 230, 242]. However, due to a lot of technicalities, we decided here to omit this part and refer the interested reader to the original full work [157].

# Locally Convex Topologies **A**

Since we are using locally convex topologies throughout this book we will recall some basis aspects of locally convex spaces. In particular we will show that there is a correspondence between locally convex topologies and separating families of seminorms. In particular: for every locally convex space there exists a separating family of continuous seminorms. Vice versa, every separating family of seminorms induces a locally convex topology such that every seminorm in continuous. Most of the results here are taken from [206]. First of all we have to recall some definitions.

**Definition A.1** Let $X$ be a vector space over $\mathbb{C}$ and $\tau$ a topology on $X$. We say that $(X, \tau)$ is a **topological vector space** if

(a)  every point of $X$ is a closed set, and
(b)  the vector space operations are continuous with respect to $\tau$.

We will not recall the basic definitions of point-set topology here. For that we refer again to [206] or to any other introductory book on topology. A subset $C \subseteq X$ is called **convex** if $tC + (1 - t)C \subseteq C$ for $0 \leq t \leq 1$, i.e., $tx + (1 - t)y \in C$ for all $x, y \in C$ and $0 \leq t \leq 1$. Having this in mind we can make the following definition.

**Definition A.2** A topological vector space $(X, \tau)$ is called **locally convex** if there is a local base consisting of convex sets. A convex subset $A \subseteq X$ is called **absorbing** if for each $x \in X$ there exists $t > 0$ such that $x \in tA$. A subset $B \subseteq X$ is called **balanced** if $\alpha B \subseteq B$ for every $\alpha \in \mathbb{C}$ with $|\alpha| \leq 1$. A subset $E \subseteq X$ is called bounded if for every neighborhood $V$ of $0$ in $X$ there exists $s > 0$ such that $E \subseteq tV$ for each $t > s$.

Now we will take also seminorms into account. We start again with some definitions.

© The Editor(s) (if applicable) and The Author(s), under exclusive license to Springer Nature Switzerland AG 2026
C. Budde, *Bi-Continuous Operator Semigroups*, Frontiers in Mathematics,
https://doi.org/10.1007/978-3-032-12948-2

**Definition A.3** A **seminorm** on a vector space $X$ is a real-valued function $p : X \to \mathbb{R}$ such that

(a) $p(x + y) \leq p(x) + p(y)$,
(b) $p(\alpha x) = |\alpha|\, p(x)$

for each $x, y \in X$ and $\alpha \in \mathbb{C}$.

**Definition A.4** A family $\mathcal{P}$ of seminorms on $X$ is called **separating** if for each $x \in X \setminus \{0\}$ there exists $p \in \mathcal{P}$ with $p(x) \neq 0$.

We show now how to get from a locally convex space $X$ to a family $\mathcal{P}$ of seminorms and back. For that define for any absorbing subset $A \subseteq X$ the so-called **Minkowski functional** by

$$\mu_A(x) = \inf \left\{ t > 0 : t^{-1}x \in A \right\}, \quad x \in X. \tag{A.0.1}$$

With the help of these functionals one obtains the following theorem.

**Theorem A.5** *Let $\mathcal{B}$ be a convex balanced local base in a topological vector space $X$. Associate to every $V \in \mathcal{B}$ its Minkowski functional $\mu_V$ as defined in (A.0.1). Then:*

*(a) $V = \{x \in X : \mu_V(x) < 1\}$ for each $V \in \mathcal{B}$, and*
*(b) $\{\mu_V : V \in \mathcal{B}\}$ is a separating family of continuous seminorms on $X$.*

The following theorem gives the possibility to go from families of seminorms to locally convex spaces.

**Theorem A.6** *Suppose $\mathcal{P}$ is a separating family of seminorms on a vector space $X$. Associate to each $p \in \mathcal{P}$ and to each $n \in \mathbb{N}$ the set*

$$V(n, p) := \left\{ x \in X : p(x) < \frac{1}{n} \right\}.$$

*Let $\mathcal{B}$ be the collection of all finite intersections of the sets $V(n, p)$. Then $\mathcal{B}$ is a convex balanced local base for a topology $\tau$ on $X$, which turns $X$ into a locally convex space such that every $p \in \mathcal{P}$ is continuous and $E \subseteq X$ is bounded if and only if every $p \in \mathcal{P}$ is bounded on $E$.*

Now we will formulate the Hahn–Banach theorems in the general context of locally convex spaces, without any proof.

**Theorem A.7** *Suppose $M$ is a subspace of a vector space $X$, $p$ is a seminorm on $X$ and $\varphi$ a linear functional on $M$ such that $|\varphi(x)| \le p(x)$ for each $x \in M$. Then $\varphi$ extends to a linear functional $\Phi$ on $X$ such that $|\Phi(x)| \le p(x)$ for each $x \in X$.*

**Corollary A.8** *If $X$ is a normed space and $x_0 \in X$, there exists $\Phi \in X'$ such that $\Phi(x_0) = \|x_0\|$ and $|\Phi(x)| \le \|x\|$ for each $x \in X$.*

**Theorem A.9** *Suppose $M$ is a subspace of a locally convex space $X$ and $x_0 \in X$. If $x_0 \notin \overline{M}$, then there exists $\Phi \in X'$ such that $\Phi(x_0) = 1$ but $\Phi(x) = 0$ for each $x \in M$.*

**Theorem A.10** *If $\varphi$ is a continuous linear functional on a subspace $M$ of a locally convex space $X$, then there exists $\Phi \in X'$ such that $\Phi = \varphi$ on $M$.*

The following criterion for locally convex topologies on a vector space $X$ induced by a family of seminorms $\mathcal{P}$ tells us how we translate convergence in $X$ to a condition on the seminorms.

**Theorem A.11** *Let $X$ be a vector space whose topology is induced by a family of seminorms $\mathcal{P}$. Given any net $(x_\iota)_{\iota \in I}$ and any $x \in X$ we have*

$$x_\iota \to x \iff p(x_\iota - x) \to 0 \ \ \forall p \in \mathcal{P}.$$

# The Mixed Topology **B**

The mixed topology becomes interesting whenever one deals with a vector space with two different topologies on it. Especially for our case this is interesting when the space is a Banach space with the norm topology and an additional locally convex topology. The idea came up when Alexiewicz started to study spaces with two norms [11–13]. Later on, Wiweger generalized this to vector spaces with locally convex topologies, cf. [239]. For unexplained topological notations in the context of locally convex topologies we refer to [73] or [239]. Let us summarize some important definitions before we can define the mixed topology.

**Definition B.1** Let $E$ be a vector space.

(a) A *ball* in $E$ is an absolutely convex subset of $E$ which does not contain a non-trivial subspace.

(b) A *(convex) bornology* on $E$ is a family $\mathcal{B}$ of balls such that

  (i) $E = \bigcup_{B \in \mathcal{B}} B$.

  (ii) If $B \in \mathcal{B}$ and $\lambda > 0$, then $\lambda B \in \mathcal{B}$.

  (iii) $\mathcal{B}$ is directed on the right by inclusion, i.e., if $B, C \in \mathcal{B}$ then there exists $D \in \mathcal{B}$ with $B \cup C \subseteq D$.

  (iv) If $B \in \mathcal{B}$ and $C$ is a ball contained in $B$, then $C \in \mathcal{B}$.

(c) A subset $B \subseteq E$ is $\mathcal{B}$-bounded if it is contained in some ball in $\mathcal{B}$.

(d) A *basis* for $\mathcal{B}$ is a subfamily $\mathcal{B}_1$ of $\mathcal{B}$ such that each $B \in \mathcal{B}$ is a subset of some $B_1 \in \mathcal{B}_1$.

(e) $\mathcal{B}$ is of *countable type* if $\mathcal{B}$ has a countable basis.

In what follows, let $E$ be a vector space with a locally convex topology $\tau$ and bornology $\mathcal{B}$ of countable type. Observe that the collection $\mathcal{B}_\tau$ of $\tau$-bounded and absolutely convex subsets of $E$ is a bornology for $E$ as well. We say that $\mathcal{B}$ and $\tau$ are *compatible* if $\mathcal{B} \subseteq \mathcal{B}_\tau$

© The Editor(s) (if applicable) and The Author(s), under exclusive license to Springer Nature Switzerland AG 2026

C. Budde, *Bi-Continuous Operator Semigroups*, Frontiers in Mathematics, https://doi.org/10.1007/978-3-032-12948-2

and $\mathcal{B}$ has a basis of $\tau$-closed sets. In the case of compatibility, one is able to choose a basis $(B_n)_{n\in\mathbb{N}}$ of $\mathcal{B}$ such that $B_n + B_n \subseteq B_{n+1}$ and $B_n$ is $\tau$-closed for all $n \in \mathbb{N}$. In fact, one can assume that a given basis of $\mathcal{B}$ has these properties.

**Definition B.2** Let $E$ be a vector space, $\mathcal{B}$ a bornology of countable type with basis $(B_n)_{n\in\mathbb{N}}$ and $\tau$ a locally convex topology on $E$. Assume that $\mathcal{B}$ and $\tau$ are compatible. For a sequence $(U_n)_{n\in\mathbb{N}}$ of absolutely convex $\tau$-neighborhoods of zero we set

$$\gamma((U_n)_{n\in\mathbb{N}}) := \bigcup_{n\in\mathbb{N}} (U_1 \cap B_1 + \cdots + U_n \cap B_n).$$

The set of all such sets $\gamma((U_n)_{n\in\mathbb{N}})$ is a basis of neighborhoods for zero for a locally convex topology on $E$. We will call this new locally convex topology the *mixed topology* and will denote this topology by $\gamma := \gamma(\mathcal{B}, \tau)$. If $\mathcal{B}$ is a bornology induced by a norm $\|\cdot\|$ on $E$ we write $\gamma(\|\cdot\|, \tau)$.

The following properties are direct consequences of the construction in Definition B.2, see for example [73, Chap. I, Proposition 1.5].

**Proposition B.3** *Let $E$ be a vector space, $\mathcal{B}$ a bornology of countable type and $\tau$ a locally convex topology on $E$. Assume that $\mathcal{B}$ and $\tau$ are compatible. Denote by $\gamma$ the corresponding mixed topology on $E$.*

(i) *$\gamma$ is finer than $\tau$.*
(ii) *$\gamma$ and $\tau$ coincide on the sets of $\mathcal{B}$.*
(iii) *$\gamma$ is the finest linear topology on $E$ which coincides with $\tau$ on the sets of $\mathcal{B}$.*

Some authors use Proposition B.3 as the definition of the mixed topology. The following result is important to characterize convergence with respect to the mixed topology, cf. [239, Theorem 2.3.1] or [73, Chap. I, Proposition 1.10].

**Proposition B.4** *Let $E$ be a vector space and $(x_n)_{n\in\mathbb{N}}$ be a sequence in $E$. The following assertions are equivalent:*

(a) *$x_n \to x$ as $n \to \infty$ with respect to the mixed topology $\gamma$.*
(b) *$(x_n)_{n\in\mathbb{N}}$ is $\mathcal{B}$ bounded and $x_n \to x$ as $n \to \infty$ with respect to $\tau$.*

Locally convex topologies can either be characterized by a neighborhood basis of zero or by continuous seminorms, which sometimes is more convenient, see for example [206]. As it is pointed out in [73, p. 41] the mixed topology can also be characterized by seminorms. To do so, choose for each $n \in \mathbb{N}$ a seminorm $p_n \in \mathcal{P}$ and define

$$p(x) := \inf \left\{ \sum_{k=0}^{n} p_k(x_k) : \ x = x_1 + \cdots + x_n, \ x_k \in B_k, \ 1 \le k \le n \right\}, \quad x \in X.$$

Then the set of all such seminorms determines the mixed topology, see also [82]. This characterization is not very useful for applications, but in certain cases a much simpler representation for the seminorms can be given, cf. [73, p. 41] and [106, Definition A.1.1].

**Definition B.5** Let $(X, \|\cdot\|, \tau)$ be a bi-admissible space according to Assumption 2.1. For a sequence $(p_n)_{n\in\mathbb{N}}$ in $\mathcal{P}$ and a sequence $(a_n)_{n\in\mathbb{N}} \in c_0$ with $a_n \ge 0$ for all $n \in \mathbb{N}$ define a seminorm

$$\widetilde{p}_{(p_n,a_n)_{n\in\mathbb{N}}}(x) := \sup_{n\in\mathbb{N}} a_n p_n(x), \quad x \in X.$$

By taking all those seminorms together, say

$$\widetilde{\mathcal{P}} := \left\{ \widetilde{p}_{(p_n,a_n)_{n\in\mathbb{N}}} : \ (p_n)_{n\in\mathbb{N}} \subseteq \mathcal{P}, \ (a_n)_{n\in\mathbb{N}} \in c_0, \ a_n \ge 0 \right\}, \tag{B.0.1}$$

this determines another locally convex topology, the so-called *submixed topology* $\widetilde{\gamma}$.

By construction, one has that $\tau \subseteq \widetilde{\gamma} \subseteq \gamma$ and $\widetilde{\gamma}$ has the same convergent sequences as $\gamma$. In some situations, the mixed topology and the submixed topology even coincide, cf. [73, Chap. I, Proposition 4.5] and [239, Theorem 3.1.1].

**Proposition B.6** *Let* $(X, \|\cdot\|, \tau)$ *be a bi-admissible space according to Assumption 2.1. Assume that either*

(i) *The closed unit ball* $B_{\|\cdot\|} := \{x \in X : \ \|x\| \le 1\}$ *is* $\tau$-*compact, or*
(ii) *for every* $x \in X$, $\varepsilon > 0$ *and* $p \in \mathcal{P}$ *there exist* $y, z \in X$ *such that* $x = y + z$, $p(z) = 0$ *and* $\|y\| \le p(x) + \varepsilon$.

*Then* $\gamma = \widetilde{\gamma}$.

**Example B.7** Let $\Omega$ be a completely regular Hausdorff space and consider the space $X := C_b(\Omega)$ of bounded continuous functions over $\mathbb{R}$ equipped with the supremum norm $\|\cdot\|_\infty$ and the compact-open topology $\tau_{co}$. The space $C_b(\Omega)$ satisfies Assumption 2.1, see also Sect. 2.1.1. Recall that $\tau_{co}$ is induced by the family of seminorms given by

$$p_K(f) := \sup_{x\in K} |f(x)|, \quad K \subseteq \Omega, \ f \in C_b(\Omega),$$

where $K$ is a compact subset of $\Omega$. It was shown by Wiweger [239, Example D)] that $C_b(\Omega)$ satisfies Proposition B.6(ii) so that the mixed and the submixed topology coincide on $C_b(\mathbb{R})$. This fact was also used by Goldys et al. when they studied diffusion semigroups on spaces of continuous functions or Markov processes, cf. [121, 122].

**Example B.8** Consider the space $X := \mathscr{L}(E, F)$ of bounded linear operators from a Banach space $E$ to another Banach space $F$ equipped with the operator norm $\|\cdot\|$ and the strong operator topology $\tau_{\mathrm{sot}}$. Then $\mathscr{L}(E, F)$ satisfies Assumption 2.1, see also Sect. 2.1.3. The strong operator topology $\tau_{\mathrm{sot}}$ on $\mathscr{L}(E, F)$ is induced by the family of seminorms given by

$$p_x(S) := \|Sx\|_F, \quad x \in E, \ S \in \mathscr{L}(E, F),$$

Let us comment on the mixed topology $\gamma\,(\|\cdot\|, \tau_{\mathrm{sot}})$. From [154, Remark 3.19(d)], we obtain that $\gamma\,(\|\cdot\|, \tau_{\mathrm{sot}})$ is $C$-sequential if $E$ is separable. Moreover, if $F$ is finite-dimensional, then the closed unit ball of $\mathscr{L}(E, F)$ is $\tau_{\mathrm{sot}}$-compact. Under this assumption, the condition of Proposition B.6(i) is satisfied, and we may conclude that $\gamma\,(\|\cdot\|, \tau_{\mathrm{sot}})$ coincides with the submixed topology $\widetilde{\gamma}$. Furthermore, if $F$ is finite-dimensional, then $\mathscr{L}(E, F)$ equipped with the mixed topology $\gamma\,(\|\cdot\|, \tau_{\mathrm{sot}})$ is semi-Montel, cf. [152, Remark 2.5(d)].

# Riesz Spaces and Banach Lattices  $\qquad$ C

We now recall some basic definitions of ordered topological vector spaces and operators acting on such spaces, see [207] or [124] for further background material. First of all, we recall the notion of vector and Banach lattices. Notice that for our purpose only real vector spaces are of interest.

**Definition C.1** A *vector lattice* or *Riesz space* $(V, \leq)$ is a vector space $V$ equipped with a partial order $\leq$ such that for each $x, y, z \in V$:

(a) $x \leq y \Rightarrow x + z \leq y + z$.
(b) $x \leq y \Rightarrow \alpha x \leq \alpha y$ for all scalars $\alpha \geq 0$.
(c) For any pair $x, y \in V$ there exists a supremum, denoted by $x \vee y$, and an infimum, denoted by $x \wedge y$, in $V$ with respect to the partial order $\leq$.

An element $x \in V$ is called *positive* if $x \geq 0$. The set of all positive elements of $V$ is denoted by $V_+$. Furthermore, the *absolute value* of an element $x \in X$ is defined by $|x| := x \vee (-x)$. Furthermore, $x_+ := x \vee 0$ and $x_- := x \wedge 0$ are called the *positive part* and *negative part*, respectively.

**Definition C.2** A *Banach lattice* is a triple $(X, \|\cdot\|, \leq)$ where $(X, \|\cdot\|)$ is a Banach space equipped with a partial order $\leq$ such that $(X, \leq)$ is a Riesz space satisfying $|x| \leq |y| \Rightarrow \|x\| \leq \|y\|$ for all $x, y \in X$.

The following important definition is due to Arendt [16].

**Definition C.3** A linear operator $(A, D(A))$ on a Banach lattice $(X, \|\cdot\|, \leq)$ is called *resolvent positive* if there exists $\omega \in \mathbb{R}$ such that $(\omega, \infty) \subseteq \rho(A)$ and such that $R(\lambda, A) \geq 0$ for each $\lambda > \omega$.

**Definition C.4**  Let $(X, \|\cdot\|)$ be a Banach lattice.

(a)  $X$ is called an AL-*space* if for all $x, y \in X_+$ the equality $\|x + y\| = \|x\| + \|y\|$ holds.

(b)  $X$ is called an AM-*space* if for all $x, y \in X_+$ the equality $\sup\{\|x\|, \|y\|\} = \|x \vee y\|$ holds.

# Bochner $L^p$-spaces

D

## D.1 Basic Definitions

Let $(\Omega, \Sigma, \mu)$ be a $\sigma$-finite, complete measure space. The well-known spaces $L^p(\Omega)$ are defined by

$$L^p(\Omega) := \left\{ f : \Omega \to \mathbb{C} : f \text{ is measurable and } \|f\|_p := \left( \int_\Omega |f|^p \, d\mu \right)^{\frac{1}{p}} < \infty \right\}, \quad 1 \le p < \infty,$$

$$L^\infty(\Omega) := \left\{ f : \Omega \to \mathbb{C} : f \text{ is measurable and } \|f\|_\infty := \operatorname*{ess\,sup}_{\omega \in \Omega} |f(\omega)| < \infty \right\}.$$

These function spaces have generalizations to vector-valued functions. Let $X$ be a Banach space. We introduce simple function in this vector-valued context. A function $s : \Omega \to X$ is called a *simple function* if it can be written as

$$s = \sum_{i=1}^{n} x_i \cdot \mathbf{1}_{A_i},$$

where $x_i \in X$ and $A_i \in \Sigma$ with $\mu(A_i) < \infty$ for each $1 \le i \le n$. The integral of such a simple function is defined by

$$\int_\Omega s \, d\mu := \sum_{i=1}^{n} x_i \mu(A_i).$$

A function $f : \Omega \to X$ is called *measurable* if there exists a sequence $(s_n)_{n \in \mathbb{N}}$ of simple functions such that $\|f - s_n\| \to 0$ almost everywhere for $n \to 0$. A measurable function $f : \Omega \to X$ is called *Bochner-integrable* if there exists a sequence $(s_n)_{n \in \mathbb{N}}$ of simple functions which converges to $f$ almost everywhere such that

© The Editor(s) (if applicable) and The Author(s), under exclusive license to Springer Nature Switzerland AG 2026
C. Budde, *Bi-Continuous Operator Semigroups*, Frontiers in Mathematics,
https://doi.org/10.1007/978-3-032-12948-2

$$\lim_{n \to \infty} \int_{\Omega} \|f - s_n\| \, d\mu = 0.$$

The integral of such an integrable function is defined by

$$\int_{\Omega} f \, d\mu := \lim_{n \to \infty} \int_{\Omega} s_n \, d\mu.$$

One observes that the definition of the integral is independent of the choice of the sequence $(s_n)_{n \in \mathbb{N}}$. The following theorem by Bochner connects the vector-valued integral with the ordinary scalar-valued integral.

**Theorem D.1** *Let $f : \Omega \to X$ be a measurable function, then $f$ is Bochner-integrable if and only if $\|f\| : \Omega \to \mathbb{R}$ is integrable. Furthermore we have*

$$\left\| \int_{\Omega} f \, d\mu \right\| \leq \int_{\Omega} \|f\| \, d\mu.$$

We remark that the Beppo–Levi theorem, the Fubini–Tonelli theorem as well as the Lebesgue theorem hold also for the vector-valued case. Observe that Theorem D.1 tells us that $f \in L^1(\Omega, X)$ if and only if $\|f\| \in L^1(\Omega, \mathbb{R})$. Hence one defines the vector-valued $L^p$ spaces for $1 \leq p \leq \infty$ by

$$L^p(\Omega, X) := \left\{ f : \Omega \to X : f \text{ is Bochner-measurable and } \|f\| \in L^p(\Omega, \mathbb{R}) \right\}.$$

with norm

$$\|f\|_{L^p(\Omega, X)} := \|\|f\|\|_{L^p(\Omega, \mathbb{R})}, \quad f \in L^p(\Omega, X).$$

Having this definition in mind, we remark that the Hölder inequality also holds for vector-valued function. In particular, let $1 \leq p \leq \infty$ and let $f \in L^p(\Omega, X)$ and $g \in L^q(\Omega, X)$ where $\frac{1}{p} + \frac{1}{q} = 1$. Then $fg \in L^1(\Omega, X)$ and

$$\|fg\|_{L^1(\Omega, X)} \leq \|f\|_{L^p(\Omega, X)} \cdot \|g\|_{L^q(\Omega, X)}.$$

## D.2    The Dual of $L^p(\Omega, X)$

Let $1 \leq p < \infty$ and $1 < q \leq \infty$ such that $\frac{1}{p} + \frac{1}{q} = 1$. We recall that the following duality result for scalar-valued functions holds

$$L^p(\Omega)' = L^q(\Omega).$$

For vector-valued functions we need a little bit more effort to conclude a result like this. Due to the Hölder inequality one has a continuous embedding $L^q(\Omega, X') \hookrightarrow L^p(\Omega, X)'$ which is actually an isometry. To continue we need the following definition.

**Definition D.2** A Banach space $X$ has the *Radon–Nikodym property* (with respect to $\mu$) if for every bounded variation, countable additive $\mu$-continuous vector measure $\nu : \Sigma \to X$ there is a Bochner-integrable function $g : \Omega \to X$ such that $\nu(A) = \int_A g \, d\mu$ for each $A \in \Sigma$.

As examples of spaces having this Radon–Nikdoym property we mention the Banach space $\mathbb{R}$ (since we have the Radon–Nikodym theorem), as well as every $\ell^p$ for $1 \leq p < \infty$ and every Hilbert space. Banach spaces which fail to have the Radon–Nikodym property are $\ell^\infty$, $L^1([0, 1])$ and $C([0, 1])$. Another characterization of Banach spaces having the Radon–Nikodym property is the following.

**Lemma D.3** *A Banach space $X$ has the Radon–Nikodym property if and only if every Lipschitz-continuous function $f : \mathbb{R} \to X$ is differentiable almost everywhere. In that case $f' \in L^\infty(\mathbb{R}, X)$ and*

$$f(t) = f(0) + \int_0^t f'(s) \, ds, \quad t \in \mathbb{R},$$

*and* $\left\| f' \right\|_\infty \leq \| f \|_{\mathrm{Lip}} := \sup_{x \neq y} \frac{\| f(x) - f(y) \|}{|x - y|}.$

**Theorem D.4** *Let $X$ be a Banach space such that $X'$ has the Radon–Nikodym property, then* $L^q(\Omega, X') \cong L^p(\Omega, X)'.$

# References


1. Adler, M., Bombieri, M., Engel, K.-J.: On perturbations of generators of $C_0$-semigroups. Abstr. Appl. Anal. **213020**(13) (2014). https://doi.org/10.1155/2014/213020
2. Albanese, A., Kühnemund, F.: Trotter-Kato approximation theorems for locally equicontinuous semigroups. Riv. Mat. Univ. Parma **7**(1), 19–53 (2002)
3. Albanese, A.A., Bonet, J., Ricker, W.J.: Grothendieck spaces with the Dunford-Pettis property. Positivity **14**(1), 145–164 (2010). https://doi.org/10.1007/s11117-009-0011-x
4. Albanese, A.A., Bonet, J., Ricker, W.J.: Mean ergodic semigroups of operators. Rev. R. Acad. Cienc. Exactas Fís. Nat., Ser. A Mat., RACSAM **106**(1), 299–319 (2012) https://doi.org/10.1007/s13398-011-0054-2
5. Albanese, A.A., Bonet, J., Ricker, W.J.: Montel resolvents and uniformly mean ergodic semigroups of linear operators. Quaest. Math. **36**(2), 253–290 (2013). https://doi.org/10.2989/16073606.2013.779978
6. Albanese, A.A., Jornet, D.: Dissipative operators and additive perturbations in locally convex spaces. Math. Nachr. **289**(8–9), 920–949 (2016). https://doi.org/10.1002/mana.201500150
7. Albanese, A.A., Lorenzi, L., Manco, V.: Mean ergodic theorems for bi-continuous semigroups. Semigroup Forum **82**(1), 141–171 (Feb2011). https://doi.org/10.1007/s00233-010-9260-z
8. Albanese, A.A., Mangino, E.: Trotter-Kato theorems for bi-continuous semigroups and applications to Feller semigroups. J. Math. Anal. Appl. **289**(2), 477–492 (2004). https://doi.org/10.1016/j.jmaa.2003.08.032
9. Alber, J.: On implemented semigroups. Semigroup Forum **63**(3), 371–386 (2001). https://doi.org/10.1007/s002330010082
10. Alexandroff, A.D.: Additive set-functions in abstract spaces. Mat. Sb., Nov. Ser. **13**, 169–238 (1943). https://eudml.org/doc/65241
11. Alexiewicz, A.: On the two-norm convergence. Stud. Math. **14**, 49–56 (1953). https://doi.org/10.4064/sm-14-1-49-56
12. Alexiewicz, A., Semadeni, Z.: A generalization of two norm spaces. Linear functionals. Bull. Acad. Pol. Sci., Sér. Sci. Math. Astron. Phys. **6**, 135–139 (1958)
13. Alexiewicz, A., Semadeni, Z.: Linear functionals on two-norm spaces. Stud. Math. **17**, 121–140 (1958b). https://doi.org/10.4064/sm-17-2-121-140
14. Amann, H.: Parabolic evolution equations in interpolation and extrapolation spaces. J. Funct. Anal. **78**(2), 233–270 (1988). https://doi.org/10.1016/0022-1236(88)90120-6





C. Budde, *Bi-Continuous Operator Semigroups*, Frontiers in Mathematics, https://doi.org/10.1007/978-3-032-12948-2

15. Ameziane Hassani, R., Blali, A., El Amrani, A., Moussa Ouja, K.: On cosine families. Rend. Circ. Mat. Palermo (2) **71**(2), 725–735 (2022) https://doi.org/10.1007/s12215-021-00635-5

16. Arendt, W.: Vector-valued Laplace transforms and Cauchy problems. Isr. J. Math. **59**, 327–352 (1987). https://doi.org/10.1007/BF02774144

17. Arendt, W., Batty, C.J.K., Hieber, M., Neubrander, F.: Vector-valued Laplace transforms and Cauchy problems, volume 96 of Monogr. Math., Basel. Basel: Birkhäuser (2001) https://doi.org/10.1007/978-3-0348-0087-7

18. Arendt, W., El-Mennaoui, O., Kéyantuo, V.: Local integrated semigroups: evolution with jumps of regularity. J. Math. Anal. Appl. **186**(2), 572–595 (1994). https://doi.org/10.1006/jmaa.1994.1318

19. Arendt, W., Grabosch, A., Greiner, G., Groh, U., Lotz, H.P., Moustakas, U., Nagel, R., Neubrander, F., Schlotterbeck, U.: One-parameter semigroups of positive operators, volume 1184 of Lecture Notes in Mathematics. Springer-Verlag, Berlin (1986). https://doi.org/10.1007/BFb0074922

20. Arendt, W., Räbiger, F., Sourour, A.: Spectral properties of the operator equation $AX + XB = Y$. Quart. J. Math. Oxford Ser. (2) **45**(178), 133–149 (1994) https://doi.org/10.1093/qmath/45.2.133

21. Arendt, W., Rhandi, A.: Perturbation of positive semigroups. Arch. Math. (Basel) **56**(2), 107–119 (1991). https://doi.org/10.1007/BF01200341

22. Arora, S., Glück, J.: Spectrum and convergence of eventually positive operator semigroups. Semigroup Forum **103**(3), 791–811 (2021). https://doi.org/10.1007/s00233-021-10204-y

23. Banasiak, J., Arlotti, L.: Perturbations of positive semigroups with applications. Springer Monogr. Math. London: Springer (2006). https://doi.org/10.1007/1-84628-153-9

24. Banasiak, J., Falkiewicz, A.: Some transport and diffusion processes on networks and their graph realizability. Appl. Math. Lett. **45**, 25–30 (2015). https://doi.org/10.1016/j.aml.2015.01.006

25. Banasiak, J., Falkiewicz, A.: A singular limit for an age structured mutation problem. Math. Biosci. Eng. **14**(1), 17–30 (2017). https://doi.org/10.3934/mbe.2017002

26. Banasiak, J., Namayanja, P.: Asymptotic behaviour of flows on reducible networks. Netw. Heterog. Media **9**(2), 197–216 (2014). https://doi.org/10.3934/nhm.2014.9.197

27. Banasiak, J., Puchalska, A.: Generalized network transport and Euler-Hille formula. Discrete Contin. Dyn. Syst., Ser. B **23**(5), 1873–1893 (2018) https://doi.org/10.3934/dcdsb.2018185

28. Baroun, M., Jacob, B.: Admissibility and observability of observation operators for semilinear problems. Integral Equations Operator Theory **64**(1), 1–20 (2009). https://doi.org/10.1007/s00020-009-1675-0

29. Bátkai, A., Eisner, T., Latushkin, Y.: The spectral mapping property of delay semigroups. Complex Anal. Oper. Theory **2**(2), 273–283 (2008). https://doi.org/10.1007/s11785-008-0052-3

30. Bátkai, A., Engel, K.-J.: Abstract wave equations with generalized Wentzell boundary conditions. J. Differ. Equations **207**(1), 1–20 (2004). https://doi.org/10.1016/j.jde.2003.12.005

31. Bátkai, A., Engel, K.-J., Haase, M.: Cosine families generated by second-order differential operators on $W^{1,1}(0,1)$ with generalized Wentzell boundary conditions. Appl. Anal. **84**(9), 867–876 (2005). https://doi.org/10.1080/00036810500148952

32. Bátkai, A., Jacob, B., Voigt, J., Wintermayr, J.: Perturbations of positive semigroups on AM-spaces. Semigroup Forum **96**(2), 333–347 (2018). https://doi.org/10.1007/s00233-017-9879-0

33. Bátkai, A., Kramar Fijavž, M., Rhandi, A.: Positive operator semigroups. From finite to infinite dimensions, volume 257 of Oper. Theory: Adv. Appl. Basel: Birkhäuser/Springer (2017). https://doi.org/10.1007/978-3-319-42813-0

34. Batty, C.J.K., Chill, R., Tomilov, Y.: Strong stability of bounded evolution families and semi-groups. J. Funct. Anal. **193**(1), 116–139 (2002). https://doi.org/10.1006/jfan.2001.3917

35. Batty, C.J.K., Chill, R., van Neerven, J.: Asymptotic behaviour of $C_0$-semigroups with bounded local resolvents. Math. Nachr. **219**, 65–83 (2000). https://doi.org/10.1002/1522-2616(200011)219:1<65::AID-MANA65>3.0.CO;2-O

36. Batty, C.J.K., Duyckaerts, T.: Non-uniform stability for bounded semi-groups on Banach spaces. J. Evol. Equ. **8**(4), 765–780 (2008). https://doi.org/10.1007/s00028-008-0424-1

37. Batty, C.J.K., van Neerven, J., Räbiger, F.: Local spectra and individual stability of uniformly bounded $C_0$-semigroups. Trans. Am. Math. Soc. **350**(5), 2071–2085 (1998). https://doi.org/10.1090/S0002-9947-98-01919-9

38. Bayazit, F., Dorn, B., Fijavž, M.K.: Asymptotic periodicity of flows in time-depending networks. Netw. Heterog. Media **8**(4), 843–855 (2013). https://doi.org/10.3934/nhm.2013.8.843

39. Ben Amor, M.A., Homann, J., Kuo, W.-C., Watson, B.A.: Characterisation of conditional weak mixing via ergodicity of the tensor product in Riesz spaces. J. Math. Anal. Appl. **524**(2), 12 (2023). Id/No 127074. https://doi.org/10.1016/j.jmaa.2023.127074

40. Birkhoff, G.D.: Proof of the ergodic theorem. Proc. Natl. Acad. Sci. **17**(12), 656–660 (1931). https://doi.org/10.1073/pnas.17.2.656

41. Bobrowski, A., Gregosiewicz, A.: A general theorem on generation of moments-preserving cosine families by Laplace operators in $C[0, 1]$. Semigroup Forum **88**(3), 689–701 (2014). https://doi.org/10.1007/s00233-013-9561-0

42. Bobrowski, A., Mugnolo, D.: On moments-preserving cosine families and semigroups in $C[0, 1]$. J. Evol. Equ. **13**(4), 715–735 (2013). https://doi.org/10.1007/s00028-013-0199-x

43. Bogachev, V.I.: Differentiable measures and the Malliavin calculus, volume 164 of Math. Surv. Monogr. Providence, RI: American Mathematical Society (AMS) (2010) https://doi.org/10.1090/surv/164

44. Boltzmann, L.: Ueber die Eigenschaften monocyklischer und anderer damit verwandter Systeme. J. Reine Angew. Math. **98**, 68–94 (1885). https://eudml.org/doc/148597

45. Bombach, C., Gabel, F., Seifert, C., Tautenhahn, M.: Observability for non-autonomous systems. SIAM J. Control. Optim. **61**(1), 313–339 (2023). https://doi.org/10.1137/22M1485139

46. Bombach, C., Gallaun, D., Seifert, C., Tautenhahn, M.: Observability and null-controllability for parabolic equations in $L_p$-spaces. Math. Control Relat. Fields **13**(4), 1484–1499 (2023). https://doi.org/10.3934/mcrf.2022046

47. Bombieri, M.: On Weiss–Staffans Perturbations of Semigroup Generators. PhD thesis, Eberhard-Karls-Universität Tübingen (2015) https://doi.org/10.15496/publikation-4568

48. Bourgain, J.: New Banach space properties of the disc algebra and $H^\infty$. Acta Math. **152**, 1–48 (1984). https://doi.org/10.1007/BF02392189

49. Bratteli, O., Robinson, D.W., Operator algebras and quantum statistical mechanics. Vol. 1. Springer-Verlag, New York-Heidelberg,: $C^*$- and $W^*$-algebras, algebras, symmetry groups, decomposition of states. Texts and Monographs in Physics (1979). https://doi.org/10.1007/978-3-662-02520-8

50. Budde, C.: General extrapolation spaces and perturbations of bi-continuous semigroups. PhD thesis, Bergische Universität Wuppertal (2019). https://doi.org/10.25926/7ss7-bt33

51. Budde, C.: Perturbations of uniformly mean ergodic semigroups. J. Semigroup Theory Appl. **21**(1) (2021). https://doi.org/10.28919/jsta/5509

52. Budde, C.: Positive Desch-Schappacher perturbations of bi-continuous semigroups on AM-spaces. Acta Sci. Math. **87**(3–4), 571–594 (2021) https://doi.org/10.14232/actasm-021-914-5

53. Budde, C.: Positive Miyadera-Voigt perturbations of bi-continuous semigroups. Positivity **25**(3), 1107–1129 (2021c). https://doi.org/10.1007/s11117-020-00806-1

54. Budde, C.: Semigroups for flows on limits of graphs. Univ. Iagell. Acta Math. **58**, 7–21 (2021d). https://doi.org/10.4467/20843828AM.21.001.14982

55. Budde, C.: How to approach stability of bi-continuous semigroups? Math. Pannonica (N.S.) **29**(2), 282–290 (2023). https://doi.org/10.1556/314.2023.00029

56. Budde, C.: Another special role of $L^\infty$-spaces-evolution equations and Lotz' theorem. AIMS Math. **9**(12), 36158–36166 (2024a). https://doi.org/10.3934/math.20241716

57. Budde, C.: On bi-continuous cosine families. Quaest. Math. **47**(8), 1589–1611 (2024b). https://doi.org/10.2989/16073606.2024.2328817

58. Budde, C., Dobrick, A., Glück, J., Kunze, M.: A monotone convergence theorem for strong Feller semigroups. Ann. Mat. Pura Appl. (4) **202**(4), 1573–1589 (2023) https://doi.org/10.1007/s10231-022-01293-9

59. Budde, C., Farkas, B.: Intermediate and extrapolated spaces for bi-continuous operator semigroups. J. Evol. Equ. **19**(2), 321–359 (2019). https://doi.org/10.1007/s00028-018-0477-8

60. Budde, C., Farkas, B.: A Desch-Schappacher perturbation theorem for bi-continuous semigroups. Math. Nachr. **293**(6), 1053–1073 (2020). https://doi.org/10.1002/mana.201800534

61. Budde, C., Fijavž, M.K.: Bi-continuous semigroups for flows on infinite networks. Netw. Heterog. Media **16**(4), 553–567 (2021). https://doi.org/10.3934/nhm.2021017

62. Budde, C., Wegner, S.-A.: A Lumer-Phillips type generation theorem for bi-continuous semigroups. Z. Anal. Anwend. **41**(1–2), 65–80 (2022). https://doi.org/10.4171/ZAA/1695

63. Busch, P., Gudder, S.P.: Effects as functions on projective Hilbert space. Lett. Math. Phys. **47**(4), 329–337 (1999). https://doi.org/10.1023/A:1007573216122

64. Cannarsa, P., Da Prato, G.: On a functional analysis approach to parabolic equations in infinite dimensions. J. Funct. Anal. **118**(1), 22–42 (1993). https://doi.org/10.1006/jfan.1993.1137

65. Cârjă, O.: On constraint controllability of linear systems in Banach spaces. J. Optim. Theory Appl. **56**(2), 215–225 (1988). https://doi.org/10.1007/BF00939408

66. Casarino, V., Engel, K.-J., Nagel, R., Nickel, G.: A semigroup approach to boundary feedback systems. Integral Equations Operator Theory **47**(3), 289–306 (2003). https://doi.org/10.1007/s00020-002-1163-2

67. Cerrai, S.: A Hille-Yosida theorem for weakly continuous semigroups. Semigroup Forum **49**(3), 349–367 (1994). https://doi.org/10.1007/BF02573496

68. Chill, R., Fašangová, E., Metafune, G., Pallara, D.: The sector of analyticity of the Ornstein-Uhlenbeck semigroup on $L^p$ spaces with respect to invariant measure. J. Lond. Math. Soc., II. Ser. **71**(3), 703–722 (2005). https://doi.org/10.1112/S0024610705006344

69. Chill, R., Keyantuo, V., Warma, M.: Generation of cosine families on $L^p(0, 1)$ by elliptic operators with Robin boundary conditions. In: Functional analysis and evolution equations. The Günter Lumer volume, pp. 113–130. Basel: Birkhäuser (2008). https://doi.org/10.1007/978-3-7643-7794-6_7

70. Chojnowska-Michalik, A., Goldys, B.: Nonsymmetric Ornstein-Uhlenbeck semigroup as second quantized operator. J. Math. Kyoto Univ. **36**(3), 481–498 (1996). https://doi.org/10.1215/kjm/1250518505

71. Chojnowska-Michalik, A., Goldys, B.: Symmetric Ornstein-Uhlenbeck semigroups and their generators. Probab. Theory Relat. Fields **124**(4), 459–486 (2002). https://doi.org/10.1007/s004400200222

72. Conway, J.B.: A course in abstract analysis, volume 141 of Graduate Studies in Mathematics. American Mathematical Society, Providence, RI (2012). https://doi.org/10.1090/gsm/141

73. Cooper, J.B.: Saks spaces and applications to functional analysis, volume 139 of North-Holland Mathematics Studies. North-Holland Publishing Co., Amsterdam, second edition (1987). Notas de Matemática [Mathematical Notes], 116

74. Cristescu, R.: Vector seminorms, spaces with vector norm, and regular operators. Rev. Roumaine Math. Pures Appl. **53**(5–6), 407–418 (2008)

75. Da Prato, G.: A new regularity result for Ornstein-Uhlenbeck generators and applications. J. Evol. Equ. **3**(3), 485–498 (2003). https://doi.org/10.1007/s00028-003-0114-x

76. Da Prato, G., Grisvard, P.: Maximal regularity for evolution equations by interpolation and extrapolation. J. Funct. Anal. **58**, 107–124 (1984). https://doi.org/10.1016/0022-1236(84)90034-X

77. Da Prato, G., Lunardi, A.: On the Ornstein-Uhlenbeck operator in spaces of continuous functions. J. Funct. Anal. **131**(1), 94–114 (1995). https://doi.org/10.1006/jfan.1995.1084

78. Da Prato, G., Zabczyk, J.: Stochastic equations in infinite dimensions, volume 44 of Encycl. Math. Appl. Cambridge etc.: Cambridge University Press (1992). https://doi.org/10.1017/CBO9780511666223

79. Daners, D., Glück, J., Kennedy, J.B.: Eventually positive semigroups of linear operators. J. Math. Anal. Appl. **433**(2), 1561–1593 (2016). https://doi.org/10.1016/j.jmaa.2015.08.050

80. Davies, E.B.: One-parameter semigroups, volume 15 of Lond. Math. Soc. Monogr. Academic Press, London (1980). https://doi.org/10.1017/S0013091500028169

81. Davies, E.B.: $L^1$ properties of second order elliptic operators. Bull. Lond. Math. Soc. **17**, 417–436 (1985). https://doi.org/10.1112/blms/17.5.417

82. De Wilde, M.: Limite inductive d'espaces linéaires semi-normes. Bull. Soc. R. Sci. Liège **32**, 476–484 (1963)

83. Desch, W., Schappacher, W.: On relatively bounded perturbations of linear $C_0$-semigroups. Ann. Sc. Norm. Super. Pisa, Cl. Sci., IV. Ser. **11**, 327–341 (1984). https://eudml.org/doc/83934

84. Desch, W., Schappacher, W.: Some perturbation results for analytic semigroups. Math. Ann. **281**(1), 157–162 (1988). https://doi.org/10.1007/BF01449222

85. Diestel, J., Uhl, J.J.J.: Vector measures, volume 15 of Math. Surv. American Mathematical Society (AMS), Providence, RI (1977). https://doi.org/10.1090/surv/015

86. DiPerna, R.J., Lions, P.L.: Ordinary differential equations, transport theory and Sobolev spaces. Invent. Math. **98**(3), 511–547 (1989). https://doi.org/10.1007/BF01393835

87. Dorn, B.: Semigroups for flows in infinite networks. Semigroup Forum **76**(2), 341–356 (2008). https://doi.org/10.1007/s00233-007-9036-2

88. Dorn, B., Keicher, V., Sikolya, E.: Asymptotic periodicity of recurrent flows in infinite networks. Math. Z. **263**(1), 69–87 (2009). https://doi.org/10.1007/s00209-008-0410-x

89. Dorn, B., Kramar Fijavž, M., Nagel, R., Radl, A.: The semigroup approach to transport processes in networks. Phys. D **239**(15), 1416–1421 (2010). https://doi.org/10.1016/j.physd.2009.06.012

90. Dorroh, J.R., Neuberger, J.W.: Lie generators for semigroups of transformations on a Polish space. Electron. J. Differ. Equ. 1993/01 (1993). https://eudml.org/doc/118513

91. Dorroh, J.R., Neuberger, J.W.: A theory of strongly continuous semigroups in terms of Lie generators. J. Funct. Anal. **136**(1), 114–126 (1996). https://doi.org/10.1006/jfan.1996.0023

92. Douglas, R.G.: On majorization, factorization, and range inclusion of operators on Hilbert space. Proc. Am. Math. Soc. **17**, 413–415 (1966). https://doi.org/10.2307/2035178

93. Eisner, T.: Stability of operators and operator semigroups, volume 209 of Oper. Theory: Adv. Appl. Basel: Birkhäuser (2010). https://doi.org/10.1007/978-3-0346-0195-5

94. Eisner, T., Farkas, B., Haase, M., Nagel, R.: Operator theoretic aspects of ergodic theory, volume 272 of Graduate Texts in Mathematics. Springer, Cham (2015). https://doi.org/10.1007/978-3-319-16898-2

95. Eisner, T., Farkas, B., Nagel, R., Serény, A.: Weakly and almost weakly stable $C_0$-semigroups. Int. J. Dyn. Syst. Differ. Equ. **1**(1), 44–57 (2007). https://doi.org/10.1504/IJDSDE.2007.013744

96. Engel, K.-J.: Spectral theory and generator property for one-sided coupled operator matrices. Semigroup Forum **58**(2), 267–295 (1999). https://doi.org/10.1007/s002339900020

97. Engel, K.-J., Fijavž, M.K.: Waves and diffusion on metric graphs with general vertex conditions. Evol. Equ. Control Theory **8**(3), 633–661 (2019). https://doi.org/10.3934/eect.2019030

278

98. Engel, K.-J., Kramar Fijavž, M.: Exact and positive controllability of boundary control systems. Netw. Heterog. Media **12**(2), 319–337 (2017). https://doi.org/10.3934/nhm.2017014

99. Engel, K.-J., Kramar Fijavž, M., Klöss, B., Nagel, R., Sikolya, E.: Maximal controllability for boundary control problems. Appl. Math. Optim. **62**(2), 205–227 (2010). https://doi.org/10.1007/s00245-010-9101-1

100. Engel, K.-J., Kramar Fijavž, M., Nagel, R., Sikolya, E.: Vertex control of flows in networks. Netw. Heterog. Media **3**(4), 709–722 (2008). https://doi.org/10.3934/nhm.2008.3.709

101. Engel, K.-J., Nagel, R.: One-parameter semigroups for linear evolution equations, volume 194 of Graduate Texts in Mathematics. Springer-Verlag, New York, 2000. With contributions by S. Brendle, M. Campiti, T. Hahn, G. Metafune, G. Nickel, D. Pallara, C. Perazzoli, A. Rhandi, S. Romanelli and R. Schnaubelt

102. Es-Sarhir, A., Farkas, B.: Positivity of perturbed Ornstein-Uhlenbeck semigroups on $C_b(H)$. Semigroup Forum **70**(2), 208–224 (2005). https://doi.org/10.1007/s00233-004-0167-4

103. Es-Sarhir, A., Farkas, B.: Perturbation for a class of transition semigroups on the Hölder space $C_{b,\mathrm{loc}}^{\theta}(H)$. J. Math. Anal. Appl. **315**(2), 666–685 (2006). https://doi.org/10.1016/j.jmaa.2005.04.024

104. Esterle, J.: Mittag-Leffler methods in the theory of Banach algebras and a new approach to Michael's problem. In: Proceedings of the conference on Banach algebras and several complex variables (New Haven, Conn., 1983), volume 32 of Contemp. Math., pages 107–129. Amer. Math. Soc., Providence, RI (1984). https://doi.org/10.1090/conm/032/769501

105. Fabian, M., Habala, P., Hájek, P., Montesinos, V., Zizler, V.: Banach space theory. The basis for linear and nonlinear analysis. CMS Books Math./Ouvrages Math. SMC. Berlin: Springer (2011). https://doi.org/10.1007/978-1-4419-7515-7

106. Farkas, B.: Perturbations of Bi-Continuous Semigroups. PhD thesis, Eötvös Loránd University (2003)

107. Farkas, B.: Perturbations of bi-continuous semigroups. Studia Math. **161**(2), 147–161 (2004a). https://doi.org/10.4064/sm161-2-3

108. Farkas, B.: Perturbations of bi-continuous semigroups with applications to transition semigroups on $C_b(H)$. Semigroup Forum **68**(1), 87–107 (2004b). https://doi.org/10.1007/s00233-002-0024-2

109. Farkas, B.: Adjoint bi-continuous semigroups and semigroups on the space of measures. Czech. Math. J. **61**(2), 309–322 (2011). https://doi.org/10.1007/s10587-011-0076-0

110. Fattorini, H.O.: Ordinary differential equations in linear topological space. II. J. Differ. Equations **6**, 50–70 (1969a). https://doi.org/10.1016/0022-0396(69)90117-X

111. Fattorini, H.O.: Ordinary differential equations in linear topological spaces. I. J. Differ. Equations **5**, 72–105 (1969b). https://doi.org/10.1016/0022-0396(69)90105-3

112. Fattorini, H.O.: Second order linear differential equations in Banach spaces. North-Holland Math, vol. 108. Stud. Elsevier, Amsterdam (1985)

113. Fomin, S.V.: Differentiable measures in linear spaces. Us*p*ehi Mat. Nauk, **23**(1(139)), 221–222 (1968)

114. Gallaun, D., Meichsner, J., Seifert, C.: Final state observability in Banach spaces with applications to subordination and semigroups induced by Lévy processes. Evol. Equ. Control Theory **12**(4), 1102–1121 (2023). https://doi.org/10.3934/eect.2023002

115. Gallaun, D., Seifert, C., Tautenhahn, M.: Sufficient criteria and sharp geometric conditions for observability in Banach spaces. SIAM J. Control. Optim. **58**(4), 2639–2657 (2020). https://doi.org/10.1137/19M1266769

116. Giga, M.: Abstract elliptic operator and its associated semigroup in a locally convex space. Hokkaido Math. J. **22**(3), 225–244 (1993). https://doi.org/10.14492/hokmj/1381413176

117. Gilbarg, D., Trudinger, N.S.: Elliptic partial differential equations of second order. 2nd ed, volume 224 of Grundlehren Math. Wiss. Springer, Cham (1983). https://doi.org/10.1007/978-3-642-61798-0

118. Glück, J., Wolff, M.P.H.: Long-term analysis of positive operator semigroups via asymptotic domination. Positivity **23**(5), 1113–1146 (2019). https://doi.org/10.1007/s11117-019-00655-7

119. Goldstein, J.A.: Semigroups of linear operators and applications. Oxford Math. Monogr. Oxford University Press, Oxford (1985)

120. Goldys, B.: On analyticity of Ornstein-Uhlenbeck semigroups. Atti Accad. Naz. Lincei, Cl. Sci. Fis. Mat. Nat., IX. Ser., Rend. Lincei, Mat. Appl., 10(3):131–140 (1999). http://eudml.org/doc/252271

121. Goldys, B., Kocan, M.: Diffusion semigroups in spaces of continuous functions with mixed topology. J. Differ. Equations **173**(1), 17–39 (2001). https://doi.org/10.1006/jdeq.2000.3918

122. Goldys, B., Nendel, M., Röckner, M.: Operator semigroups in the mixed topology and the infinitesimal description of Markov processes. J. Differ. Equations **412**, 23–86 (2024). https://doi.org/10.1016/j.jde.2024.08.024

123. Greiner, G.: Perturbing the boundary conditions of a generator. Houston J. Math. **13**, 213–229 (1987)

124. Groenewegen, G.L.M., van Rooij, A.C.M.: Spaces of continuous functions, volume 4 of Atlantis Studies in Mathematics. Atlantis Press, Paris (2016). https://doi.org/10.2991/978-94-6239-201-4

125. Groh, U., Neubrander, F.: Stabilität starkstetiger, positiver Operatorhalbgruppen auf $C^*$-Algebren. Math. Ann. **256**(4), 509–516 (1981). https://doi.org/10.1007/BF01450545

126. Haase, M.: Operator-valued $H^\infty$-calculus in inter- and extrapolation spaces. Integral Equations Oper. Theory **56**(2), 197–228 (2006). https://doi.org/10.1007/s00020-006-1418-4

127. Hadd, S., Manzo, R., Rhandi, A.: Unbounded perturbations of the generator domain. Discrete Contin. Dyn. Syst. **35**(2), 703–723 (2015). https://doi.org/10.3934/dcds.2015.35.703

128. Hassani, R.A., Blali, A., Amrani, A.E., Moussaouja, K.: Cosine families of operators in a class of Fréchet spaces. Proyecciones **37**(1), 103–118 (2018). https://doi.org/10.4067/S0716-09172018000100103

129. Hennig, B., Neubrander, F.: On representations, inversions, and approximations of Laplace transforms in Banach spaces. Appl. Anal. **49**(3–4), 151–170 (1993). https://doi.org/10.1080/00036819108840171

130. Henríquez, H.R., and C. L. H.: A generation result for cosine functions of operators. Z. Anal. Anwend. **28**(1), 103–118 (2009). https://doi.org/10.4171/ZAA/1375

131. Hille, E.: Representation of one-parameter semi-groups of linear transformations. Proc. Nat. Acad. Sci. U.S.A. **28**(5), 175–178 (1942). https://doi.org/10.1073/pnas.28.5.175

132. Hille, E., Phillips, R.S.: Functional analysis and semigroups. Rev. ed, volume 31 of Colloq. Publ., Am. Math. Soc. American Mathematical Society (AMS), Providence, RI (1957). www.ams.org/online_bks/coll31/

133. Homann, J., Kuo, W.-C., Watson, B.A.: Ergodicity in Riesz spaces. In: Positivity and its applications. Proceedings from the conference Positivity X, Pretoria, South Africa, July 8–12, 2019, pp. 193–201. Cham, Birkhäuser (2021). https://doi.org/10.1007/978-3-030-70974-7_9

134. Homann, J., Kuo, W.-C., Watson, B.A.: A Koopman-von Neumann type theorem on the convergence of Cesàro means in Riesz spaces. Proc. Am. Math. Soc., Ser. B, **8**, 75–85 (2021)https://doi.org/10.1090/bproc/75

135. Ito, K., Kappel, F.: The Trotter-Kato theorem and approximation of PDEs. Math. Comput. **67**(221), 21–44 (1998). https://doi.org/10.1090/S0025-5718-98-00915-6

136. Jacob, B., Nabiullin, R., Partington, J., Schwenninger, F.: On input-to-state-stability and integral input-to-state-stability for parabolic boundary control systems. In 2016 IEEE 55th Conference on Decision and Control (CDC), pp. 2265–2269 (2016). https://doi.org/10.1109/CDC.2016.7798600

137. Jacob, B., Nabiullin, R., Partington, J.R., Schwenninger, F.L.: Infinite-dimensional input-to-state stability and Orlicz spaces. SIAM J. Control. Optim. **56**(2), 868–889 (2018). https://doi.org/10.1137/16M1099467

138. Jacob, B., Schwenninger, F.L., Zwart, H.: On continuity of solutions for parabolic control systems and input-to-state stability. Journal of Differential Equations, p. 23 (2018). https://doi.org/10.1016/j.jde.2018.11.004

139. Jacob, B., Wegner, S.-A.: Asymptotics of evolution equations beyond Banach spaces. Semigroup Forum **91**(2), 347–377 (2015). https://doi.org/10.1007/s00233-014-9659-z

140. Jara, P.: Rational approximation schemes for bi-continuous semigroups. J. Math. Anal. Appl. **344**(2), 956–968 (2008). https://doi.org/10.1016/j.jmaa.2008.02.068

141. Jara, P., Neubrander, F., Özer, K.: Rational inversion of the Laplace transform. J. Evol. Equ. **12**(2), 435–457 (2012). https://doi.org/10.1007/s00028-012-0139-1

142. Kalton, N.J.: Mackey duals and almost shrinking bases. Proc. Camb. Philos. Soc. **74**, 73–81 (1973)

143. Kato, T.: Perturbation theory for linear operators. Class. Math. Berlin: Springer-Verlag, reprint of the corr. print. of the 2nd ed. 1980 edition (1995). https://doi.org/10.1007/978-3-642-66282-9

144. Kenari, H.M., Saadati, R., Azhini, M., Cho, Y.J.: Mean ergodic theorem for semigroups of linear operators in multi-Banach spaces. J. Inequal. Appl.: 10 (2014). Id/No **402**,(2014). https://doi.org/10.1186/1029-242X-2014-402

145. Kisyński, J.: On cosine operator functions and one parameter groups of operators. Stud. Math. **44**, 93–105 (1972). https://doi.org/10.4064/sm-44-1-93-105

146. Komatsu, H.: Semi-groups of operators in locally convex spaces. J. Math. Soc. Japan **16**, 230–262 (1964). https://doi.org/10.2969/jmsj/01630230

147. Kōmura, T.: Semigroups of operators in locally convex spaces. J. Funct. Anal. **2**, 258–296 (1968). https://doi.org/10.1016/0022-1236(68)90008-6

148. Konishi, Y.: Cosine functions of operators in locally convex spaces. J. Fac. Sci., Univ. Tokyo, Sect. I A, **18**, 443–463 (1972)

149. Kovrijkine, O.: Some results related to the Logvinenko-Sereda theorem. Proc. Am. Math. Soc. **129**(10), 3037–3047 (2001). https://doi.org/10.1090/S0002-9939-01-05926-3

150. Kraaij, R.: Strongly continuous and locally equi-continuous semigroups on locally convex spaces. Semigroup Forum **92**(1), 158–185 (2016). https://doi.org/10.1007/s00233-015-9689-1

151. Kramar, M., Sikolya, E.: Spectral properties and asymptotic periodicity of flows in networks. Math. Z. **249**(1), 139–162 (2005). https://doi.org/10.1007/s00209-004-0695-3

152. Kruse, K.: Mixed topologies on Saks spaces of vector-valued functions. Topology Appl. **345**, 27 (2024). Id/No 108843. https://doi.org/10.1016/j.topol.2024.108843

153. Kruse, K., Meichsner, J., Seifert, C.: Subordination for sequentially equicontinuous equibounded $C_0$-semigroups. J. Evol. Equ. **21**(2), 2665–2690 (2021). https://doi.org/10.1007/s00028-021-00700-7

154. Kruse, K., Schwenninger, F.L.: On equicontinuity and tightness of bi-continuous semigroups. J. Math. Anal. Appl. **509**(2), 27 (2022). Id/No 125985. https://doi.org/10.1016/j.jmaa.2021.125985

155. Kruse, K., Schwenninger, F.L.: Sun dual theory for bi-continuous semigroups. Anal. Math. **50**(1), 235–280 (2024). https://doi.org/10.1007/s10476-024-00014-z

156. Kruse, K., Seifert, C.: A note on the Lumer-Phillips theorem for bi-continuous semigroups. Z. Anal. Anwend. **41**(3–4), 417–437 (2022). https://doi.org/10.4171/ZAA/1709

157. Kruse, K., Seifert, C.: Final state observability estimates and cost-uniform approximate null-controllability for bi-continuous semigroups. Semigroup Forum **106**(2), 421–443 (2023). https://doi.org/10.1007/s00233-023-10346-1

158. Kühmemund, F.: Approximation of bi-continuous semigroups. Tübinger Berichte zur Funktionalanalysis (2000/2001)

159. Kühnemund, F.: Bi-Continuous Semigroups on Spaces with Two Topologies: Theory and Applications. PhD thesis, Eberhard-Karls-Universität Tübingen (2001)

160. Kühnemund, F.: A Hille-Yosida theorem for bi-continuous semigroups. Semigroup Forum **67**(2), 205–225 (2003). https://doi.org/10.1007/s00233-002-5000-3

161. Kühnemund, F., van Neerven, J.: A Lie-Trotter product formula for Ornstein-Uhlenbeck semigroups in infinite dimensions. J. Evol. Equ. **4**(1), 53–73 (2004). https://doi.org/10.1007/s00028-003-0078-y

162. Kühnemund, F., Wacker, M.: Commutator conditions implying the convergence of the Lie-Trotter products. Proc. Am. Math. Soc. **129**(12), 3569–3582 (2001). https://doi.org/10.1090/S0002-9939-01-06034-8

163. Kunstmann, P.C.: On continuity properties of semigroups in real interpolation spaces. J. Evol. Equ. **21**(3), 3503–3520 (2021). https://doi.org/10.1007/s00028-020-00652-4

164. Kunze, M.: Continuity and equicontinuity of semigroups on norming dual pairs. Semigroup Forum **79**(3), 540–560 (2009). https://doi.org/10.1007/s00233-009-9174-9

165. Kuo, W.-C., Labuschagne, C.C.A., Watson, B.A.: Ergodic theory and the strong law of large numbers on Riesz spaces. J. Math. Anal. Appl. **325**(1), 422–437 (2007). https://doi.org/10.1016/j.jmaa.2006.01.056

166. Landsman, K.: Foundations of quantum theory, volume 188 of Fundamental Theories of Physics. Springer, Cham (2017). From classical concepts to operator algebras. https://doi.org/10.1007/978-3-319-51777-3

167. Lebeau, G., Robbiano, L.: Exact control of the heat equation. Commun. Partial Differ. Equations **20**(1–2), 335–356 (1995). https://doi.org/10.1080/03605309508821097

168. Logvinenko, V.N., Sereda, Y.F.: Äquivalente Normen in Räumen ganzer Funktionen vom Exponentialtyp. Teor. Funkts. Funkts. Anal. Prilozh. **20**, 102–111 (1974)

169. Lorenzi, L., Bertoldi, M.: Analytical methods for Markov semigroups, volume 283 of Pure and Applied Mathematics (Boca Raton). Chapman & Hall/CRC, Boca Raton, FL (2007). https://doi.org/10.1201/9781420011586

170. Lorenzi, L., Rhandi, A.: Semigroups of bounded operators and second-order elliptic and parabolic partial differential equations. Monogr. Res. Notes Math. Boca Raton, FL: CRC Press (2021). https://doi.org/10.1201/9780429262593

171. Lotz, H.P.: Uniform convergence of operators on $L^\infty$ and similar spaces. Math. Z. **190**(2), 207–220 (1985). https://doi.org/10.1007/BF01160459

172. Lovelady, D.L.: Semigroups of linear operators on locally convex spaces with Banach subspaces. Indiana Univ. Math. J. **24**, 1191–1198 (1975). https://doi.org/10.1512/iumj.1975.24.24097

173. Lumer, G., Phillips, R.S.: Dissipative operators in a Banach space. Pac. J. Math. **11**, 679–698 (1961). https://doi.org/10.2140/pjm.1961.11.679

174. Lumer, G., Weis, L, editors. Evolution equations and their applications in physical and life sciences, volume 215 of Lecture Notes in Pure and Applied Mathematics. Marcel Dekker, Inc., New York (2001). https://doi.org/10.1201/9780429187810

175. Lunardi, A.: Interpolation theory, volume 16 of Appunti, Sc. Norm. Super. Pisa (N.S.). Pisa: Edizioni della Normale, 3rd edition edition (2018). https://doi.org/10.1007/978-88-7642-638-4

176. Lutz, D.: Der infinitesimale Erzeuger für lösungen der Funktionalgleichung des Cosinus mit Werten im Bereich der abgeschlossenen linearen Operatoren auf einem Banachraum. Seminarberichte FB Mathematik (1978)

177. Luxemburg, W.A.J., Zaanen, A.C.: Riesz spaces. Vol. I. North-Holland Publishing Co., Amsterdam-London; American Elsevier Publishing Co., New York, North-Holland Mathematical Library (1971)

178. Magal, P., Ruan, S.: Theory and applications of abstract semilinear Cauchy problems, volume 201 of Applied Mathematical Sciences. With a foreword by Glenn Webb. Springer, Cham (2018). https://doi.org/10.1007/978-3-030-01506-0

179. Mátrai, T., Sikola, E.: Asymptotic behavior of flows in networks. Forum Math. **19**(3), 429–461 (2007). https://doi.org/10.1515/FORUM.2007.018

180. Meise, R., Vogt, D.: Einführung in die Funktionalanalysis, volume 62 of Vieweg Stud. Braunschweig: Friedr. Vieweg & Sohn (1992). https://doi.org/10.1007/978-3-322-80310-8

181. Meyer-Nieberg, P.: Banach lattices. Berlin etc.: Springer-Verlag (1991). https://doi.org/10.1007/978-3-642-76724-1

182. Michael, E.: On k-spaces, $k_R$-spaces and k(X). Pac. J. Math. **47**, 487–498 (1973). https://doi.org/10.2140/pjm.1973.47.487

183. Miller, L.: A direct Lebeau-Robbiano strategy for the observability of heat-like semigroups. Discrete Contin. Dyn. Syst., Ser. B, 14(4):1465–1485 (2010). https://doi.org/10.3934/dcdsb.2010.14.1465

184. Miyadera, I.: Semi-groups of operators in Fréchet space and applications to partial differential equations. Tôhoku Math. J. 2(11), 162–183 (1959). https://doi.org/10.2748/tmj/1178244580

185. Miyadera, I.: On perturbation theory for semi-groups of operators. Tôhoku Math. J. 2(18), 299–310 (1966). https://doi.org/10.2748/tmj/1178243419

186. Nagel, R.: Sobolev spaces and semigroups. Semesterbericht Funktionalanalysis, pp. 1–19 (1983)

187. Nagel, R.: Extrapolation spaces for semigroups. RIMS Kôkyûroku, **1009**, 181–191 (1997). https://doi.org/2433/61496

188. Nagel, R., Nickel, G., Romanelli, S.: Identification of extrapolation spaces for unbounded operators. Quaest. Math. **19**(1–2), 83–100 (1996). https://doi.org/10.1080/16073606.1996.9631827

189. Nagel, R., Sinestrari, E.: Inhomogeneous Volterra integrodifferential equations for Hille-Yosida operators. In Functional analysis (Essen, 1991), volume 150 of Lecture Notes in Pure and Appl. Math., pp. 51–70. Dekker, New York (1994)

190. Nagel, R., Sinestrari, E.: Extrapolation spaces and minimal regularity for evolution equations. J. Evol. Equ. **6**(2), 287–303 (2006). https://doi.org/10.1007/s00028-006-0246-y

191. Nagy, B.: On cosine operator functions in Banach spaces. Acta Sci. Math. **36**, 281–289 (1974)

192. Nakić, I., Täufer, M., Tautenhahn, M., Veselić, I.: Sharp estimates and homogenization of the control cost of the heat equation on large domains. ESAIM, Control Optim. Calc. Var., **26**, 26 (2020). Id/No 54. https://doi.org/10.1051/cocv/2019058

193. Neuberger, J.W.: Lie generators for strongly continuous equiuniformly continuous one parameter semigroups on a metric space. Indiana Univ. Math. J. **21**, 961–971 (1972). https://doi.org/10.1512/iumj.1972.21.21077

194. Neuberger, J.W.: Lie generators for one parameter semigroups of transformations. J. Reine Angew. Math. **258**, 133–136 (1973). https://doi.org/10.1515/crll.1973.258.133

195. Nickel, G.: A new look at boundary perturbations of generators. Electron. J. Differ. Equ., **2004**, 14 (2004). Id/No 95. https://eudml.org/doc/124901

196. Ōuchi, S.: Semi-groups of operators in locally convex spaces. J. Math. Soc. Japan **25**, 265–276 (1973). https://doi.org/10.2969/jmsj/02520265

197. Pazy, A.: Semigroups of linear operators and applications to partial differential equations, volume 44 of Appl. Math. Sci. Springer, Cham (1983). https://doi.org/10.1007/978-1-4612-5561-1

198. Phóng, V.Q.: The operator equation $AX - XB = C$ with unbounded operators $A$ and $B$ and related abstract Cauchy problems. Math. Z. **208**(4), 567–588 (1991). https://doi.org/10.1007/BF02571546

199. Phóng, V.Q.: On the exponential stability and dichotomy of $C_0$-semigroups. Studia Math. **132**(2), 141–149 (1999). https://doi.org/10.4064/sm-132-2-141-149

200. Phóng, V.Q., Schüler, E.: The operator equation $AX - XB = C$, admissibility, and asymptotic behavior of differential equations. J. Diff. Equ. **145**, 394–419 05 (1998). https://doi.org/10.1006/jdeq.1998.3418

201. Piskarev, S., Shaw, S.Y.: On certain operator families related to cosine operator functions. Taiwanese J. Math. **1**(4), 527–546 (1997). https://doi.org/10.11650/twjm/1500406127

202. Priola, E.: On a class of Markov type semigroups in spaces of uniformly continuous and bounded functions. Stud. Math. **136**(3), 271–295 (1999). https://eudml.org/doc/216671, https://doi.org/10.4064/sm-136-3-271-295

203. Priola, E.: On a Dirichlet problem involving an Ornstein-Uhlenbeck operator. Potential Anal. **18**(3), 251–287 (2003). https://doi.org/10.1023/A:1020933325029

204. Prüss, J., Metafune, G., Rhandi, A., Schnaubelt, R.: The domain of the Ornstein-Uhlenbeck operator on an $L^p$-space with invariant measure. Ann. Sc. Norm. Super. Pisa, Cl. Sci. (5) **1**(2), 471–485 (2002). https://eudml.org/doc/84478

205. Rainville, E.D.: Special functions. Bronx, N. Y.: Chelsea Publishing Comp. XII, p. 365 (1971)

206. Rudin, W.: Functional analysis, 2nd edn. International Series in Pure and Applied Mathematics. McGraw-Hill Inc, New York (1991)

207. Schaefer, H.H.: Banach lattices and positive operators. Springer-Verlag, New York-Heidelberg, 1974. Die Grundlehren der mathematischen Wissenschaften, Band 215. https://doi.org/10.1007/978-3-642-65970-6

208. Schaefer, H.H., Wolff, M.P.: Topological vector spaces, volume 3 of Graduate Texts in Mathematics. Springer-Verlag, New York, second edition (1999). https://doi.org/10.1007/978-1-4612-1468-7

209. Schwartz, L.: Lectures on Mixed Problems in Partial Differential Equations and Representation of Semi-groups. Number Nr. 11 in Lectures on Mixed Problems in Partial Differential Equations and Representation of Semi-groups. Tata Institute of Fundamental Research (1957)

210. Schwenninger, F.L., Zwart, H.: Zero-two law for cosine families. J. Evol. Equ. **15**(3), 559–569 (2015). https://doi.org/10.1007/s00028-015-0272-8

211. Scirrat, A.K.: Evolution Semigroups for Well-Posed, NonAutonomous Evolution Families. PhD thesis, Louisiana State University and Agricultural and Mechanical College (2016)

212. Sentilles, F.D.: Semigroups of operators in $C(S)$. Can. J. Math. **22**, 47–54 (1970). https://doi.org/10.4153/CJM-1970-006-8

213. Sentilles, F.D.: Bounded continuous functions on a completely regular space. Trans. Am. Math. Soc. **168**, 311–336 (1972). https://doi.org/10.2307/1996178

214. Shaw, S.-Y.: On $w^*$-continuous cosine operator functions. J. Funct. Anal. **66**, 73–95 (1986). https://doi.org/10.1016/0022-1236(86)90082-0

215. Shimizu, M., Miyadera, I.: Perturbation theory for cosine families on Banach spaces. Tokyo J. Math. **1**, 333–343 (1978). https://doi.org/10.3836/tjm/1270216503

216. Shiryaev, A.N.: Probability. Transl. from the Russian by R. P. Boas, volume 95 of Grad. Texts Math. Springer, Cham (1984)

217. Sinestrari, E.: Interpolation and extrapolation spaces in evolution equations. In J. Cea, D. Chenais, G. Geymonat, and J. L. Lions, editors, Partial Differential Equations and Functional

Analysis: In Memory of Pierre Grisvard, pp. 235–254. Birkhäuser Boston, Boston, MA (1996). https://doi.org/10.1007/978-1-4612-2436-5_16

218. Skorohod, A.V.: On the differentiability of measures which correspond to stochastic processes. I. Processes with independent increments. Teor. Veroyatnost. i Primenen **2**, 417–443 (1957)https://doi.org/10.1137/1102030

219. Skorohod, A.V., Integration in Hilbert space. Springer-Verlag, New York-Heidelberg,: Translated from the Russian by Kenneth Wickwire. Ergeb. Math. Grenzgeb. **79**,(1974). https://doi.org/10.1007/978-3-642-65632-3

220. Snipes, R.F.: C-sequential and S-bornological topological vector spaces. Math. Ann. **202**, 273–283 (1973). https://doi.org/10.1007/BF01433457

221. Sova, M.: Cosine operator functions. Diss. Math. **49** (1966). http://eudml.org/doc/268428

222. Staffans, O.: Well-posed linear systems, volume 103 of Encyclopedia of Mathematics and its Applications. Cambridge University Press, Cambridge (2005). https://doi.org/10.1017/CBO9780511543197

223. Staffans, O.J., Weiss, G.: Transfer functions of regular linear systems. III. Inversions and duality. Int. Equ. Oper. Theory **49**(4), 517–558 (2004). https://doi.org/10.1007/s00020-002-1214-8

224. Travis, C.C., Webb, G.F.: Second order differential equations in Banach space. Nonlinear equations in abstract spaces, Proc. int. Symp., Arlington 1977, 331–361 (1978)

225. Trotter, H.F.: Approximation of semi-groups of operators. Pac. J. Math. **8**, 887–919 (1958). https://doi.org/10.2140/pjm.1958.8.887

226. van Neerven, J.: The adjoint of a semigroup of linear operators, volume 1529 of Lecture Notes in Mathematics. Springer-Verlag, Berlin (1992). https://doi.org/10.1007/BFb0085008

227. van Neerven, J.: The asymptotic behaviour of semigroups of linear operators, volume 88 of Oper. Theory: Adv. Appl. Basel: Birkhäuser (1996). https://doi.org/10.1007/978-3-0348-9206-3

228. van Neerven, J.M.A.M., Straub, B., Weis, L.: On the asymptotic behaviour of a semigroup of linear operators. Indag. Math., New Ser. **6**(4), 453–476 (1995). https://doi.org/10.1016/0019-3577(96)81760-5

229. van Zuijlen, W.: Integration of functions with values in a Riesz space. Master's thesis, Radboud Universiteit Nijmegen (2012)

230. Vieru, A.: On null controllability of linear systems in Banach spaces. Syst. Control Lett. **54**(4), 331–337 (2005). https://doi.org/10.1016/j.sysconle.2004.09.004

231. Voigt, J.: On the perturbation theory for strongly continuous semigroups. Math. Ann. **229**(2), 163–171 (1977). https://doi.org/10.1007/BF01351602

232. Voigt, J.: On resolvent positive operators and positive $C_0$-semigroups on AL-spaces. Semigroup Forum **38**(2), 263–266 (1989). http://eudml.org/doc/134959

233. von Below, J., Lubary, J.A.: The eigenvalues of the Laplacian on locally finite networks. Results Math. **47**(3–4), 199–225 (2005). https://doi.org/10.1007/BF03323026

234. von Below, J., Lubary, J.A.: The eigenvalues of the Laplacian on locally finite networks under generalized node transition. Results Math. **54**(1–2), 15–39 (2009). https://doi.org/10.1007/s00025-009-0376-y

235. von Neumann, J.: Proof of the quasi-ergodic hypothesis. Proc. Natl. Acad. Sci. U.S.A. **18**, 70–82 (1932). https://doi.org/10.1073/pnas.18.1.70

236. Walker, R.C.: The Stone-Cech compactification. Ergeb, vol. 83. Math. Grenzgeb. Springer-Verlag, Berlin (1974)

237. Walter, T.: Störungstheorie von Generatoren und Favardklassen. Semesterbericht Funktionalanalysis, pp. 207–220 (1986)

238. Wegner, S.-A.: Universal extrapolation spaces for $C_0$-semigroups. Ann. Univ. Ferrara, Sez. VII, Sci. Mat. **60**(2), 447–463 (2014). https://doi.org/10.1007/s11565-013-0189-5

239. Wiweger, A.: Linear spaces with mixed topology. Stud. Math. **20**, 47–68 (1961). https://doi.org/10.4064/sm-20-1-47-68

240. Yosida, K.: On the differentiability and the representation of one-parameter semi-group of linear operators. J. Math. Soc. Japan **1**, 15–21 (1948). https://doi.org/10.2969/jmsj/00110015

241. Yosida, K.: Functional analysis. Springer-Verlag, New York-Heidelberg, fourth edition. Die Grundlehren der mathematischen Wissenschaften, Band 123 (1974). https://doi.org/10.1007/978-3-642-61859-8

242. Yu, X., Liu, K., Chen, P.: On null controllability of linear systems via bounded control functions. In: Proceedings of the 2006 American Control Conference, pp. 1458–1461 (2006). https://doi.org/10.1109/ACC.2006.1656423

# Index